T0296742

Computational Modeling of Infectious Disease

Computational Modeling of Infectious Disease

With Applications in Python

Chris von Csefalvay
Starschema Inc.
Arlington, VA, United States

Academic Press is an imprint of Elsevier
125 London Wall, London EC2Y 5AS, United Kingdom
525 B Street, Suite 1650, San Diego, CA 92101, United States
50 Hampshire Street, 5th Floor, Cambridge, MA 02139, United States
The Boulevard, Langford Lane, Kidlington, Oxford OX5 1GB, United Kingdom

ISBN: 978-0-323-95389-4

For information on all Academic Press publications
visit our website at https://www.elsevier.com/books-and-journals

Publisher: Stacy Masucci
Acquisitions Editor: Kattie Washington
Editorial Project Manager: Lira Faurillo
Production Project Manager: Kamatchi Madhavan
Cover Designer: Mark Rogers

Typeset by VTeX

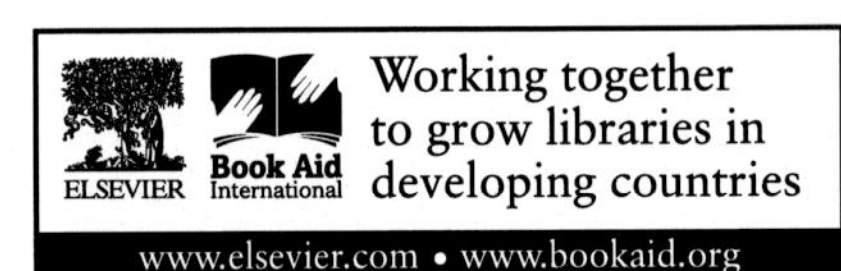

To Katie, who has been with me every step of this journey.

And to a new generation of infectious disease modelers: may they carry our hopes and dreams into the future with steadfastness, courage, and integrity.

Contents

List of figures

Biography

Chris von Csefalvay

Born in Budapest, Hungary, in 1986, Chris von Csefalvay was educated at Oxford, Leiden, and Cardiff. A data scientist by background, he has advised enterprises, NGOs, and governments on the use of computational tools and Big Data to manage the challenges of public health in a rapidly changing world. He joined Starschema Inc. in 2018, serving as Vice President for Special Projects. Board certified in public health (CPH), he is a Fellow of the Royal Society for Public Health.

Foreword

The Computational Modeling of Infectious Diseases by Chris von Csefalvay is a critical read. Chris has assembled in one place a clear and understandable guide to the development and use of models in transmissible infectious diseases. He walks through the math behind the models in a manner that facilitates understanding the precise variables that are critical to define when developing and utilizing these computational models.

Some may ask why this is important. The current COVID-19 pandemic laid bare the lack of understanding of these formulas, how to modify the variables, and how to continuously adapt the models based on the reality on the ground. Many public health officials, modelers, and epidemiologists sought solely to modify existing models that had been developed for influenza over the past decades. This resulted in predictions that did not match the reality on the ground. Why did that matter? It mattered as it resulted in some political leaders throwing up their collective hands and saying "no one knows," "the experts don't know," "the experts don't agree," "so any interpretation by anyone is potentially valid." Poorly predicted models added to the mistrust between political leaders and public health officials and, ultimately, the public.

But fortunately, there was another small group. There were modelers who were willing to start at the beginning, willing to move past dogma and perception. They were willing to move past and modify flu models. They understood early that SARS-CoV-2 was different: in its asymptomatic spread, in the persistent aerosols contributing to indoor transmission, in the need to go beyond droplet spread to aerosol spread. These modelers built a new model, a model from the evidence base of COVID-19 step by step, using the very formulas so clearly provided in this book. These modelers were willing continue to learn; when new data became available, they continuously updated the inputs to the formula, resulting in ever more precise predictions.

Every public health official, every epidemiologist, every member of the public interested in pandemics, every person working on pandemic preparedness should read this book and understand the math. Critically they must be able to evaluate the models used to predict the pathogen infectious disease spread so that they understand both the strengths and weaknesses around the data. They need to understand the importance of each variable in the formulas. Why? So that they can explain this in simple terms to their leadership and the public. So that they can clearly articulate the unknowns, what is still being learned, and how as more real-world data becomes available, the models will be refined. So that they can explain how we will learn together, and how with each refinement our model, and thus our predictions, will become more and more reliable.

Finally, the core value of this book and the need to understand each of these chapters is the ability to clearly demonstrate the impact of each element of mitigation on community spread, as well as the potential synergy of different mitigating measures that can be layered, so that the masses can understand risk reduction and the science and data behind any public health guidance. Trust is fractured when public health officials do not provide the data, the math, and the science behind their guidance. Trust is fractured when officials cannot explain what is known and what is still being discovered and how that impacts the model and the predictions. Trust is fractured when officials cannot provide the science and data and show the specific impact of each of the mitigation on the community, infection rates, and survival. I have had the privilege to work all over the world on many pandemics and the common thread throughout is that the public is smart: they can understand complex concepts when explained in clear language. They can understand what is known and what remains unknown. They can tell if you do not understand the science and data behind what you are saying, and they will question your guidance. But when you can explain things clearly, provide the evidence base in clear terms, public knowledge and understanding leads to active decision making to protect themselves, their families, and their community; that is how we will control this COVID-19 pandemic and be ready for pandemics in the future.

Deborah L. Birx, MD
Former White House COVID-19 Response Coordinator

Preface

At the time of writing, the COVID-19 pandemic has been raging for almost three years. It has cost five million people their lives. The toll of destruction, the human cost, and the economic losses remain to be counted. Few outbreaks in history leave this kind of lasting mark on society: the Plague of Athens (430 BC), the Plague of Galen (165–180 AD), the Plague of Justinian (541–549), the Black Death (1346–1353), the Spanish Flu (1918–1920), and the HIV/AIDS pandemic (1981 onwards) are the most notable exceptions. COVID-19 has now joined the ranks of these sad episodes of human history.

Yet humans are not helpless against pandemics. Amidst all the destruction and grief of the COVID-19 pandemic, science has been a bright, shining beacon showing how humanity can prevail against fearful odds.

- SARS-CoV-2, the coronavirus responsible for COVID-19, was sequenced faster than the pathogen responsible for any other major outbreak of a novel infectious disease in history.
- Within months, thousands of SARS-CoV-2 genomes were collected by researchers at the Fred Hutchinson Cancer Research Center, who built a real-time phylogenetic tree of COVID-19.
- Tests were developed in record time. Rapid point-of-care tests and lateral flow assays were deployed faster than at any time in the past.
- Viable and safe vaccines were administered to the general public within a year of the pathogen's first detection.
- Quantitative models of disease, largely resting on computational resources, took centre stage in the efforts to respond to the pandemic.

In *The Coming Plague*, Laurie Garrett calls for multidisciplinary perspectives, writing that to "comprehend the interactions between *Homo sapiens* and the vast and diverse microbial world, perspectives must be forged that meld such disparate fields as medicine, environmentalism, public health, basic ecology, primate biology, human behavior, economic development, cultural anthropology, human rights law, entomology, parasitology, virology, bacteriology, evolutionary biology, and epidemiology" [1]. This book seeks to introduce computational methods, computer science, and high-performance computing into Garrett's list of perspectives.

Computational models of infectious disease can make all the difference in our response to pandemics. As habitat loss and climate change make zoonotic spillover events increasingly more likely, COVID-19 is almost certainly not the last major pandemic of the 21st century. In fact, it is reasonable to assume that such outbreaks will become increasingly frequent. Computational models can be powerful weapons in our fight against pandemics. The increasing availability and decreasing cost of computing resources means that large-scale computation has never been this affordable in human history. The average high school student's graphing calculator packs more computing power than has been available for the first several decades of computer science. Massively distributed architectures can simulate populations of millions in agent-based models and examine the effectiveness of increasingly complex public health policies in a data-driven and evidence-based manner. And slowly but surely, a new generation of epidemiologists is coming to a better understanding and greater acceptance of computational methods.

There is a kind of magic to the human ability to predict health and disease in large populations. As a teenager, I was introduced—by my father—to the work of Isaac Asimov, whose *Foundation* series described the fictional discipline of psychohistory, a form of what we would today call social behavioral data science. Whether the approaches to behavioral analytics have reached the level of Asimov's fictional discipline at this point is up for debate. What is not is that the dynamics of infectious disease in populations as small as thousands and as large as billions have been laid open to exploration over the last decades, purely the way of mathematics and, where that tool fails, computational science. Asimov's psychohistory might be fiction, but the mathematical dynamics of the ways large populations behave rests firmly in the realm of science. This text is but one in a long series of works attempting to harness the quantitative understanding of the dynamics of populations for the purpose of preserving and protecting life, human and animal. It follows a long history, going back at least a century and a half, of quantitative analyses of infectious disease, but seeks to supplement it with the much more recent developments in computational science.

This book is designed to be a "first port of call" for computational epidemiologists and disease modelers, in two senses of the word. First, I hope that this book will be a springboard into a complex yet incredibly exciting field for many who wish to gain an understanding of modeling infectious disease. As such, it is not a book for any one audience; it is not epidemiology for computer scientists or computational modeling for epidemiologists, but rather a unified approach that should be digestible (hopefully with no adverse effects) to a wide audience. There are trade-offs that have to be made when writing such a book, and one trade-off I chose to adopt was to orient the book towards practical use. Consequently, some of the more in-depth mathematical content, including proofs and derivations, as well as detailed discussion of analytical solutions, had to give way. Where these are omitted, I have sought to provide references to works of other authors where interesting readers may find more details.

A disease modeler needs to know more than what model to use. She needs to also know what assumptions that model makes, and judge how well that fits the reality of the circumstances. All models are wrong, as the saying attributed to English statistician George Box goes [2], but some are useful. I hope that this book will be a valuable addition to every infectious disease modeler's bookshelf in helping them to find the least wrong and most useful model for their envisaged applications.

My name may be on the cover, but no single book is truly a work of a single individual. I have the good fortune to stand on the capable shoulders of extremely patient giants, who have freely lent their time, assistance, and expertise in preparing this manuscript. I am, more than anything, grateful to my wife Katie, to whom this volume is dedicated. Her companionship and encouragement have rescued this book more than once from landing on a discarded pile of half-finished manuscripts. I am also grateful to my parents, H.E. Dr. Zoltan Csefalvay, who instilled in me a love for asking difficult questions, and Dr. Anna Maria Bartal, who taught me to never stop looking for answers to them. Their example as humans and as academics alike has been an enduring inspiration. I have been extremely fortunate to study under many shining lights at the University of Oxford, who have shaped my outlook on many things, including data, evidence, and truth: the late John Gardner, as well as John M. Finnis, Andrew Burrows, Stefan Vogenauer, and Adrian Zuckerman. I am grateful to all their efforts, and their extraordinary patience in putting up with me. Their legacy, even though it is in a rather different field, lives on in me to this day.

I am grateful to H.E. Dr. Deborah Birx for agreeing to contribute a fascinating foreword to this title. Her example, along with that of thousands of public health physicians and scientists taking up the fight against COVID-19, has been one of extraordinary courage under fire. From first responders to public health directors, from emergency department physicians to laboratory scientists, the threat of a pandemic was once again met by a resolute and steadfast alliance, and showed that even in the darkest of times, humanity is capable of tapping into secret sources of strength, courage, and kindness. It has been an extraordinary privilege to see millions of nameless heroes all across the world respond to the worst that COVID-19 could throw at us with the best humankind has to offer. Their unheralded efforts and quiet sacrifices have been an enduring source of inspiration throughout writing this book.

My editors at Elsevier, Ms. Kattie Washington and Ms. Lira Faurillo, have been patient and capable guides in shepherding along this work, and I cannot overstate my gratitude for them.

I am also greatly indebted to my colleagues at Starschema Inc., especially Tamas Foldi, Eszter Windhager-Pokol, Balazs Zempleni, and Chimed Altandush. CDR Tom Van Gilder, MD, USPHS (Ret.) helped me spread my wings in computational epidemiology many years ago, for which I will be forever grateful. Many of the ideas in this book have benefited from conversations I have had with my long-term collaborator, Dr. Victor Mutua. Last—but definitely not least—I am deeply grateful for the late night conversations with John Abdo, who has encouraged me to think about new realms of complexity.

Countless other friends, acquaintances, and family members have lent their encouraging presence, kind words, and the occasional patient ear to me, and I cannot thank them enough, even if I cannot list all of them without risking omissions.

Figs. 1.1, 4.1, 4.7, and 9.3 were created using BioRender and are used under license. Figs. 8.7, 8.8, 8.9, 8.10, 9.11, and 9.12 use data from the OpenStreetMap project, whose diligent work in making global street network topologies available is greatly acknowledged.

Finally, I am grateful to my students at the Budapest University of Technology and Economics, past and present, whose spirit of enthusiasm and inquiry have been an inspiration throughout.

All errors and omissions are, of course, the author's own.

Chris von Csefalvay
Great Falls, VA, United States
September 2022

Introduction
Why and how we model infectious disease

For ere this the tribes of men lived on earth remote and free from ills and hard toil and heavy sickness which bring the Fates upon men; for in misery men grow old quickly.

But the woman took off the great lid of the jar with her hands and scattered all these and her thought caused sorrow and mischief to men.

*Only **Hope** remained there in an unbreakable home within under the rim of the great jar.*

Hesiod, Works and Days, tr. H.G. Evelyn-White [3]

1.1 Why we model infectious disease

This is a book on the computational modeling of infectious disease. Thus if you have picked it up for any reason other than to judge its suitability as a doorstop, one might well presume you have an understanding of why we model infectious disease. Nevertheless, it might be useful to briefly summarize the reasons behind infectious disease modeling at large. For other than academic interest, infectious disease modeling has significant ramifications in the "real world." By that, I mean that models of infectious disease are used to answer practical questions, many of which are quite significant, even vital.

The recent COVID-19 pandemic has offered a nearly inexhaustible wealth of examples for such questions. What follow are but a few representative examples, but they should give you an idea of the sheer breadth of the queries that computational epidemiology is called upon to answer.

- By July 2021, approximately 60 percent of all Americans were vaccinated. Should parents expect their children to return to full-time in-person schooling come September?
- There are several variants of SARS-CoV-2, the virus responsible for COVID-19. Some of these, called variants of concern (VOCs), are known to exhibit a higher propensity to cause symptomatic disease. If the vaccine is a given percentage less effective against a certain variant, what is the overall ratio of effectively protected persons given a certain seroprevalence of a VOC?
- According to the Kaiser Family Foundation and the 2008 AHA survey, there are 2007 staffed ICU beds in Virginia. Given a base occupancy of approximately 70 per cent, what is the maximum number of new cases per unit time before healthcare capacity is overloaded?

Computational Modeling of Infectious Disease. https://doi.org/10.1016/B978-0-32-395389-4.00010-4

Infectious disease modeling is exciting, challenging and, at least for some people, fun. But it is also a matter of life and death, and therefore it is vital that we keep in mind throughout the importance of doing things the right way.

Though many of our tools are of recent design, the notion of using mathematical methods to model health and disease is not. John Snow (1813–1858) is often lauded as one of the first to deploy this in practice. In 1854, during a raging cholera epidemic in London, he charted the cases by place of residence. From those, he inferred that the focal point of the infection was the pump at Broad Street [4,5]. Today, we would use kernel density estimates to predict the maximum locus of likelihood, but in the early days of epidemiology, guesswork was as much part of the trade as scientific reasoning.

The 1854 London cholera epidemic shows one of what would become a facet of computational epidemiology: the investigation of disease and the identification of its causes using quantitative means. But mathematical modeling of infectious disease does not only look at causal interactions (in fact, Snow-esque detective work is rarely the province of the infectious disease modeler). Rather, we look at the four defining factors of an infectious disease process:

- the pathogen,
- the people and/or animals it affects,
- the infectious process in time, and
- the infectious process in space.

This book reflects these four preoccupations, although since they are rarely neatly separable, not in any specific order. We will look at quantitative models of different pathogens, and how differences in those pathogens affect our methodology for charting the course of an outbreak. We will also look at time. Time is the main variable over which infection takes place, which is to blame for the abundance of differential equations across this field. And finally, we will also look at the infectious disease process in space. All of this, of course, takes place within various populations of humans and non-human animals.

As we venture across these areas, we will journey through the main competences of public health: disease investigation, disease control, and disease prevention. There are numerous ways in which the computational modeling of infectious disease interacts with other fields in the process. There is a growing recognition that the computational approach brings significant new dividends to many scientific fields [6]. The COVID-19 pandemic has highlighted the enormous value that quantitative modeling can bring to analyzing a dynamic process amidst considerable uncertainty. Quantitative modeling has moved from the province of academic interest into the mainstream of public health discourse, and it is set to occupy the fourth point of the compass rose—together with infectious disease medicine, microbiology, and public health policy—of responding to novel pathogenic threats for a long time to come. While mathematical modeling of infectious disease is hardly new, the increasing ability to leverage vast computing power in search of answers to questions of public health significance is creating a cross-disciplinary approach to epidemiology.

Hesiod's description of Pandora's box reminds us that though we live in a world of danger, where infectious diseases continue to maim and kill millions, especially across the developing world, we are not without hope. Part of that hope is our ability as humans to bring mathematics, genomics, data science, statistics, and computational science to bear on this problem and call these altogether rather disparate disciplines into humanity's service against disease. Infectious disease modeling is part of that wider story of hope.

Infectious disease modelers may thus come from a medical background, a background in computer science or mathematics, traditional education in epidemiology or other disciplines. The purpose of this book is to give the reader, their background notwithstanding, a solid grounding in the principles of quantitative and computational infectious disease modeling.

1.2 A brief history of the discipline

In a 1675 letter to Robert Hooke, an unusually self-effacing Isaac Newton described his accomplishments as "standing on the shoulders of giants" [7]. Computational epidemiology as a discipline has a relatively short history, but stands on the shoulders of gigantic achievements indeed. Most of the tools we use are the products of the last three decades, which saw a considerable development in fast, efficient computational tools to solve ordinary differential equations, agent-based models, simulations, and numerical optimization. In parallel with this, the rapidly decreasing cost of computational power (Moore's law) [8] and the rise of cloud computing services, which afford individuals the ability to deploy massive computing clusters have democratized data analysis. The discipline itself, however, owes much to the giants on whose shoulders it stands.

As Brauer [9] notes, the first mathematical model of an infectious disease, properly so called, owes its existence to Bernoulli [10]. Two major events over the century and a half that followed Bernoulli's paper have contributed to the emergence of mathematical epidemiology in the early 20th century:

- John Snow's work in identifying the source of the 1854 London cholera epidemic has created epidemiology as a discipline in its own right.
- The development of calculus and differential equations, from Newton on through Frobenius, Lie, and Liouville, has created mathematical tools to represent and analyze rates of flows and quantities over time [11].

Thus when in the early 20th century, the fundamental pathological mechanisms behind infectious disease came to be understood with some more accuracy, the mathematical tools to express them were already present. In 1915, the erstwhile discoverer of the plasmodial lifecycle that underlies malarial infection, Sir Ronald Ross, expanded on his earlier work with what he referred to as a work of "pathometry," the use of quantitative means to understand an infectious disease [12]. Shortly thereafter, Kermack and McKendrick [13] created the first true compartmental models, which to this day serve as our line of departure towards more complicated models. The genius of Kermack and McKendrick's work lay in creating an expandable structural framework for infectious diseases, in which transitions between populations are represented as population flows, the values of which are comprised in a system of ordinary differential equations.

At this point, the evolution of what eventually became computational epidemiology became, for a time being, intrinsically intertwined with the history of solving ordinary differential equations. All but the simplest models of infectious diseases were not quite amenable to analytical solution. Therefore a way to solve ordinary differential equations numerically, and preferably in an automated manner (i.e., computationally), was required.

The quest for numerical solutions for ordinary differential equations goes back to the 18th century, when Leonhard Euler first presented what is known today as the forward Euler method [14]. The methods described by Runge and later expanded on by Kutta in 1895 and 1901, respectively, have laid the groundwork for future solution algorithms, many of which are used to this day [15].

The next impulse to find better ways to solve ordinary differential equations came from a somewhat unexpected corner. Warfare in the late 19th century saw the role of artillery both expanded in ambit and expanded in reach. Because long-range indirect-fire artillery necessarily involved longer times of flight, new variables had to be accounted for, such as lower drag once the projectile reached thinner air at altitude, or the inertial effect of the Earth's rotation (Coriolis force). In addition, long range naval gunfire required tools that could accurately adjust a naval gun's azimuth and elevation rapidly enough to keep track with the ship's chaotic motion and speeds above 20 knots, as was becoming common in the era of dreadnoughts. The first "computers" were entirely mechanical, such as the ball-and-disk integrator developed by Lord Kelvin. By the outbreak of World War I, many vessels of the Royal Navy had Dreyer fire control tables, and a decade and a half later, the improved admiralty fire control table was introduced [16]. World War II also saw the increasing use of electro-mechanical analog computers. Perhaps the most influential among these was the Norden Bombsight, which among others played a part in the atomic bombings of Hiroshima and Nagasaki [17]. These mechanical and electro-mechanical devices took differential equations from the realm of pen-and-paper mathematics into the contemporary precursor of the computational field.

At the same time, by the mid-1940s, Alan Turing's pioneering work at Bletchley Park has laid the groundwork for the first true general-purpose programmable computers. Just how intrinsically linked the development of computing has been with the problem of solving differential equations is evident from the name given to the first such computer: Electronic numerical integrator and computer or ENIAC, for short. The room-sized multi-panel ENIAC eventually gave way to smaller and more efficient computational hardware. The development of microprocessors that began in the 1960s has made computing affordable, widely available, and much, much smaller: in 1995, the entire circuitry of ENIAC was replicated on a single chip approximately the size of a dime (a U.S. 10¢ coin, with a diameter of 17.91 mm), and at the time of writing (2022), the entire architecture can be replicated on programmable logic devices (FPGAs) available for less than US$ 100 [18].

Computational biology emerged as the result of broader introduction of computational methods into the sciences in the 1960s and 1970s. The rapid advances in genomics and proteomics following Crick and Watson's discovery of DNA's structure in 1953 created a demand for efficient ways to search for DNA and protein sequence alignments between samples. Around the same time, the first computational models were developed that used individual agent-level simulation to explain larger-scale processes. The most notable among these was Schelling [19]'s model of segregation in the presence of a surprisingly low threshold value: Schelling's simulation proved that even quite weak preferences to be surrounded by the same group (homophily) can result in rather significant spatial segregation between the groups.

The rise of Big Data, cloud computing, and the slow and arduous trek towards using the internet to collect and disseminate information on ongoing epidemic processes is the final thread that makes up the complex fabric of computational epidemiology [20–22].

Any subject whose history ranges from pump handles on London's Broad Street, tide tables, naval gunfire, and models of social segregation is bound to have rich parentage. It took "a village" to beget computational epidemiology: as a true multidisciplinary subject, it evolved at the crossroads of mathematics, computation, statistics, and medicine, with some contributions from systems biology, virology, microbiology, game theory, geography, and perhaps even the social sciences. Of course, each of those fields have histories of their own, lengthy enough to fill a bookshelf. This short historical sketch is nevertheless a useful overview that elucidates the rich historical heritage of the subject matter at hand, as well as the diverse intellectual influences that have converged to create modern computational epidemiology.

1.3 What this book is about

Clinical computational epidemiology, which I sometimes treat as synonymous with infectious disease modeling, is the use of mathematical models to represent the dynamics of infectious disease, and the use of computational tools to analyze those models. In this book, we are concerned with the interaction of quantitative methods, computation, and infectious disease. For our purposes, the term "infectious disease" shall encompass any disease that broadly fulfills Koch's postulates, most importantly the criterion that introduction of the causal organism or agent causes disease in a healthy individual [23]. As such, our examination will span a wide range of causal agents of human and animal disease, including bacteria, eukaryotes, parasites, and viruses.

Modeling, by necessity, ties reality to a Procrustean bed of mathematics. No model is perfect, but a good model is one from which we can derive practical use. What is a good, or even adequate, model is of course strongly context-dependent. For purposes of prediction, a good model would be capable of identifying, with reasonable accuracy, some parts of the epidemic curve (e.g., number of cases) over a reasonable time. Other models might be concerned with the effect of a particular intervention, including non-pharmacological interventions (NPIs), such as social distancing or travel restrictions. For such models, adequacy is largely determined by how well the model predicted the differential effect of the intervention. Arguably, it is often difficult to identify these effect sizes accurately. Nonetheless, our purpose should be to try to reflect reality, or a well-defined possible future, using quantitative means.

Thus, by the end of this book, an attentive reader ought to feel confident in their ability to tackle questions of infectious disease propagation in time and in space, even where the pathogen's dynamics and compartmental structure might be different. This book seeks to provide the practitioner with a toolkit and with an equivalent of the Basic School in infectious disease modeling. Practice makes perfect, and the book has been structured with a wealth of practical examples and code listings to allow readers to try their hands on various models. Computational epidemiology is not a spectator sport, and the more a reader devotes to the use and practice of the techniques in this book, the more rewarding the entire experience will be.

The focus of this book is primarily on human infectious disease. There are significant differences between human populations and non-human animals. However, we will discuss disease in non-human animals in at least two contexts. First, animals are relevant for human infectious disease as part of host-vector and host-host interactions (e.g., zoonotic diseases). Second, animal disease often provides illuminating examples that may help us understand human disease better. For this reason, a primary focus on human disease notwithstanding, there will be sporadic reference to animal health, too.

There are applications to computational epidemiology that share methods but not subjects. For instance, network diffusion models that are frequently used to model disease spread in communities (see Subsection 9.1.2) have been used to model the spread of information in the context of social networks [24,25]. It is not by accident that when an item of information, be it as harmless as a cute cat video or as harmful as misinformation about important matters, is, once widely shared by individuals, described as having "gone viral." The notion of "thought contagion" [26] or Dawkins's original notion of the "meme" echoes the close and self-evident relationship between socially spread information and socially spread pathogens. Though this book does not explicitly discuss the application of computational infectious disease modeling to the modeling of information in a social context at length, much of the theoretical framework we lay down for infectious disease can be used, *mutatis mutandis*, for analogous social processes of propagating sentiments, opinions, or information.

1.4 The computational mindset

This book is written from the perspective of the computational mindset. What I mean by this is that we are interested in any problem we can solve using any computational means. What I also mean is that our horizons are not limited by what can be analytically solved. As computationalists, any quantitative method that gets us on the target is a suitable solution. If we can integrate a differential equation numerically, we shall do so in preference to searching for an analytical solution that might prove elusive. The results of a well-designed numerical simulation are, in the computational mindset, just as valid as those arrived at analytically.

In this text, we will be dealing with considerable complexity and non-linearity at times. Even the simplest dynamic processes can become quite complex. As we progress towards modeling the spatial and temporal dynamics of epidemics and the complex choices individuals make in these scenarios, we move increasingly away from comfortable abstractions, such as compartmental models, and towards solving increasingly complex problems with computational methods.

For this reason, this book treats computation on an equal footing with analysis. The challenges of public health rarely afford one the opportunity to come up with a solution of sublime mathematical elegance. This is not to say that to do so is not a worthwhile pursuit. Indeed, much of that work been instrumental in furthering our understanding of infectious dynamics. Rather, the purpose is to highlight the subtle difference between mathematical and computational epidemiology. The latter, which is the subject of this book, shares its kinship with the former in that we, too, shall rely extensively on mathematics to describe systems and processes of infectious disease. We shall, however, take a more expansive view of solutions, embracing computational techniques and numerical methods as first-class citizens.

1.5 Who this book is for

This book is primarily for the practising infectious disease modeler, and those who wish to be so. As such, it seeks to convey just enough knowledge to be effective and get started tackling a problem, rather than be the exhaustive scholarly work on all matters in infectious disease modeling. In particular, I am dispensing with complex discussions of analytical solutions, for reasons explained in Section 1.4.

This book is also meant for those who consume quantitative models from epidemiology: public health officials, physicians, decision-makers, and indeed the wider public. All too often, the results of mathematical models are communicated without a critical reflection on the model itself, its suitability, and its fundamental assumptions. All models operate on the basis of some fundamental assumptions, and this book is rather more explicit in highlighting them than many others. As we journey through different and increasingly complex ways of modeling infectious disease, we should do so with an understanding of when one model is preferable to another, and indeed when modeling is entirely inappropriate.

Finally, this book is also for anyone who wants to understand the mathematical modeling of health and disease. In 1993, Roberts and Heesterbeek published a wonderful article titled "Bluff your way in epidemic models" [27]. In it, they sought to provide microbiologists in particular with a painless introduction to the mathematical modeling of infectious disease. This book is, of course, much longer and much more extensive than their manual. It is also much more in depth, and I would not presume the reader to wish to bluff their way in epidemic models. By the time one has worked one's way through this book, one will have acquired a thorough enough understanding of epidemic models to be able to apply them, comment on them and, most importantly, use them.

1.6 What this book is not about

Epidemiology and public health is an expansive field, and its reach broadens with time as we understand the increasingly complex interactions between health and other social factors. It would be folly to attempt to survey the entire breadth of the field in a single monograph. As such, there are several worthy areas we are nonetheless not concerned with in this text.

- First, we are not concerned with **non-infectious causes**, such as non-infectious chronic diseases. That is not to discount the applicability of our techniques to non-infectious phenomena. In his classic study of von Economo's disease (*encephalitis lethargica*), neurologist Oliver Sacks points out that response to treatment with L-DOPA had, at times, an infectious nature: when some patients did well, others improved (regardless of changes to their treatment regimen), while setbacks afflicted more than the index patient [28]. In a more recent contribution, Sánchez et al. used techniques from compartmental infectious disease modeling to mathematically describe the risk of relapsing among persons with problematic alcohol use [29]. Though such techniques are often helpful, especially where there is an element of social contagion involved, we will largely stick to infectious disease.
- We are also not primarily concerned with **the underlying mathematics**, there are plenty of textbooks on that subject, and the assumption is that the reader has digested at least one of these previously, or would do so. As such, from time to time, an understanding of certain basic mathematical methods will be taken for granted.
- We are **not concerned with efficiency in general**; much of the code exists for readability over speed. Python, which is used for the computational parts of this text, is an incredibly versatile language, and understanding how we computationally approach a problem can easily translate to other languages. This benefit would be lost altogether if we aimed for the most streamlined and most efficient code. That being said, we will not eschew the occasional opportunity to optimize our code.
- Finally, we are **generally concerned with animal disease, and particularly with human disease**. The epidemiology of plant diseases is a fascinating field [30,31], which poses the unique challenge that most of its subjects are sessile. Plants, absent rather extraordinary circumstances, do not move around in the Brownian-like motion that much of mathematical epidemiology assumes for human and non-human animals. While computational methods, such as diffusion cellular automata [32], have been applied to plant diseases, the differences in the infectious process put plant disease outside the scope of this text.

Other exciting fields we are not concerned with in this book include trial and experimental design and the analysis of clinical studies. This is a field that merits its own literature, and several comprehensive, thorough, and thoughtful works are available on the subject (e.g., Meinert [33], Carter et al. [34] and Yin [35], to name a few). Furthermore, this book largely bypasses the nascent field of genomic infectious disease epidemiology; the domain of this field is large and rapidly changing, and it would be much too difficult to do it adequate justice as well as Tibayrenc [36] and Ginsburg et al. [37] had.

1.7 How to use this book

You may use this book either as a course or as a reference. Unless you are participating in a course where this book is a set text, it is up to you whether you wish to follow a chronological path or use it to look up what you need *pro re nata* (as your needs arise). This book was written with both uses in mind, and as such, you should feel comfortable with it either way.

Appendix A contains a crash course in Python, which should be sufficient to help you hit the road running. Python is a simple but enormously versatile programming language, and the best way to learn it is largely by experimenting with it. If you are interested in additional resources to deepen your knowledge of Python, Section A.12 of Appendix A outlines useful starting points for beginners.

1.7.1 Prerequisites

This book assumes that the reader has at least some familiarity with

- calculus (including differential equations) and elementary linear algebra,
- Python (basic programming, use of packages, syntax, control flow, plotting), and
- the fundamentals of infectious disease epidemiology.

Though this is a book intended for practising professionals as well as students, there is no assumption of deep expertise in any of these fields, nor of any previous exposure to infectious disease modeling. Students with two semesters of college-level calculus and a semester's worth of Python will be more than comfortable with the concepts involved. Readers with less extensive training in mathematics may benefit from the concomitant use of a mathematical techniques textbook, such as Jordan and Smith [38]. For readers who wish to reflect more deeply on the mathematical concepts and models involved, Keeling and Rohani [39]'s excellent book (which this text partially follows in terms of structure and chapter order) is a good starting point. As far as infectious disease epidemiology is concerned, the assumption is that the reader comes to this field from a primarily quantitative point of view, and for this reason, relatively little knowledge of epidemiology or infectious disease medicine is presumed, beyond fundamental biology (the basic nature of different pathogens) and elementary concepts of epidemiology (transmission, immunity and reproduction).

1.7.2 Software specifications

As the subtitle suggests, this book makes extensive use of Python as a fast, agile, and easy to learn language that has extensive support for the numerical constructs that are the infectious disease modeler's daily bread. It is accompanied by a set of

Jupyter notebooks, which allow for interactive experimentation with Python code. The code listings in the book are often not exhaustive, omitting some boilerplate code, and code snippets in Computational Notes are more akin to explanations of the notebooks. In short, code should always be read in conjunction with the notebook. The notebooks are available on the book's companion Github repository, which can be found under github.com/chrisvoncsefalvay/computational-infectious-disease/. The book's companion website (https://chrisvoncsefalvay.github.io/computational-infectious-disease/) also contains useful content, including code updates wherever necessary.

All code throughout this book was written for compatibility against Python 3.8, but should run perfectly well on more recent versions. Older versions of Python might not support the type annotation syntax (PEP 484), of which we will make (sparing, but judicious) use. For this reason, you might want to run at least Python 3.6 or more recent. The notebooks in the book's companion Github repository have been executed under Python 3.10. At the beginning of each notebook, a "watermark" segment identifies the version of both the Python runtime and the versions of major imported packages. More details about Python are to be found in Appendix A.

1.7.2.1 Ellipses and omissions

In code, the ellipsis (...) symbol refers to omission or incompleteness. There exists a use for the same symbol in NumPy for a form of advanced indexing. Where the ellipsis (which in NumPy is usually called "dots") is used to denote index expansion, this will be noted specifically. Otherwise, it is to be assumed that an ellipsis refers to an omission. As a general rule, an ellipsis is on the level of indent, where the omission occurs.

```
def foo(arg1, arg2):
    if arg1 == arg2:
        ...
    else:
        raise()
```

In the code example above, the ellipsis is located under the scope of the first branch of the conditional statement, meaning that contents there are omitted, but the rest of the function goes on. Contrast this with

```
def foo(arg1, arg2):
    if arg1 == arg2:
    ...
```

In this case, the ellipsis is at the indent level of the function body, meaning that some or all of the rest of the function definition will be omitted.

1.7.2.2 Line lengths

Python is an indentation-sensitive language. For this reason, there are strict rules as to when a line might be broken. This makes displaying code somewhat difficult in the limited width allotted to text.

The explicit line join operator in Python, `\`, more commonly known as the backslash, tells the interpreter that the input goes on in the next line. Notably, since it is considered the same line by Python, one can indent the continuation line as much as one wishes. This is, indeed, good practice to indicate that there is a logical continuation happening here rather than the start of a new line at the indentation level.

1.7.2.3 Typing

Python has, for much of its existence, worked perfectly well without type checking and type specification. Typing, however, makes for better, clearer, and more intelligible code. For this reason, it is good practice to specify types.

In this text, we will do so where appropriate, where types are relevant, and where specifying them makes sense. Their omission in many examples is a matter of saving space rather than a suggestion that type specification should be foregone.

1.7.2.4 Standard imports

In general, most examples assume that certain standard imports have been made. These are as follows:

```
import numpy as np
import pandas as pd
from matplotlib import pyplot as plt
```

In most cases, we will not explicitly discuss imports. To see the exact imports required, please consult the relevant notebooks linked to each computational example.

1.7.2.5 Plotting and visualization

With the exception of flow diagrams, every illustration in this book was generated by a notebook available on the book's companion Github repository. Creating compelling graphics for public health is an art and a science of its own [40], combining the sensory psychology of visual representations, design philosophies, and information architecture. It would be impossible to do the subject of visualization justice in a monograph dealing primarily with the construction and analysis of models. For this reason, the Computational Notes typically omit the visualization stage. However, interested readers will find the visualization code in the notebooks, with extensive comments.

1.7.3 A note on structure

This book was written with the practising epidemiologist or infectious disease modeler in mind. As such, it treats particular scenarios, types of models and questions on one hand and their mathematical solutions on the other as a united whole. This follows texts, such as Keeling and Rohani [39], which is structurally quite similar to this book, and contrasts with others, such as Li [41], which devotes specific parts to the

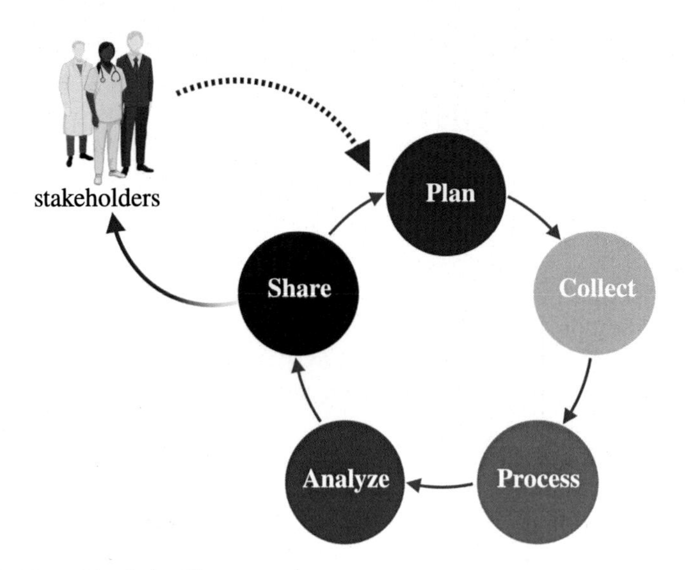

Figure 1.1 The epidemic intelligence cycle.

mathematical methods before discussing applications. This structure is largely a compromise to allow users to refer to the book as both a reference manual and a guided tour through infectious disease modeling.

The approach adopted by this text for solving problems is a five-step intelligence cycle (Fig. 1.1):

1. **Plan**: Identify the question.
2. **Collect**: Obtain data that characterizes the infectious process, including ground truths against which we can calibrate our models.
3. **Process**: Create a mathematical description capable of being computationally/quantitatively ascertained. This is your analytical instrument, akin to an instrument with which you can probe the question by computational means.
4. **Analyze**: Develop a computational answer to the question and obtain a model, a prediction, or any other product that answers your question meaningfully.
5. **Share**: Represent and communicate the information, creating actionable and audience-appropriate visual representations to bring your findings to public health stakeholders.

Similar to the intelligence cycle in the national security context [42], this is typically iterative: each finding raises a host of new questions.

This text introduces each section introduced using a case study. The purpose of case studies is to give a tangible introduction to the subject matter and discuss the use of the computational methods the chapter will go on to deal with, using a real-world example.

Throughout the book, reference is made to the relevant Jupyter notebooks. The notebooks are available in the book's Github companion repository. You may run these on your own computer, or on a cloud-based Jupyter implementation, such as Google's Colaboratory or JetBrains's Datalore.

1.7.4 Language and terminology

Though computational sciences are generally close cousins of mathematics, there are some differences in terminology.

- In NumPy, our main tool for representing various mathematical objects computationally, an **array** is essentially a homogeneous array of arbitrary dimensions, where each element is indexable by a tuple of non-negative integers. Note that **array** refers to the mathematical concept, whereas `np.array` (identical to `np.ndarray`) refers to the function implementing it in Numpy. On the whole, an array can have any positive nonzero integer number of dimensions, although NumPy does not support more than 32 (which is more than enough for our purposes). A NumPy `array` may represent a scalar, a vector, a matrix, or any higher dimensional construct.
- The **shape** of an array is an array of its size in respect of each of its dimensions. Thus an $m \times n$ matrix has the shape (m, n). The shape of a vector, sometimes also referred to as its **length**, is the number of elements in the vector, and somewhat inconveniently, its **size** is a single integer valued count of its elements, i.e., $m\ n$.
- The individual dimensions of an array are commonly referred to as its **axes**.
- A **subset** of an array is a part of an array and, contrary to the connotations of its name, not a set but an array of its own. In this text, we will strictly use the terminology of subsetting to refer to instances where the subset is defined by indices. A related operation is **Boolean indexing**, which is discussed in detail in Subsection A.11.1.3 of Appendix A.
- We will, from time to time, do things to arrays of different sizes that is not necessarily supported by ordinary mathematics. The magic that enables this is called **broadcasting**, and is explained in Appendix A.
- A **vectorized function** is a function that is applied to a Numpy array object without explicit iteration. It is, notably, not the same as a vector-valued function, as it can return any array-like of any size or dimensionality.
- From time to time, the terminology of **vector**s may be used to refer generically to one-dimensional multi-element arrays.

1.7.5 Conventions

The field of mathematical modeling of infectious disease has a number of conventions as to notation and representation. Unless otherwise noted, symbols throughout are to be understood to refer to the variables as described in Table 1.1 and Table 1.2.

1.7.5.1 Compartment definitions

An important formalism, explained in detail in Subsection 2.1.2, is that we do not differentiate between compartments and their sizes. I often denotes both the set of all

Table 1.1 Conventional symbols for compartments used throughout this book.

Symbol	Definition
S	Compartment of all susceptible individuals in a compartmental model.
I	Compartment of all infectious individuals in a compartmental model.
E	Compartment of all exposed individuals in a compartmental model.
R	Compartment of all removed in a compartmental model.
D	Compartment of all deceased in a compartmental model.
C	Compartment of all asymptomatic infected carriers in a compartmental model.
V	Compartment of all vaccinated in a compartmental model.
Q	Compartment of all individuals quarantined in a compartmental model.
M	Compartment of all with maternal immunity in a compartmental model.
T	Compartment of all treated in a compartmental model.
N	Number of individuals in a compartmental model without demographic change.
$N(t)$	Number of individuals in a compartmental model with demographic change, at time t.

infectious individuals and the cardinality of that set (i.e., their number). Because most often, infectious processes are related to proportions rather than absolute numbers, we will make our lives a little easier and simply use these symbols as fractions of the total population. This permits us to dispense with having to write $\frac{I}{N}$ any time we want to refer to the fraction of infectious individuals within the population. Some texts use the terms X, Y, Z, and W to describe proportions, but we shall leave this mainly to context. To differentiate between the size of a compartment as a fraction of all compartments and the absolute size (cardinality) of a compartment, we will use C for the former and $\mathcal{C}$ for the latter where this is relevant, but otherwise simply use C if context suffices to explain. Note, however, that N is *not* a compartment, but the total number of individuals within the system or within a compartment.

1.7.5.2 A note on symbols

It is common in statistics texts to see the symbols μ and σ used for the mean and standard deviation of a distribution, respectively. In the modeling of infectious disease, μ commonly denotes mortality rates. For this reason, I shall use "bar notation" for means, e.g., $\overline{x}$ denotes the mean of the vector (or variable) x.

In keeping with computer science literature, especially in the fields of machine learning, hats—unless othervise specified—denote predicted values of the "hatted" variable, thus $\hat{y}$ is the predicted or estimated value of y, which itself is the observed value. Matrices are generally set in bold font, and where they pertain to a parameter, the parameter is romanized (e.g., a matrix of values of β would ordinarily be $\mathbf{b}$).

For avoidance of doubt, the maximize operator is always spelled out, whereas $\max(x, y)$ denotes the maximum of the values or, where the argument is a set, the maximum of the set's members.

Table 1.2 Other conventional symbols used throughout this book.

Symbol	Definition
$\mathfrak{R}_0$	Basic reproduction number.
$\mathfrak{R}_t$	Reproduction number at time t.
ϕ	Birth rate (live births).
μ	(for constant population models) Mortality and birth rate.
β	Transmission coefficient.
φ	Rate of funerary transmission.
γ	Recovery rate.
ω	(in SIR models) Rate of waning immunity, (in temporal forcing) period of forcing.
ν	Vaccination rate.
λ	Force of infection.
p_c	Quarantine capture fraction.
$T = \{0 \dots t_{\max}\}$	T denotes the overall time span of a process and typically begins at 0 and ends at $t_{\max}$.
τ_c	(in SIRC models) Clearance time.
τ_l	Latency time or latency period.
τ	Mean infectious period.
τ'	Serial interval.
τ_m	Duration of maternal immunity.
τ_q	Quarantine period.
τ_T	Length of treatment period.
ϱ	Encounter rate.
θ	(in SIRC models) Fraction of recovery into carrier state, (in SI*D models) mortality fraction of removals, (in SIRT models) treatment rate, (in optimization) parameter vector.
T_d	Doubling time.

Some texts in mathematical epidemiology use the Newton or "flyspeck" notation for differentiation by time, i.e., $\dot{R}$ *in lieu* of $\frac{dR}{dt}$. My preference is strongly with the Leibniz notation, for two reasons. First, even if context would normally make this a non-issue, it is easy to overlook a differentiation dot. Second, even though we are almost always differentiating by time, it is good to be reminded of what exactly it is we are doing. Where the prime symbol (x') is used, it also refers to a modified or derived quantity from x, not the first derivative.

Some authors use a curly brace to denote that a number of differential equations together constitute a system of differential equations. In the interest of saving space, I have dispensed with this throughout this book. Systems of differential equations always appear together and share an index number, which should make identifying what belongs into the same system.

Another formalism relates to flow diagrams, which we use extensively in the mathematical modeling of infectious disease. There are numerous philosophies as to what goes onto the arrows between compartments. My philosophy, which I seek to adhere to in this book, is to align flow diagrams with the corresponding systems of differential equations. Consequently, in this book, arrows between compartments show the rate of flow.

1.7.5.3 Symbols in code

In the code examples, I have tried to use the same symbols as were used in the text. Some divergences are necessary due to reserved words in Python. Of these, most notably, λ is written as `blamda`, to avoid collision with the `lambda` keyword in Python (on which also see Appendix A). Because Python does not permit parentheses in variable names, the common notation of $C(t)$ for the size of a compartment C at time t is transliterated as `C_t`. This is to be distinguished from $\mathfrak{R}_0$ and $\mathfrak{R}_t$, which are respectively rendered in code as `R0` and `Rt`.

In theory, Python supports Unicode characters in variable names, as long as they represent a letter-like symbol in a foreign language. Thus `{\theta}` is technically a perfectly valid variable name. In the examples in this book, we will limit ourselves to the letters of the English alphabet, numbers, and underscores.

1.7.5.4 $\mathfrak{R}$ versus R

In what is somewhat unfortunate, computational epidemiology uses some confusingly similar terms, especially with the symbol R, used both for the basic reproductive number and the recovered/removed compartment. To disambiguate, we are going to use functional notation to denote the parameters of a value. Thus, in particular, $\mathfrak{R}_0$ and $\mathfrak{R}_t$ denote the basic and the time-dependent reproduction number, whereas $R(0)$ and $R(t)$ are the size of the removed compartment at time 0 and t, respectively.

To distinguish from $\mathfrak{R}$ is Re, the real-part operator, which sometimes uses the same symbol. In this text, real and imaginary parts are denoted by the Re and Im symbols, respectively.

1.7.6 Flows and multiple infection

In some cases, we wish to denote a flow between compartments, or a compartment's status with respect to multiple pathogens. Context should typically elucidate what is meant, e.g., $\beta_{H,L}$ is to be generally understood to refer to the transmission coefficient from subpopulation H to subpopulation L.

In multi-pathogen models, $N_{i_1,i_2,\ldots,i_n}$ describes status with respect to each of multiple pathogens, e.g., $N_{S,I,S}$ would correspond to the size of the compartment susceptible to pathogens 1 and 3, and infectious with respect to pathogen 2.

1.7.7 A note on matrix multiplication

In computation, the multiplication operation over tensors, matrices, vectors, and other arraylike objects is typically not the matrix multiplication as known to and loved by linear algebra, but rather the elementwise or Hadamard product. The elementwise product of two identical sized matrices is a matrix of the same size, where

$$(A \odot B)_{i,j} = A_{i,j} B_{i,j}.$$

In NumPy, which provides our main abstraction of tensors, matrices, and vectors, the multiplication operator `*` provides the Hadamard product for equal-sized arguments, as long as they are `array` objects. It also does so for unequal-sized `array` arguments, which is the feature referred to above as broadcasting and explained in more detail in Subsubsection A.11.1.3.

1.8 Definitions, computational examples, and practice notes

The book's main text is interspersed with definitions, computational examples, and practice notes.

Definition 1.1 (Definition). A definition lays out the meaning of a particular term, concept, or value. Terms following the definition are used with the understanding that their meaning is "as defined."

Computational Note 1.1 Computational notes

Computational notes explain aspects of devising and implementing calculations in Python. The Computational Notes are a good starting point for implementing your own models and building up an understanding of computational problem-solving in the infectious disease modeling domain. Note that computational notes often only include the key parts of the code. For the whole code, you should always consult the book's Github companion repository.

Practice Note 1.1 Practice notes

Practice Notes describe aspects of professional practice, ethics, and responsible use of data. The social importance of computational epidemiology means that ethical rules are more important than ever. And with the growing public awareness of the risk of pandemics, epidemiologists are often called upon to interpret quite complicated scientific matters for lay audiences. Practice Notes provide guidance on the ethical and professional conduct of research and practice in computational epidemiology.

Simple compartmental models
The bedrock of mathematical epidemiology

The statistician knows ... that in nature there never was a normal distribution, there never was a straight line, yet with normal and linear assumptions, known to be false, he can often derive results which match, to a useful approximation, those found in the real world.

Box, "Science and Statistics", 1976 [2]

2.1 The intuition of compartmental models

Compartmental models are the bread and butter of infectious disease modeling. The development of compartmental models by Kermack and McKendrick, building on earlier work by Sir Ronald Ross (1857–1932) and Hilda Phoebe Hudson (1881–1965), represented the first rigorous mathematical treatment of infectious disease dynamics, transforming what was a mostly qualitative inquiry into a scientifically rigorous discipline. This section examines basic compartmental models as the building blocks of more complex structures in the modeling of infectious disease.

2.1.1 Case study: influenza A outbreak on Tristan da Cunha

The small British Crown dependency of Tristan da Cunha is one of the most isolated points on the planet. Over 1700 miles off the coast of South Africa, it is home to fewer than 300 permanent inhabitants. The island's capital, Edinburgh of the Seven Seas, is commonly considered to be the most remote permanent human settlement. Without suitable ground for a paved airstrip, Tristan da Cunha's only lifeline to the outside world is by sea.

In 1971, the supply ship *Tristania* brought an unwelcome stowaway to the island: the alphainfluenzavirus H3N2, known as the Hong Kong flu at the time. By winter's end, 96% of the population would contract influenza, with almost a third experiencing at least two distinct episodes of illness [43].

Isolated settings, such as Tristan da Cunha, are the closest reality ever tends to come to the simple compartmental models that take no account of migration, demographic changes (births and deaths) and other more complex processes. Though reality tends to be much more complicated than compartmental models give it credit for, such models are mathematically convenient and close enough to the real world to be valuable sources of illumination. We shall therefore use them to begin our journey through the mathematical and computational modeling of infectious disease.

In this section, we will be looking at the basic notions of a compartmental model. Such models are enticingly simple without being overly simplistic, and offer the

Computational Modeling of Infectious Disease. https://doi.org/10.1016/B978-0-32-395389-4.00011-6

benefit of being quite easy to represent mathematically using systems of ordinary differential equations.

Compartmental models are in many ways the Lego bricks of quantitative epidemiology: they can be combined and almost endlessly expanded. For instance, a model that has a single compartment for "removed" cases can be expanded to include separate compartments for survival and mortality. This in turn can be broken down to characterize various forms of survival: survival with sequelae, survival without sequelae, and so on. The variety and complexity of compartmental models is limited only by imagination; SIDARTHE, for instance, is a model of COVID-19 transmission composed of eight compartments [44]. Yet no matter the complexity, compartmental models all follow the same basic mathematical notions, so that exploring an adequate but nonexhaustive sample of compartmental models suffices to provide infectious disease modelers with blueprints from which they can adapt and construct ever more ambitious models to fit specifications.

2.1.2 Defining compartments

The fundamental intuition of compartmental models is that if we define a number of mutually exclusive but jointly exhaustive states, we can characterize every individual's course through the disease process as state transitions at given times (e.g., from susceptible to infectious and from infectious to recovered) [13]. Given a population, then, we can estimate the rate at which these state transitions occur.

Definition 2.1 (Partition). A set of n nonempty subsets $\mathcal{S}_{1\ldots n}$ is a **partition** of a set if every element of that set belongs to exactly one subset.

This gives us a very convenient mean-field model, so that instead of having to track each person individually, we may simply conceive of "flows" between compartments (states) at a given rate that reflects the approximate likelihood of subjects in one compartment to transition into another. In this way, we can characterize an entire pathogenic process in a population by the size of the compartments at the start (commonly denoted as t_0) and the volume of flow between those compartments. We can then represent the entire system as a set of ordinary differential equations (ODEs), which in turn lays the dynamics governing the system open to mathematical investigation and modeling. This is the "skeleton key" to much of the mathematical—and with that, computational—reasoning about infectious disease.

Compartments solve the thorny problem of relatively large populations, which would otherwise make such computations more difficult. The total population in a model may be a rather large figure; it was not until relatively recent days that the computing power required to operate large-scale agent-based models (which are discussed at length in Chapter 9) has become ubiquitous, and even so, there are limits to the size of agent-based models that can be feasibly run. It is often more efficient to look not at individuals but at populations and their mean-field governing statistical dynamics, just as statistical mechanics can make generalized statements about the whereabouts and behaviors of particles without having to account for each individual

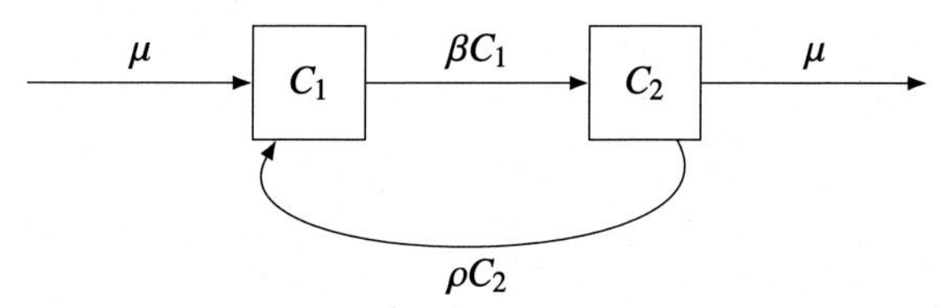

Figure 2.1 Flow diagram of a generic compartmental model with two compartments (C_1 and C_2), a constant-rate influx and outflow (μ), and a loopback from C_1 to C_2 at the rate of ρ.

particle. The benefit of the compartmental formalism is that it is robust to the total population size; subject to the issue of value overflows, the time complexity of solving a system of ODEs is not computationally dependent on the number of individuals or the total quantities.

A compartment, in the context of infectious disease modeling, denotes both a state and the set of all individuals from the entire population (which by convention we shall refer to as N) that are in a particular state. Rather than tracking each member individually, we consider clinically coherent groups of patients—the compartments in question—and look at the sum total of "state changes" by patients through the lens of flows between compartments.

Definition 2.2 (Compartment). A **compartment** is a set of individuals in the population, who share a characteristic relevant for the modeling of infectious disease (e.g., all recovered alive persons). Every compartment is a subset of the set N of all individuals in the model. The set of all compartments is a partition of N.

Fig. 2.1 is an example of a generic compartmental model. There are two compartments, C_1 and C_2. The rate of flow between C_1 and C_2 is governed by the parameter β, the rate of reflux is ρ. This example also hints at the fact that not all compartmental models are closed systems; the number of individuals within the system changes, through the parameters of μ and γ (e.g., immigration/emigration, births and deaths). The representation adopted in Fig. 2.1 is called a flow diagram, in which we denote compartments by rectangles, flows by arrows, and the rate of flow by labels over the arrows.

Over the course of our journey through various models of infectious disease propagation, we will encounter a range of increasingly intricate flow diagrams of compartments. They will be our *mappa mundi* to help us understand anything from simple SIR models to multilayered host-pathogen interactions. At the heart of compartmental models lie three fundamental axioms:

Definition 2.3 (Closed system axiom). Subject to demographic change (immigration, emigration, deaths, and births), the sum of each compartment's cardinality at any given time t is N. Subject to demographic change (immigration, emigration, deaths, and births), the **sum of all flows is zero**.

This follows from the fact that every patient in N is accounted for in exactly one compartment at any given time, which in turn is a consequence of our definition of compartments in a way that the set of all nonempty compartments is a partition of N.

Definition 2.4 (Partition property axiom). The set of all nonempty compartments at time t is a **partition** of $N(t)$, the population at time t.

From the partition property follows that every element of N is in exactly one—no more, no fewer—compartment at any given time (or inversely, the sum of the cardinalities of the compartments is N). As such, we can comfortably represent individuals by flows between compartments without the risk of information loss (as would be the case if a person could be in multiple compartments at the same time).

Definition 2.5 (Markov property axiom). The cardinality of any compartment at any time $t \geq t_0$ can be determined from the state of the system at t_0 (its initial conditions) if the system can be described by an autonomous and deterministic ODE.

Definition 2.6 (Autonomous and nonautonomous systems of ODEs). A system of ODEs is **autonomous** if and only if its right-hand side is independent of t.

This can be proven by recursion. For discrete time, the Markov property means that the state of the system at $t + 1$ is a function of the state of the system at t. That state, in turn, is a function of the state of the system at $t - 1$, and so on until t_0. The corresponding proof for continuous time is left to the reader, but should not prove difficult. Compartmental models that fit the first two criteria but incorporate a Wiener process or some other source of stochasticity are known as stochastic compartmental models. Those that integrate a time-dependent function on the right-hand side are known as nonautonomous.

Definition 2.7 (Stochastic versus deterministic compartmental models). A compartmental model is **deterministic** if it strictly complies with the Markov property axiom, and otherwise **stochastic**.

We will generally not discuss stochastic compartmental models, because other highly stochastic compartmental models, such as agent-based models (see Chapter 9) often reflect stochasticity better. Keeling and Rohani [39] provide an outstanding survey of stochastic models that an interested reader might enjoy.

Together, these three axioms underlie our ability to represent the entire model as a system of ordinary differential equations. In the remainder of this chapter, we will be making use of these axioms in practice as we examine some basic compartmental models.

Computational Note 2.1 ODE solvers

We reason about flows using ordinary differential equations (ODEs) and, in some cases, delay differential equations (DDEs) and partial differential equations (PDEs). The numerical solutions for ODEs are performed by the use of integrating ODE solvers. These solve a class of problems for ODEs known as initial value problems (IVPs): given some initial conditions (e.g., the sizes of the compartments at time zero), a set of parameters, and a system described by some differential equations, what will the state of the system be at an arbitrary time $t > 0$?

Until quite recently, the workhorse of such operations in Python was the `scipy.integrate.odeint` function, which wrapped around the venerable ODEPACK library written in FORTRAN. This has been significantly overhauled by the `solve_ivp` API, which incidentally also allows a number of different methods (mainly of the Runge-Kutta family, but also including backward differentiation) to be employed. In this text, we shall predominantly use the new API. An important practical difference is that whereas `odeint` takes a differential function of the form $\frac{dy}{dt} = f(y, t, \ldots)$, i.e., specifying y first, `solve_ivp` takes a "time-first" specification $\frac{dy}{dt} = f(t, y, \ldots)$. Framing the derivative function with the arguments in the right order for the respective API is thus essential.

A notebook implementing the contents of this Computational Note is available on the book's companion Github repository in the folder `/ch02/sir_ models`.

2.1.3 The principle of mass action

The fundamental assumption of compartmental models is that a fundamentally homogeneous mixing of the populations is occurring in a manner resembling Brownian motion, i.e., that every member of the population is in contact with every other one, and the likelihood of any person in any compartment interacting with any other is equal. Consequently, infection depends on the likelihood of transmission upon an encounter, and the likelihood of the encounter itself. The former is typically described as a parameter β. The latter results from the product of the proportion of susceptibles and the proportion of infectious individuals in the population. Together, they give us the transmission term or mass action term, βSI, which we will encounter repeatedly through our journeys across the realm of compartmental models.

Definition 2.8 (Mass action). By **mass action**, we denote the probabilistic nature of infectious disease transmission that depends on both the number of infected and the number of infectible individuals. Mathematically expressed, $\frac{dI}{dt} \propto S, I$. The proportionality is conditioned by the transmission coefficient β, which describes the likelihood of transmission upon an encounter eligible to create a transmission event, whereas SI describes the likelihood of such events occurring. The overall term βSI represents the rate at which infectious (I) and susceptible (S) individuals encounter each other and transmission takes place. We refer to this as the **mass action term**.

It may be helpful to derive this term due to its fundamental nature. Let us assume a population of $\mathcal{N}$, of whom $\mathcal{S}$ are susceptible and $\mathcal{I}$ are infected (with S and I, respectively, being the proportion of these compartments within N). Recall that we use the notation $\mathcal{C}$ for the absolute number of members in a compartment and C for the fraction $\frac{\mathcal{C}}{N}$, the proportion of that compartment in the total model population. If a susceptible individual meets n individuals in unit time, some of these will be infectious. We know that $I = \frac{\mathcal{I}}{N}$ describes the proportion of individuals who are infectious, which under the assumption of homogeneous mixing equates to the likelihood that any randomly encountered individual will be infectious. For every encounter, the likelihood that the other party was infectious is I, and for n encounters, it is nI. We may generalize this for any timespan Δ_t as $nI\Delta_t$.

Then, we assume that following a contact, there is a p_i probability of infection. The likelihood thus of a contact not resulting in an infection is $1 - p_i$. The likelihood that our susceptible person is infected after Δ_t time is

$$p_{S,I} = 1 - (1 - p_i)^{nIt}. \tag{2.1}$$

If we substitute β for $-n\log(1 - p_i)$ and insert it into (2.1), we get

$$p_{S,I} = 1 - e^{-\beta It}. \tag{2.2}$$

The limit of this expression is then

$$\lim_{t \to 0} \frac{1 - e^{-\beta It}}{t} = \beta I. \tag{2.3}$$

The above is often referred to as the force of infection, λ.

Definition 2.9 (Force of infection). The **force of infection** λ is the likelihood of infection per unit time. It is often also expressed as the rate per capita of acquiring the infection.

Since the force of infection describes the *per capita* rate of acquiring the infection, we must multiply this by the amount of susceptibles S to obtain the rate of new infections. This term will have a negative sign, since it is a flow out of the pool of susceptible individuals and into the pool of infectious ones:

$$\frac{dS}{dt} = -\underbrace{\lambda}_{\text{force of infection}}\underbrace{S}_{\text{susceptibles}}. \tag{2.4}$$

Since $\lambda = \beta I$, (2.4) becomes the mass action term

$$\frac{dS}{dt} = -\underbrace{\underbrace{\beta I}_{\text{force of infection}}\underbrace{S}_{\text{susceptibles}}}_{\text{mass action}}. \tag{2.5}$$

By convention, we write this in the order βSI. Beyond its mathematical importance, the mass action term tells us something important about infectious diseases, namely that the likelihood of their spread, under the assumption of homogeneous mixing, depends on the relative sizes of the infectious and the susceptible populations.

- There can be no epidemic without any infectious individuals; this is the idea at the heart of eradication and quarantines.

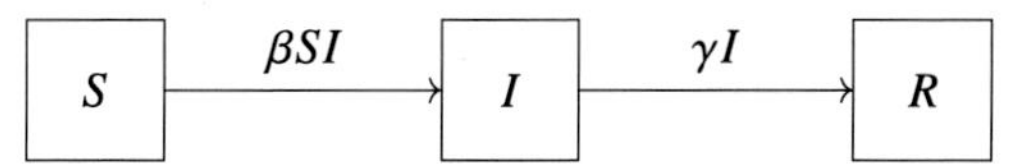

Figure 2.2 Flow diagram of a basic SIR model. β is the transmission coefficient, a term that accounts for the frequency of interactions between susceptible and infectious individuals. γ is the removal rate; it reflects the rate at which infectious individuals become no longer infectious.

- There can also be no epidemic without any susceptible individuals, that is, the basic driving idea of vaccination and pre-exposure prophylaxis.
- Finally, there can be no epidemic without mixing of populations, which is the principle that underlies social distancing.

The principle of mass action provides a mathematical way of interpreting and analyzing the touchpoint between infectious and susceptible populations. As such, it is the cornerstone on which much of our subsequent compartmental modeling is built.

2.1.4 A basic SIR model

The simplest of all compartmental models is the SIR model for a closed population (i.e., a population with no demographic change). A flow diagram of such a model is presented in Fig. 2.2.

Definition 2.10 (Constant and closed populations). In a model with a **constant population**, the total population is a constant and always equal to the sum of the population of all compartments. Formally, in a closed population with n compartments $C_{1...n}$, it holds for all t that

$$\frac{dN}{dt} = \sum_{i=1}^{n} \frac{dC_i}{dt} = 0.$$

A population is closed if there are no inflows and no outflows.

This model is conditioned by two key parameters: the transmission coefficient β, which reflects the likelihood that an encounter between a susceptible and an infectious individual results in transmission, and the removal rate γ, which accounts for the rate at which infectious individuals are removed, i.e., become immune, deceased or otherwise inaccessible to the pathogen. Their relationship is encapsulated in the system of differential equations that represents SIR models:

$$\begin{aligned}
\frac{dS}{dt} &= -\underbrace{\beta S I}_{\text{mass action}}, \\
\frac{dI}{dt} &= \underbrace{\beta S I}_{\text{mass action}} - \underbrace{\gamma I}_{\text{recovery}},
\end{aligned} \tag{2.6}$$

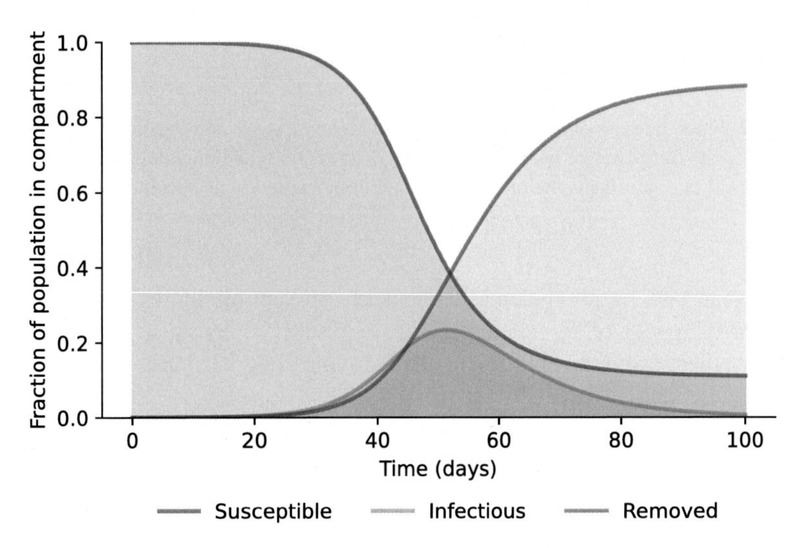

Figure 2.3 Solution of a typical SIR model with $\mathfrak{R}_0 = 2.5$ and $\tau = 8$.

$$\frac{dR}{dt} = \underbrace{\gamma I}_{\text{recovery}} .$$

In the following, we are going to examine, first, the assumptions underlying the SIR model, followed by its key relationships, and concluding in analytical and numerical approaches.

Computational Note 2.2 Solving ODEs in Python

Solving an initial value problem for a system of ODEs in Python generally proceeds in two stages. First, we need to describe the ODE as a function, then pass it onto the solver. The way of specifying a system of ODEs as a function might appear somewhat idiosyncratic at first. The overarching idea is that the function defining the ODEs will be called repeatedly with its own output passed to it as a tuple. For this reason, we will have to destructure the tuple input from the previous iteration.

Let us consider the system of differential equations from (2.6). As we transpose this to a Python function (by convention, the function describing the ODE is called `deriv`), we must consider this destructuring. The solver passes at least two arguments to the function describing the ODE. The first of these (again by convention, denoted `y`), is a tuple containing the results of the previous iteration, which we will have to break down to be reused. The differential equation for a simple SIR model would thus be described as follows:

```
def deriv(t, y, beta, gamma):
    S, I, R = y
    dSdt = -beta * S * I
    dIdt = beta * S * I - gamma * I
    dRdt = gamma * I
    return dSdt, dIdt, dRdt
```

A notebook implementing the contents of this Computational Note is available on the book's companion Github repository in the folder `/ch02/sir_models`.

The simple SIR model makes four fundamental assumptions:

1. **Static (or closed) population**: $\mathcal{S}+\mathcal{I}+\mathcal{R}=N$ and consequently, $\frac{dS}{dt}+\frac{dI}{dt}+\frac{dR}{dt}=0$. This holds for $\frac{d\mathcal{S}}{dt}$, $\frac{d\mathcal{I}}{dt}$ and $\frac{d\mathcal{R}}{dt}$, too.
2. **Homogeneous mixing**: every member of S has the same likelihood (namely, SI) to encounter someone in I.
3. **Equal outcomes**: there is a single outcome of infection, and every infected person is equally likely to experience that outcome. This outcome is represented by the γ term.
4. **Survival confers immunity**: it is assumed in this model that a removed person gains life-long immunity (or dies, which from the perspective of viral dynamics is altogether not all that different).

Definition 2.11 (Homogeneous mixing). Under **homogeneous mixing**, the likelihood of infection is identical for every infectious and every susceptible individual.

Let S be the set of susceptible individuals $S_{1\ldots m}$ and I the set of all infectious individuals. S and I are said to be homogeneously mixed if the likelihood of encounter between any randomly selected members of I and S are equal, namely SI.

These assumptions are, of course, broad and sweeping generalizations that do not quite align with reality. Mixing, in practice, is far from homogeneous; we do not, in any meaningful sense, encounter every other individual in our population with equal likelihood. However, at a large enough scale, the actual patterns of encounters we do experience are adequately approximated by the homogeneity assumption. Thus the SIR model is a good first approximation of complex realities, and a good first step towards mathematically interpretable models of infectious disease.

Computational Note 2.3 Phase portraits

A useful tool to characterize the population approaching equilibrium is the phase portrait. A phase portrait is a quite simple plot of two compartment values, typically S and I, against each other. The following snippet plots a phase portrait from the output of the ODE solver:

```
fig = plt.figure(facecolor="w", figsize=(4, 4))
ax = fig.add_subplot(111, axisbelow=True)

ax.plot(S, I)
ax.set_ylabel("Fraction of infectious")
ax.set_xlabel("Fraction of susceptible")

plt.show()
```

The phase portrait can be understood as a view of how a dampened oscillation eventually converges upon an equilibrium. Fig. 2.4 shows the phase portrait of the SIR model we solved in Computational Note 2.2. The infectious process starts from the right lower corner and gradually converges on the equilibrium point. This type of phase portrait, with a gradually decreasing radius from a stable equilibrium point, is sometimes referred to as a spiral sink and suggests a process gradually converging on an equilibrium.

A notebook implementing the contents of this Computational Note is available on the book's companion Github repository in the folder `/ch02/phase_space`.

2.1.5 The basic reproduction number $\mathfrak{R}_0$

As we have noted above, our basic SIR model has two main parameters: β, which is associated with the rate at which new cases are created, and γ, which is associated with the rate at which individuals cease to be infectious.

- If $\beta > \gamma$ ($\frac{\beta}{\gamma} > 1$), the number of infections will increase.
- If $\beta = \gamma$ ($\frac{\beta}{\gamma} = 1$), a case is lost by way of recovery for each new case gained by way of infection. The number of infectious individuals remains constant.
- If $\beta < \gamma$ ($\frac{\beta}{\gamma} < 1$), more cases are lost by recovery than are gained through infection. The number of infectious will decrease and the infection will die out.

Thus the ratio between β and γ characterizes the long-term destiny of the pathogenic process. The ratio of β to γ describes the number of new cases that an infectious case produces, under the assumption that the entire population is immunologically naive (i.e., everyone except the index infectious patient is susceptible), it is a dimensionless quantity that characterizes the infectiousness of a pathogen. We term this fraction the basic reproduction number, represented by $\mathfrak{R}_0$ (pronounced "r-nought" and not to be confused with $R(0)$). This relationship between β, γ (by way of its inverse, τ) and $\mathfrak{R}_0$ is illustrated in Fig. 2.5. Table 2.1 lays out the estimated $\mathfrak{R}_0$ values for some common pathogens.

Definition 2.12 (The basic reproduction number $\mathfrak{R}_0$). The **basic reproduction number** $\mathfrak{R}_0$ is the expectation of cases produced by an infectious case in a population, where every other individual is assumed to be susceptible.

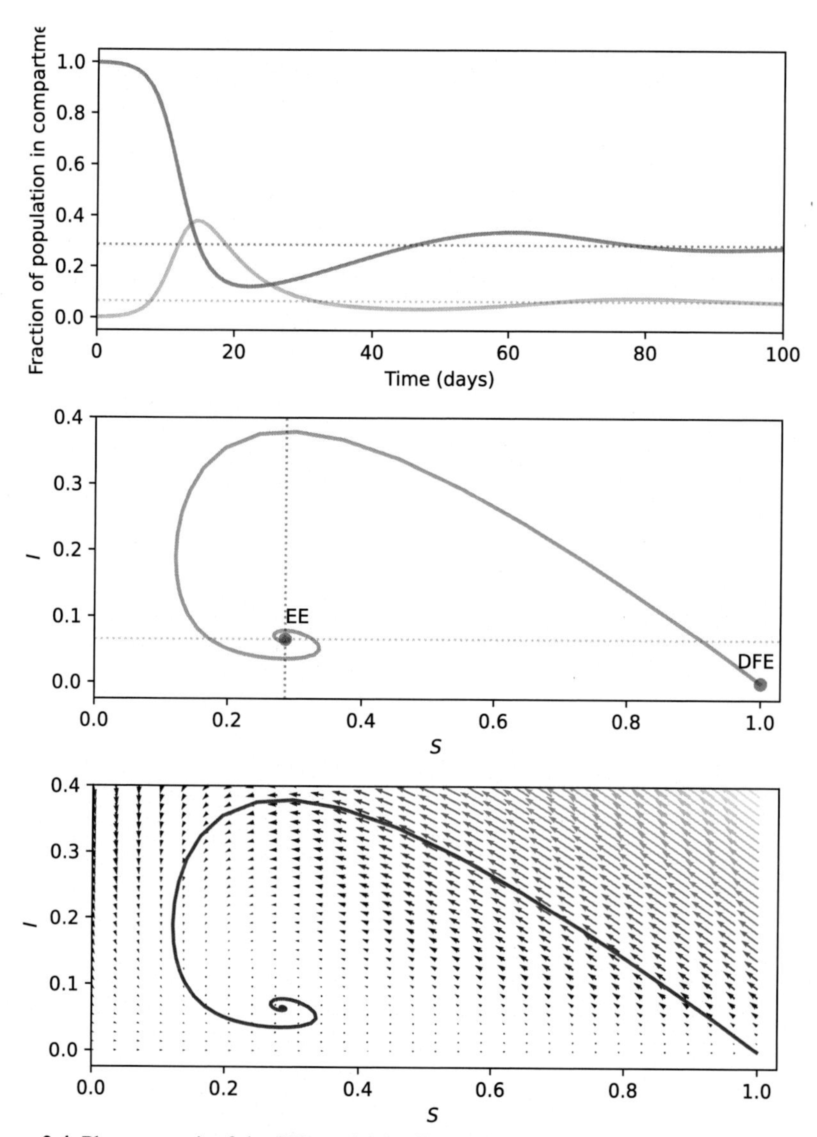

Figure 2.4 Phase portrait of the SIR model for $\mathfrak{R}_0 = 3.5$, $\tau = 5$ and $\mu = 0.02/\text{year}$. The red and blue dotted lines denote the S and I nullclines, respectively.

As a first approximation, it may be helpful to consider the forces governing $\mathfrak{R}_0$. Simply put, $\mathfrak{R}_0$ is the number of infections created during a person's infectious state We may thus conceptualize $\mathfrak{R}_0$ as proportional to the product of the encounter rate (encounters per unit time), the transmission ratio (transmissions per encounter), and the mean infectious period τ:

$$\mathfrak{R}_0 \propto \frac{\text{encounters}}{\text{time}} \times \frac{\text{transmission}}{\text{encounters}} \times \tau. \tag{2.7}$$

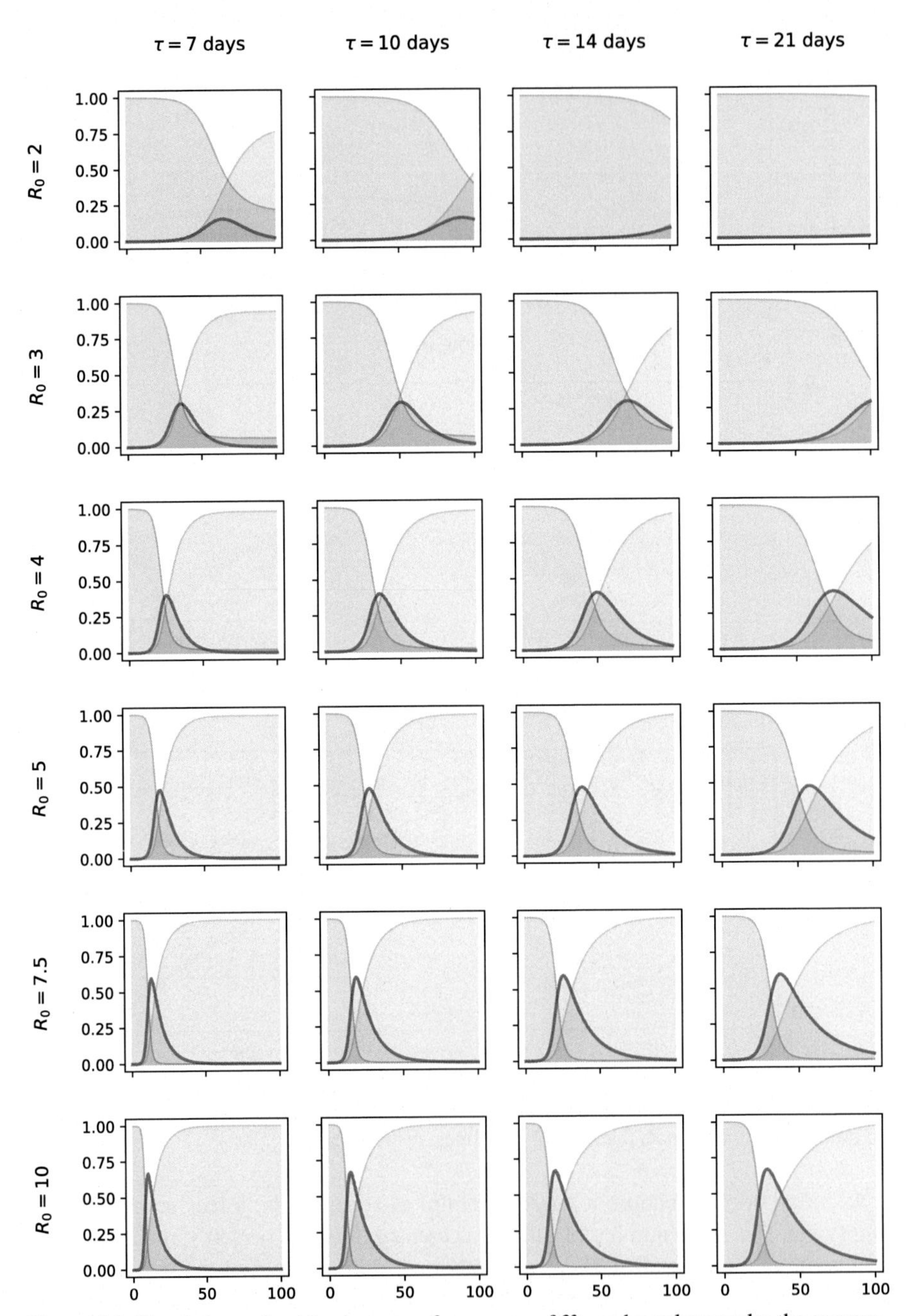

Figure 2.5 Simulations of epidemic curves for a range of $\mathfrak{R}_0$ and τ values under the assumption of a wholly susceptible population with 0.01 percent initial infected.

Table 2.1 $\mathfrak{R}_0$ values of some well-known pathogens.

Pathogen	$\mathfrak{R}_0$	
Measles	12.0–18.0	Guerra et al. [45]
Mumps	4.0–10.0	Deeks et al. [46]
Smallpox	3.0–5.0	Costantino et al. [47]
HIV	3.6–4.7	Nishiura [48]
COVID-19, Delta variant	2.3–7.0	Liu and Rocklöv [49]
COVID-19, ancestral strain	1.4–3.1	Billah, Miah, and Khan [50], Mallela et al. [51]
Zaire ebolavirus	1.4–1.8	Khan et al. [52]
Influenza A, H1N1, recent	1.4–1.6	Coburn, Wagner, and Blower [53]
Influenza A, H1N1, 1918–19	1.4–2.8	Coburn, Wagner, and Blower [53]
Nipah virus	0.4–0.5	Royce and Fu [54]
Nipah virus, swine	5.0	Wongnak et al. [55]

We may best derive $\mathfrak{R}_0$ mathematically by looking at a single infected individual. This person has, of course, the likelihood β to create new infections (once again, assuming a naive population). Since we are interested not in a likelihood at a point in time to create new cases, but rather in the overall likelihood of a person creating new infections throughout their infectious state, we must multiply this figure by τ, the mean infectious period. This, we note, is the inverse of γ, since γ is the rate at which people are removed from the infectious (I) compartment. Consequently,

$$\mathfrak{R}_0 = \beta\,\tau = \beta\gamma^{-1} = \frac{\beta}{\gamma}. \tag{2.8}$$

2.1.6 The time-dependent reproduction number $\mathfrak{R}_t$

One of the urban legends of epidemiology is that the zero-index in $\mathfrak{R}_0$ exists to differentiate it from the compartment R. This, alas, is indeed an urban legend, but what exactly *is* zero in $\mathfrak{R}_0$?

There are two schools of thought on $\mathfrak{R}_0$ and $\mathfrak{R}_t$. One holds that $\mathfrak{R}_0$ is a theoretical figure, whereas $\mathfrak{R}_t$ is empirically ascertained. The other, perhaps somewhat more accurate, notion is that $\mathfrak{R}_0$ is the reproduction number at "time zero," that is, in the presence of a large pool of susceptibles, a single infectious individual and no recovered/resistant population.

$\mathfrak{R}_0$ is often explained as the number of secondary infections that flow from each case on average. Strictly speaking, that is not true. $\mathfrak{R}_0$ is the number of secondary infections that *would* flow from a case if the whole population were susceptible, except for the lone infectious index case. Because in the initial phases of an outbreak, the population of susceptible individuals is vastly larger than that of the infected, we may dispense with this strict formalism. However, the dynamics of an outbreak begin to follow more complex trajectories over time, and the $\mathfrak{R}_0$ notion becomes less viable.

Enter $\mathfrak{R}_t$, the time-dependent reproductive number. This figure is almost always used in the context of the empirical ascertainment of the reproduction number (on which see Section 2.5 in more detail).

Definition 2.13 (The time-dependent reproductive number $\mathfrak{R}_t$). The **time-dependent reproductive number** $\mathfrak{R}_t$ (also known as the **effective reproductive number** and sometimes abbreviated R_e) is the expectation of cases produced by an infectious case in a population at time t.

One might helpfully conceive of $\mathfrak{R}_t$ as the reproduction rate given the state of the system at time t. It thus reflects the system "as things are" at t, given $S(t)$ and $I(t)$. If a system has $I(t)$ infectious individuals at time t and $I(t+\epsilon)$ after a short period of time ϵ, then the number of cases created by the cases in existence at t is $I(t+\epsilon) - I(t)$. From this, it follows that

$$\mathfrak{R}_t = \lim_{\epsilon \to 0} \frac{\overbrace{I(t+\epsilon) - I(t)}^{\text{new cases over } \epsilon}}{R(t+\epsilon) - R(t)}. \tag{2.9}$$

Contact tracing efforts can often provide relatively accurate estimates of cases by generation, which allow for accurate calculation of $\mathfrak{R}_t$. Note, however, that such figures often do not take reporting lags into consideration, and the actual time to which a value of $\mathfrak{R}_t$ is related may be at variance with the time it was calculated. However, if the serial interval τ is known, $\mathfrak{R}_t$ can be calculated from a simple time series of infectious cases. The practical application of this is discussed in Subsection 2.5.4.

Computational Note 2.4 Setting initial parameters

After defining the function, we will need to set the initial parameters. For this example, we will set the initial proportion of infectious persons at 0.01 percent, and assume that everyone who is not infected is susceptible. Then, we define $\mathfrak{R}_0$ at 2.5 and τ at 8 days. From this, we obtain γ as τ^{-1} and β as $\mathfrak{R}_0\gamma$.

```
I_0 = 0.0001
S_0 = 1 - I_0
R_0 = 0

y_0 = (S_0, I_0, R_0)

R0 = 2.5
tau = 8
gamma = 1/tau
beta = R0 * gamma
```

Now, we can invoke the solver. For the purposes of this demonstration, we will set the interval of integration (`t_span`) at $0 \to 100$ and the maximum step size (`max_step`) to 1; in other words, we want a calculation for at least every day.

You do not have to specify `max_step`; the solver will try to guess an appropriate value. However, this typically is fairly parsimonious and the resulting curves can look quite choppy.

```
res = solve_ivp(fun=deriv,
                t_span=(0, 100),
                y0=y_0,
                args=(beta, gamma),
                max_step=1)
```

The solver returns an `OdeResult` object. This provides several attributes, the most important of which are the following:

1. `success: True` indicates that the solver managed to resolve the integration.
2. `t` is an array with the points at which integration was performed.
3. `y` is a nested array (array of arrays), containing one array each for S, I and R.

We may now plot this result using our plotting tool of choice. Fig. 2.3 displays the result. This representation is, of course, population-agnostic. It is expressed in terms of a fraction of N, rather than real values of susceptible, infectious, and recovered patients. It is not difficult, however, to integrate those figures at the population level or at the level of plotting.

A notebook implementing the contents of this Computational Note is available on the book's companion Github repository in the folder `/ch02/sir_ models`*.*

2.1.7 Threshold conditions of epidemics

Fig. 2.3 shows us the size (specifically, in terms of proportion of the entire population) of each compartment as plotted against time, for a given value of $\mathfrak{R}_0$ and τ. In Fig. 2.4, we represent each value of $S(t)$ against the corresponding value of $I(t)$, a representation known as a phase portrait; it is discussed in more detail in Computational Note 2.3. Each of the lines refers to the same values of $\mathfrak{R}_0$ and γ, but different starting conditions in terms of starting values $S(0)$ and $I(0)$. We denote this family of curves as the characteristic curves of the SIR model with $\mathfrak{R}_0 = 2.5$ and $\tau = 8$.

Definition 2.14 (Characteristic curves). The characteristic curves of a model refer to the family of curves for given values of $\{\beta, \gamma, \cdots\}$ in dependence of different values of $\{S(0), I(0)\}$.

Note that regardless of the starting conditions, as long as the dynamic parameters are identical, the maxima of the family of curves will lie at $S = \frac{1}{\mathfrak{R}_0} = \frac{\gamma}{\beta}$. Behind this intuition is a crucial mathematical relationship that governs the shape of an epidemic.

The phase plane (the S/I plane in which the phase portrait is drawn) allows us to express $\frac{dS}{dt}$ and $\frac{dI}{dt}$ as a single equation, namely

$$\frac{dI}{dS} = \frac{-\overbrace{\beta SI}^{\text{new infections}}}{\underbrace{\beta SI}_{\text{new infections}} - \underbrace{\gamma I}_{\text{recoveries}}} = \frac{\gamma}{\beta S} - 1 = \frac{\mathfrak{R}_0^{-1}}{S} - 1. \tag{2.10}$$

In this equation, which is sometimes described as the phase portrait of the model, $\mathfrak{R}_0^{-1}$ is the threshold number for the parameters governing the model.

What we are in particular interested in is what happens around $\mathfrak{R}_0^{-1}$. Beginning with an almost entirely susceptible population ($S(0) \approx 1$), the ratio of infected versus susceptibles rises rapidly as individuals transfer from the susceptible to the infectious population. This persists until a point in time t_n, when $S(t_n) = \mathfrak{R}_0^{-1}$. Thereafter, the number of infectious individuals begins to decline sharply, leading to the eventual extinction of the infection.

Characteristic curves convey three fundamental truths about infectious processes. First, it explains a phenomenon called "epidemic burnout." Recall that we derived the mass action term from the fact that new infections require an infectious and a susceptible individual to meet. This is governed by the product of their relative proportions, and a constant (the coefficient of transmission, β). For a SIR model or some other model, where infectious cases do not revert to susceptibility (or do so with a significant delay), the infection eventually runs out of susceptible individuals and dies out. This receding stage of the infection begins once $S(t)$ reaches the critical value of $\mathfrak{R}_0^{-1}$.

Second, it indicates that an infection past that critical point is necessarily in decline. A consequence is that $S(0) < \mathfrak{R}_0^{-1}$, $I(t)$ will monotonically converge on zero. In other words, if $S(0)\mathfrak{R}_0 < 1$, then the corresponding $I(0)$ will be a global maximum, and the infection will monotonically decline.

Finally, and perhaps most importantly to us in our role as epidemiologists, is the fact that the time t_n at which the maximum is reached is a function of $S(0)$, β and γ:

- Interventions that reduce $S(0)$, such as vaccination, reduce both the duration of the overall outbreak and, if it is reduced below $\mathfrak{R}_0^{-1}$ (the threshold of collective immunity, on which Chapter 6 provides detail), outbreaks cannot sustain themselves.
- Interventions that reduce β, such as quarantine or social distancing, shift the critical point to the right, reducing the time during which the outbreak can grow.
- Interventions that increase γ by shortening the mean recovery time τ, such as early treatment with antimicrobials, reduces the overall extent of the disease.

An elegant analytical derivation of this is to be found in Li [41], but for our purposes, the principal takeaway is that $\mathfrak{R}_0$ does not only deal with the progress of an outbreak but also with the very fact of whether it will ever come to pass, in its relationship to $S(0)$. Thus an outbreak requires two indispensable factors:

- $I(0) > 0$, that is, an initial (or "seed") population of infectious individuals, and
- $S(0) > \mathfrak{R}_0^{-1}$, that is, a suitable susceptible population through which to spread.

Together, these constitute the threshold criteria of epidemics, and the most remarkable results of Kermack and McKendrick's model of infectious disease.

2.1.8 Density-dependence, frequency-dependence, and other factors

The standard form of the SIR model discussed in Subsection 2.1.4 assumes that the rate of encounters ϱ is governed by the relative densities of susceptible and infectious individuals. This assumption holds for many diseases, but not all of them:

- In **frequency-dependent** transmission, the encounter rate is not governed by population density, i.e., population density does not principally govern the encounter rate.
- In **hybrid** transmission, the interactions are nonlinear, i.e., neither purely frequency-dependent nor purely density-dependent [56].

Definition 2.15 (Frequency and density dependence). **Frequency-dependent transmission** obtains where $\varrho \propto \mathcal{S}\ \mathcal{I}$. **Density-dependent transmission** obtains where $\varrho \propto S\ I$. **Hybrid** (or **intermediate**) **transmission** occurs where $\varrho \propto f(\mathcal{S}, \mathcal{I}, \mathcal{N})$, where f may be a nonlinear function of relative densities.

It is worth noting that even for frequency-dependent transmission, density is often a secondary but nonetheless influential factor. Frequency dependence suggests that encounters are at a constant rate. This recapitulates the kind of dynamics where individuals seek out encounters from an available supply. For instance, for sexually transmitted infections, there is generally a large available supply and the number of encounters is governed by an individual's rate of sexual encounters over time. For this reason, many STIs, such as HIV, are considered frequency-dependent. However, at and beyond a critical low level of density, frequency-based processes become governed nonetheless by density, where the density is so low as to make it impossible for an individual to seek out the preferred volume of encounters.

Definition 2.16 (Hard critical point of frequency dependence). A frequency-dependent process becomes governed by density if and only if $S\ I < (S + I)\varrho^{-1}$, that is, where the density of available contacts makes it impossible for a frequency-dependent agent to maintain the specific constant number of encounters.

It is helpful to expand this to the situation where $S\ I > (S + I)\varrho^{-1}$, but approaches it. The cost of seeking out another individual to contact is a monotonously increasing function $J(S, I)$ of S and I. As SI converges to $(S + I)\varrho^{-1}$, the cost of obtaining ϱ^{-1} contacts reaches a level beyond what an individual may be able to afford. For instance, an individual's attempts to find ϱ^{-1} sexual partners will be frustrated if finding each partner requires a significant investment of time, energy, or risk from predators. Thus there is a soft critical point of frequency dependence that lies above the hard critical point.

Definition 2.17 (Soft critical point of frequency dependence). A frequency-dependent process tends towards being governed by density as $\varrho^{-1}J(S, I)$ approaches the individual's capacity to bear the costs of pursuing the encounters.

Thus pure frequency-dependent transmission does not really exist; after a critical point, density governs even ordinarily constant-frequency processes. In most cases, frequency-dependent dynamics converge towards a hybrid model that transitions from frequency dependence to density dependence as SI converges the critical point.

2.1.9 Peak severity of epidemics

To those in charge of responding to an epidemic, from disaster management professionals through hospitals to governmental authorities, it is not merely the overall number of cases over the epidemic's lifetime that matters, but also how that number is distributed over time. Rapid epidemic spikes can overwhelm healthcare capacities and cause excess mortality from lack of care, both among the victims of the epidemic and from unaffected individuals who cannot avail themselves of adequate care due to acute overload of the hospital system. For this reason, we are interested in the maxima that we have explored previously in Subsection 2.1.10. We noted there that in an epidemic, $I(t)$ will approach a maximum, where $S = \mathfrak{R}_0^{-1}$. The value of this maximum will be

$$\max I = I(0) + S(0) + \mathfrak{R}_0^{-1}(\ln \mathfrak{R}_0^{-1} - 1 - \ln S(0)). \tag{2.11}$$

Since $\mathfrak{R}_0 = \beta S(0)\gamma^{-1}$, we can estimate maxima based on initial conditions and $\mathfrak{R}_0$. In practice, it may often be quite difficult to know the number of infectious populations at the outset. However, in many cases, $I(0)$ will be vanishingly small compared to $S(0)$. In such situations, well-informed guesswork yielding results that are accurate to an order of magnitude tend to be sufficient. Alternatively, as long as it is quite early in the outbreak (i.e., the number of infectious cases appears to follow an exponential increase), it is possible to simply set the first available data point to represent $S(0)$ and $I(0)$.

Practice Note 2.1 A caveat about counting cases

It may be tempting to assume that the number of infected individuals is a knowable fact, and thanks to the advances of electronic medical records and integrated public health reporting systems, we might be closer to this than ever. At the same time, not all infectious patients come to the attention of healthcare services. This may be because of socio-economic factors, or because their symptoms are nonspecific or subclinical. For the former, it is often possible to compare local or regional presentation rates with the national average to estimate the relative likelihood of a symptomatic individual not seeking medical care. For the latter, serosurveys can be correlated against self-reported symptoms to calculate the likelihood of asymptomatic illness. Since asymptomatic individuals may be infectious, but also do not constitute part of the susceptible population, it is crucial to ascertain the ratio of asymptomatic infection vis-a-vis all infections. This may not always be a trivial exercise, as the intense controversy around the proportion of asymptomatic infections during the early days of the COVID-19 pandemic has demonstrated. Nevertheless, estimates that might not be perfectly accurate are much better than no estimates at all.

2.1.10 The final size of epidemics

The final size of epidemics depends on the epidemic model. It is typical to specify these in an implicit manner, which makes them analytically unsolvable, but which can be numerically estimated. Recall that the phase portrait of I and S (see Eq. (2.10)) can be described as

$$\frac{dI}{dS} = \frac{\mathfrak{R}_0^{-1}}{S} - 1. \tag{2.12}$$

Consequently, we can take I as a function of S by taking the indefinite integral of the above, which yields

$$\begin{aligned} I(S) &= \int \frac{\mathfrak{R}_0^{-1}}{S} - 1. \\ &= \mathfrak{R}_0^{-1} \ln S - S + c \end{aligned} \tag{2.13}$$

We can express the constant of integration as a function of S and I, namely

$$\psi(S, I) = S + I - \mathfrak{R}_0^{-1} \ln S. \tag{2.14}$$

Because N is constant, this has the same solution for every pair of S and I at any time. Consequently, for any $S(t)$ and $I(t)$,

$$\psi(S(t), I(t)) = \psi(S(0), I(0)). \tag{2.15}$$

We assume that $I(0)$ is vastly smaller than $S(0)$, and can be ignored. Therefore

$$S(\infty) - \mathfrak{R}_0^{-1} \ln S(\infty) = S(0) - \mathfrak{R}_0^{-1} \ln S(0), \tag{2.16}$$

which simplifies to the implicit transcendental equation for the final size of the epidemic,

$$\mathfrak{R}_0\left(1 - \frac{S(\infty)}{S(0)}\right) = \ln S(0) - \ln S(\infty). \tag{2.17}$$

The same logic can be applied to any more complex models. An alternative is the model proposed by Heesterbeek and Dietz [57], which we also use in Subsubsection 2.5.3.3 in the inverse to estimate $\mathfrak{R}_0$. Consider a random member of the population. The likelihood that this individual will be susceptible at $t = 0$ is, of course, $S(0)$. At any given time, the likelihood that this individual sustains infection is $\frac{\mathfrak{R}_0}{N}$.

Then, the likelihood that this individual will be infected at some point over the course of the epidemic,

$$I(\infty) = S(0)\left(1 - \frac{\mathfrak{R}_0}{N}\right)^{I(\infty)N}. \tag{2.18}$$

For a sufficiently large N,

$$I(\infty) = \lim_{N\to\infty} S(0)\left(1 - \frac{\mathfrak{R}_0}{N}\right)^{I(\infty)N} = 1 - S(0)e^{-\mathfrak{R}_0 I(\infty)}. \tag{2.19}$$

Assuming $I(0)$ is quite small, $S(0)$ is approximately 1, and consequently, Eq. (2.19) becomes

$$I(\infty) = 1 - e^{-\mathfrak{R}_0 I(\infty)}. \tag{2.20}$$

In more complex models, deducing $I(\infty)$ from the likelihood of a notional individual's infection by the time the infectious process has concluded is often easier than the classical approach using the phase portrait.

Practice Note 2.2 Population turnover

An inconvenient result of reality's staunch refusal to conform to the best of mathematical models is population turnover. People die and others are born, and statistically, a person being born is much more likely to be susceptible to a disease than a person dying, partly because absent maternal (vertical) immunity, newborns are generally immunologically naive, and partly because people are likely to die at older ages, which also correlates to the (time-dependent) lifetime likelihood of exposure to the pathogen. For this reason, counting on epidemic burnout to end an outbreak is rarely a wise idea. For very low population turnover and low likelihood of death from disease, counting on epidemic burnout to exhaust the pathogen's host population may be part of a multilimb strategy. However, population turnover makes reliance on epidemic burnout alone to be infeasible above an $\mathfrak{R}_0$ of approximately 5 in human populations, or where the disease has serious mortality risks or economic or social costs.

2.1.11 *Epidemic burnout*

The phenomenon of epidemic burnout is by far one of the most perplexing features of epidemic processes. We commonly talk of susceptibles as a vital "resource" for a pathogen, without which it cannot persist. This, and intuition, may give rise to the perception that epidemics "burn through" a population. This is correct, insofar as the absence of any susceptibles is a sufficient condition for the end of an infectious process. It is, however, incorrect in that it is not a necessary condition.

Let us consider Eq. (2.6), and divide the first and third equations:

$$\frac{dS}{dR} = \frac{-\beta S}{\gamma} = -\mathfrak{R}_0 S. \tag{2.21}$$

Integrating this expression, we obtain

$$S(t) = S(0)e^{-R(t)R(0)}. \tag{2.22}$$

We know that $e^{-R(t)\mathfrak{R}_0}$ is always positive. Since R may, due to the partition property ($S + I + R = 1$), assume only values in the domain [0, 1], $S > e^{-\mathfrak{R}_0}$, which itself is above zero.

In biological terms, this means that for any t, there will be, at the end of the epidemic, $e^{-\mathfrak{R}_0}$ individuals who escape infection. Or, in other terms, epidemics cease, because they run out not of susceptibles but infected individuals. This is the phenomenon of epidemic burnout.

2.2 Modeling mortality and vital dynamics

So far, we have looked at models where the sum of all compartment cardinalities equaled a fixed quantity N. In this section, we are beginning to look at models in the space of vital dynamics: births and deaths.

Definition 2.18 (Vital dynamics). The **vital dynamics** (also termed population dynamics) of a population comprise changes in the population size, namely

- births,
- deaths,
- immigration, and
- emigration.

2.2.1 Case study: births and deaths in HIV

Where an infection lasts for a long time, births and disease-unrelated deaths become relevant for our calculations. As we move away from the closed-system models of the previous sections, we encounter the issue of allocation: we may know something about the growth or decrease of a population (births and deaths, correspondingly), but we need to identify which compartments these each fall into.

Definition 2.19 (Vertical transmission). Vertical transmission refers to the transmission of a pathogen to an offspring *in utero* (transplacental infection) or at time of birth.

Given that vertical transmission is more the exception than the rule, it is particularly prevalent in the case of certain viral infections. Vertical transmission is thus the reason why we cannot always assume that all births will fall into the immunologically naive

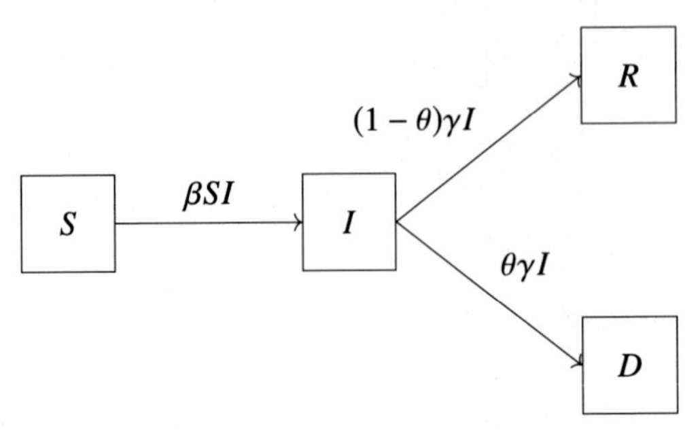

Figure 2.6 A multioutcome SIR model with mortality. θ is the mortality fraction, which separates the deceased (D) from the recovered (R).

susceptible category. The reverse of this situation is, of course, maternal antibody-based immunity, which for structure's sake will be discussed with other models of immunity, in Subsection 2.3.4.

Human immunodeficiency virus infection is an example for a long-term infectious condition that is affected both by births and deaths. Thanks to modern antiretroviral therapy, the lifespan of patients with HIV has increased significantly, with many of them never progressing to full-blown AIDS. Consequently, a good proportion of them die with, rather than from, the disease, and population dynamics allows us to model that process. At the same time, vertical transmission of HIV is documented. This may be transplacental or occur at the time of birth. A recent survey by Forbes et al. found that vertical transmission occurs in about 3% of HIV positive mothers, and highly active antiretroviral therapy (HAART) reduces this to 1% or below [58].

In this section, we will examine the way vital dynamics affect a population. We will differentiate between the ordinary case of immunologically naive births, vertical transmission, and the effects of disease on fecundity; the special case of being born with a degree of immunity (maternal immunity) is discussed in Subsection 2.3.4.

2.2.2 A multioutcome SIR model: SIRD

Our first refinement of the SIR model will be by introducing differential outcomes (Fig. 2.6). This will be the foundation of later models, where these differential outcomes feed back into the system. For instance, in models that account for lapsing immunity, it is important to keep the deceased (who will never return to the pool of susceptibles) separate from the recovered (who will).

By far the most widely accounted-for differential outcome is short-term mortality. Fig. 2.7 shows a potential multioutcome model, in which after the course of illness, a fraction of patients recover, whereas others succumb to their illness. The mortality fraction θ denotes the fraction of removed cases that are allocated to the deceased (D) compartment. This is from time to time referred to as the “mortality rate”; it is incorrectly so, since a rate is a time-denominated value. In the long run, θ equals the case-fatality ratio (CFR) of the infection.

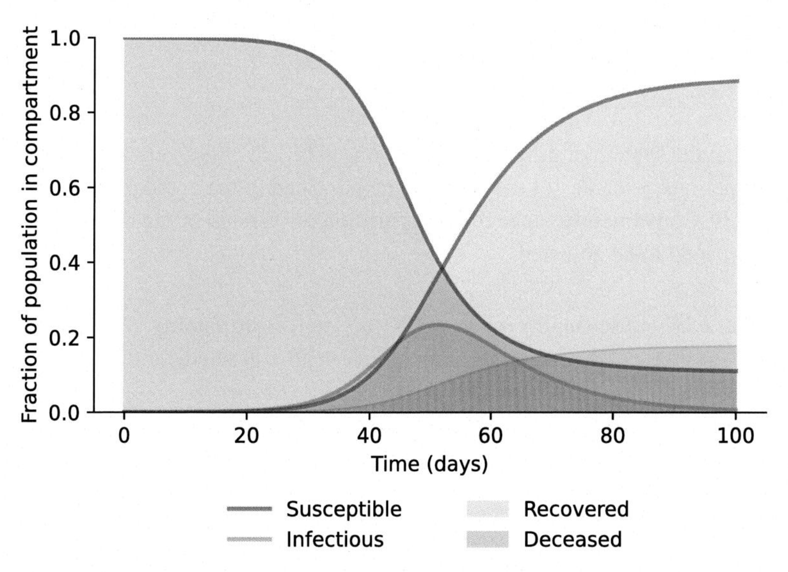

Figure 2.7 Solutions of a typical SIRD model with $\mathfrak{R}_0 = 2.5$, $\tau = 8$ and $\theta = 0.2$.

The practical utility of a SIRD model is primarily as a building block for more complex models, in which recovered populations exhibit different behaviors from the deceased (typically, by virtue of being alive). A classical example of this is waning immunity; decedents cannot revert to the susceptible pool, but recovered immune patients can. Similarly, SIRD models are useful in modeling differential infection-fatality or case-fatality ratios between subpopulations, e.g., the characteristically higher mortality of the elderly in most infectious diseases. Finally, in models of funerary transmission (see Subsection 2.2.7), decedents are—for a certain period —sources of infection, and thus part of the mass action term.

2.2.2.1 Governing equation

Just as the flow diagram of the SIRD model is quite similar to that of the simple SIR model outlined in Subsection 2.1.4, the system of differential equations that govern such circumstances is quite similar, too.

$$
\begin{aligned}
\frac{dS}{dt} &= -\underbrace{\beta S I}_{\text{mass action}}, \\
\frac{dI}{dt} &= \underbrace{\beta S I}_{\text{mass action}} - \underbrace{\gamma I}_{\text{removals to } R \text{ and } D}, \\
\frac{dR}{dt} &= \gamma \underbrace{(1-\theta) I}_{\text{surviving part}},
\end{aligned}
\tag{2.23}
$$

$$\frac{dD}{dt} = \gamma \underbrace{\theta I}_{\text{decedent part}} .$$

Note that the system of differential equations in (2.23) effectively apportions γ into the two fractions, $\gamma\theta$ and $\gamma(1-\theta)$, which are allocated to the compartments D and R, respectively. Conveniently, since this is a partition of γ, none of the other equations in the system need to be adjusted.

Definition 2.20 (Case-fatality ratio (CFR) and infection-fatality ratio (IFR)). The **infection-fatality ratio** IFR_t is the number of deaths divided by the number of infections at time t.

$$\text{IFR}_t = \frac{D(t)}{\int_0^t I(u)\, du} \tag{2.24}$$

The idealized value of IFR is θ.

The **case-fatality ratio** CFR_t is the number of deaths divided by the number of recognized cases.

$$\text{CFR}_t = \frac{D(t)}{\int_0^t I(u) r\, du}, \tag{2.25}$$

where r is the ratio of recognized cases divided by all infections. Therefore $\text{IFR}_t = \text{CFR}_t\ r$.

2.2.2.2 Blighted lives

Survivors of infectious diseases may escape short-term mortality but experience an elevated mortality risk, vis-a-vis their healthy counterparts. A 2021 study by Al-Aly, Xie, and Bowe [59] looked at survivors of COVID-19, and found that after a mean follow-up of 4 months, users of the Department of Veterans Affairs health system who were diagnosed with COVID-19 and survived for 30 or more days after their initial diagnosis still had an increased risk of mortality [59]. Similar results have emerged from long-term studies of survivors of the West African ebolavirus outbreak [60], and post-viral syndromes are widely recognized in the aftermath of a range of other infectious diseases.

In a "blighted lives" model (illustrated in Fig. 2.8), θ_s determines the fraction of deceased *versus* recovered in the short term, much as θ *simpliciter* did for the simple SIRD model. However, in addition, we have the quantity μ_l to account for long-term mortality. We do so by subtracting the term $R\theta_l$ from $\frac{dR}{dt}$ and adding it to $\frac{dD}{dt}$, maintaining the closed system axiom's requirements. This gives us the following system of

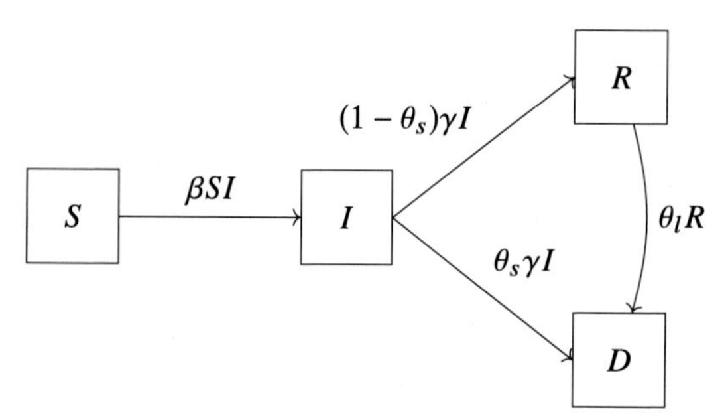

Figure 2.8 In this model, we are accounting for short-term mortality (θ_s) and long-term mortality (θ_l). Survivors experience increased mortality, in this case, for life. The long-term mortality coefficient θ_l reflects this by allocating a fraction of survivors to the deceased (D) compartment.

differential equations:

$$
\begin{aligned}
\frac{dS}{dt} &= - \underbrace{\beta SI}_{\text{mass action}} , \\
\frac{dI}{dt} &= \underbrace{\beta SI}_{\text{mass action}} - \underbrace{\gamma I}_{\text{total removed}} , \\
\frac{dR}{dt} &= \underbrace{(1-\theta_s)\gamma I}_{\text{removed recovered}} - \underbrace{\theta_l R}_{\text{removed deceased}} , \\
\frac{dD}{dt} &= \underbrace{\gamma \theta_s I}_{\text{deceased from } I} + \underbrace{\theta_l R}_{\text{deceased from } R} .
\end{aligned}
\tag{2.26}
$$

2.2.3 Modeling births

Births and deaths affect the overall population under examination. Until now, this population has been steady, which was a convenient mathematical formalism but of course flew in the face of reality. In this chapter, notably, we are not concerned with disease-related deaths, but rather natural mortality. For this reason, infectious disease modelers should keep in mind that accounting for births and deaths is not necessary nor suitable for every disease. For short, fast-moving infections, births and natural mortality are generally negligible for anything beyond a trivial population. Longer-term models, however, may usefully take account of these variables. As a rule of thumb, for human populations, including births and deaths makes sense if the model is expected to reflect a process that runs over more than twice the human gestational interval, i.e., 1.5 years. One may wish to include births and deaths at shorter time periods if they are significantly unequal, such as where population growth vastly outpaces natural mortality or vice versa.

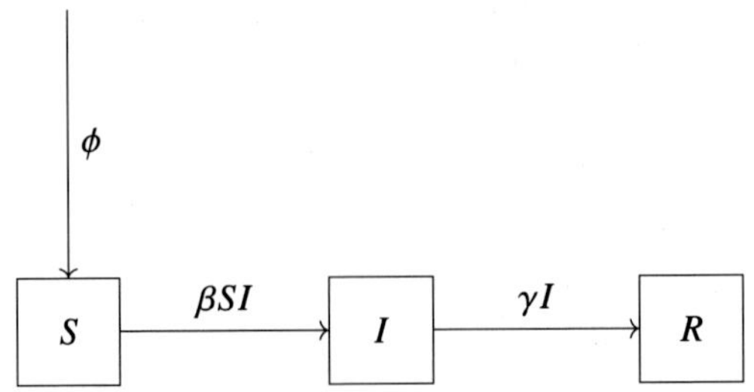

Figure 2.9 In the naive births model, we assume all births to be susceptible, and thus credit them to the S compartment. The rate of births over time is defined by the birth rate per population of N per unit time, symbolized by ϕ.

2.2.3.1 *Naive births*

In the naive birth model, we assume (as is in fact mostly the case) that anyone born is by definition susceptible at time of birth. We account for this by introducing the birth term, and assigning it entirely to S. For this, we assume that anyone who is in N is equally fecund, and consequently the rate of change of S attributable to new births is a function of the entire population size.

Definition 2.21 (Immunologically naive birth condition). The **immunologically naive birth condition** assumes that any individual born to the population will be neither infected nor immune, i.e., all births accrue to the susceptible (S) compartment.

In the naive birth model (Fig. 2.9), we account for births by crediting them to the S compartment. Thus for a rate of live births of ϕ,

$$\begin{aligned}
\frac{dS}{dt} &= -\underbrace{\beta SI}_{\text{mass action}} + \underbrace{\phi}_{\text{births}}, \\
\frac{dI}{dt} &= \underbrace{\beta SI}_{\text{mass action}} - \underbrace{\gamma I}_{\text{recovery}}, \\
\frac{dR}{dt} &= \underbrace{\gamma I}_{\text{recovery}}.
\end{aligned} \tag{2.27}$$

This model makes two significant assumptions:

1. All births are immunologically naive, i.e., neither infectious nor immune.
2. The population is equifecund, i.e., the compartment in which someone is unrelated to fertility.

We know that in reality, these assumptions do not hold. Some infections can be passed on from mother to child, by way of vertical transmission (Definition 2.19). Notable examples for the latter are the pathogens that make up the "TORCH complex," a group of infections that account for most transplacental transmission of infectious disease in humans, and are associated with perinatal infectious presentations. It includes the following diseases:

- **t**oxoplasmosis
- **o**ther infections (HIV, chlamydia, parvovirus B19)
- **r**ubella
- **c**ytomegalovirus (CMV), and
- **h**erpes simplex virus 2 (HSV-2)

Subsubsection 2.2.3.2 will discuss accounting for vertical transmission, whereas Subsubsection 2.2.3.3 shall deal with disease-related effects on fecundity.

Practice Note 2.3 When to account for maternal immunity

Accounting for maternal immunity may be useful for populations (especially nonhuman populations), where the duration of maternal immunity is significant in relation to the overall lifespan or the period under examination. For a human life, the duration of maternal immunity is quite short. This may be different for short-lived animal species.

The duration of maternal immunity may also be more significant if we are only looking at a relatively short period of time. For instance, while maternal immunity to pertussis is relatively brief, it is a significant protective factor, because most severe cases of pertussis occur in very young children.

The naive birth model is a better, and mathematically more convenient, approximation when considering longer timespans in relation to the duration of maternal immunity.

2.2.3.2 Modeling vertical transmission of disease

In vertical transmission, offspring born to infected individuals have a certain probability p_I of being infected themselves (see Definition 2.19). For this reason, we must account for population flows separately. We split the birth term, which until now we conceptualized as a single variable ϕ, into separate birth terms:

- All births accruing to susceptible or recovered individuals are susceptible.
- p_I of the births accruing to infected individuals are infected.
- $1 - p_I$ of the births accruing to infected individuals are susceptible.

Therefore the birth term ϕ is separated into susceptible births (ϕ_S) and vertically infected births (ϕ_I) as follows:

$$\begin{aligned} \phi_S &= \overbrace{(S+R)\phi}^{\text{births to } S \text{ and } R} + \overbrace{I(1-p_I)\phi}^{\text{uninfected births to } I}, \\ \phi_I &= \underbrace{I p_I \phi}_{\text{vertically infected births}}. \end{aligned} \tag{2.28}$$

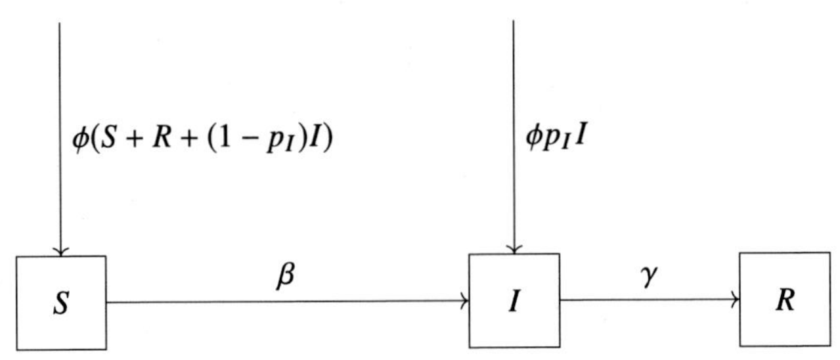

Figure 2.10 In the vertical transmission model, births in respect of S and R will be deemed susceptible, and credited to S. A fraction of births to infected individuals, p_I, will be born infected and be credited to I, whereas the remainder will be credited to S.

With that, we can adjust (2.27) to reflect vertical transmission to incorporate Eq. (2.28) (see Fig. 2.10):

$$
\begin{aligned}
\frac{dS}{dt} &= -\underbrace{\beta SI}_{\text{mass action}} + \underbrace{\overbrace{(S+R)\phi}^{\text{births to } S \text{ and } R} + \overbrace{I(1-p_I)\phi}^{\text{uninfected births to } I}}_{\text{naive births}}, \\
\frac{dI}{dt} &= \underbrace{\beta SI}_{\text{mass action}} - \underbrace{\gamma I}_{\text{recovery}} + \underbrace{I p_I \phi}_{\text{vertically infected births}}, \\
\frac{dR}{dt} &= \underbrace{\gamma I}_{\text{recovery}}.
\end{aligned}
\tag{2.29}
$$

2.2.3.3 *Disease-related effects on fecundity (DREF)*

Some infectious diseases cause a depression in the rate of live births, often by increasing the rate of miscarriages and stillbirths. We refer to these as disease-related effects on fecundity. Such effects are particularly often observed for infections of long duration, even if they are otherwise almost commensal (asymptomatic). A study of foragers in Bolivia by Blackwell et al. [61], for instance, found that gastrointestinal hookworm infection (*Ancylostoma duodenale*) reduced fertility by approximately 30% [61]. In the animal world, tsetse flies, who carry *Trypanosoma brucei rhodesiense*, have been found to suffer a 30% decrease in the number of offspring [62]. Disease-related effects on fecundity describe the cost, in terms of reduced fecundity, of an infectious disease.

Definition 2.22 (Disease-related effects on fecundity). A **disease-related effect on fecundity** is a temporary depression of the birth rate for individuals in an infected compartment in comparison to similar individuals in the susceptible compartment.

From time to time, such infections are described as "sterilizing," but this is a somewhat clumsy formulation. First, it is somewhat inaccurate in that often there's a depression of fecundity, rather than an outright sterilization. More importantly, it is liable to cause confusion with sterilizing treatments and vaccines, which describe treatments and vaccines that terminate infectious status upon administration.

We may account for DREF in our models by separating the birth rate attributable to infectious individuals and applying a discount factor ρ_i that reflects this depression. Thus the number of total births will be

$$\frac{dN}{dt} = \underbrace{\phi}_{\text{birth rate}} (S + \underbrace{\underbrace{(1-\rho_i)}_{\text{discount factor}} I}_{\text{births to infectious individuals}} + R) = \phi(N - \rho_i I). \tag{2.30}$$

We may conceptualize $\phi \rho_i I$ as the number of births attributable to individuals in I that would take place but for the disease-related reduction in fecundity. We may insert (2.30) into (2.27) for a naive births model that accounts for disease-related effects on fecundity:

$$\begin{aligned} \frac{dS}{dt} &= -\underbrace{\beta SI}_{\text{mass action}} + \underbrace{\phi}_{\text{birth rate}} (N - \underbrace{\rho_i I}_{\text{DREF}}), \\ \frac{dI}{dt} &= \underbrace{\beta SI}_{\text{mass action}} - \underbrace{\gamma I}_{\text{recovery}}, \\ \frac{dR}{dt} &= \underbrace{\gamma I}_{\text{recovery}}. \end{aligned} \tag{2.31}$$

This assumes, of course, that the reduction in fecundity and infectiousness are coterminous. Where that is not the case, such as where there is an enduring depression of fecundity after resolution of symptoms and the end of infectiousness, the appropriate approach is to model this as a separate compartment F, with a rate of reversion to fertile immune recovery or fertile susceptibility of $\frac{1}{\overline{\tau_F}}$, where $\overline{\tau_F}$ is the mean duration of suppressed fertility. In that scenario, for the discount factor ρ_F of the long-term suppressed fecundity compartment, population growth would be calculated not on the basis of N but $N - \rho_i I - \rho_F F$.

2.2.4 Modeling natural mortality

We use the term natural mortality to distinguish the "background" rate of mortality (e.g., accidental deaths, suicide/homicide, deaths from other unconnected illnesses) from disease-related mortality. A good first approximation is that the natural death rate is the inverse of the host organism's average lifespan. Where the variance of this figure is relatively low, this assumption tends to hold quite well.

Definition 2.23 (Natural mortality). **Natural mortality** refers to mortality from all causes extraneous to the infectious disease under consideration. Notably, natural mortality does not coincide with the legal concept of death from natural causes. Homicides, suicides, and deaths from warfare and violence are, for the purposes of epidemiological modeling, considered "natural." Insofar as epidemiological modeling is concerned, any births from any cause other than the infection under consideration is deemed "natural."

Natural mortality also assumes that recovered, infectious, and susceptible cases have the same mortality rate. We have seen in Subsection 2.2.2.2 that survivors of serious infectious diseases might often experience a depression in overall lifespans. However, for diseases that generally affect a large percentage of the population, such as influenza, we may consider these to be "factored into" life expectancy, and thus omit accounting for this term specifically.

Thus under the assumption that the risk of natural mortality is isotropic (equal across all compartments), we can break down the mortality term

$$\frac{dN}{dt} = \theta N \tag{2.32}$$

in proportion to the size of each compartment:

$$\begin{aligned}
\frac{dS}{dt} &= \underbrace{\theta}_{\text{birth}} - \underbrace{\beta SI}_{\text{mass action}} - \underbrace{\theta S}_{\text{deaths from } S}, \\
\frac{dI}{dt} &= \underbrace{\beta SI}_{\text{mass action}} - \underbrace{\gamma I}_{\text{recovery}} - \underbrace{\theta I}_{\text{deaths from } I}, \\
\frac{dR}{dt} &= \underbrace{\gamma I}_{\text{recovery}} - \underbrace{\theta R}_{\text{deaths from } R}.
\end{aligned} \tag{2.33}$$

This is an isotropic model in which all births accrue to S, but deaths are distributed across the compartments according to their respective sizes. Since $S + I + R = 1$, $\theta(S + I + R) = \theta$.

Definition 2.24 (Isotropy). A process or value is said to be **isotropic** over a compartmental model if it affects every compartment to the same degree.

Practice Note 2.4 Estimating natural mortality

The inverse lifespan estimator ($\theta = \overline{\tau_L}^{-1}$, where $\overline{\tau_L}$ is the average life expectancy) is a convenient first guess of the mortality rate, but not always accurate. In particular, it is much less accurate where the risk of death is unevenly distributed.

In many developing countries, the likelihood of mortality is bimodal due to high infant mortality. Additionally, age-differential mortality factors, such as prolonged warfare (which affects young men more than older men and women), often interfere with this estimation. In affluent countries, where infant mortality is low and natural mortality is more or less a monotonously increasing function of age, the inverse lifespan estimator is likely to be adequate for most purposes.

Nevertheless, when modeling infectious disease over a longer period of time, we must be sensitive to the local circumstances, including how the population's demographics are distributed. Life tables are often available for populations, from which one may then infer precise population distributions.

2.2.5 Constant-population naive-birth isotropic mortality model

Many epidemiological models, especially those executed over longer timespans, benefit from introducing a term to reflect population turnover. This creates a supply of new susceptible individuals, while keeping the population constant. The constant-population naive-birth isotropic mortality model is a useful formalism to allow the long-term oscillatory dynamics that flow from a depletion of immune individuals (through waning or death) and an influx of susceptibles to manifest. Fig. 2.11 describes this distribution. By convention, since the birth rate and the death rate are equal, they are represented by the same symbol, μ, representing the population's turnover rate.

Given a model of n compartments, all births accrue to S, at a rate of μ, and deaths occur in each compartment in proportion to its size, i.e., the mortality term for the i-th compartment is $C_i\mu$. Since $\sum_{i=1}^{n} C_i\mu = \mu$, the total mortality term will be the same as the birth term, and consequently, $\frac{dN}{dt} = 0$.

$$\begin{aligned}
\frac{dS}{dt} &= \overbrace{\mu}^{\text{births}} - \ldots - \overbrace{\mu S}^{\text{deaths from } S}, \\
\frac{dC_i}{dt} &= \ldots - \underbrace{\mu C_i}_{\text{deaths from } C_i}.
\end{aligned} \tag{2.34}$$

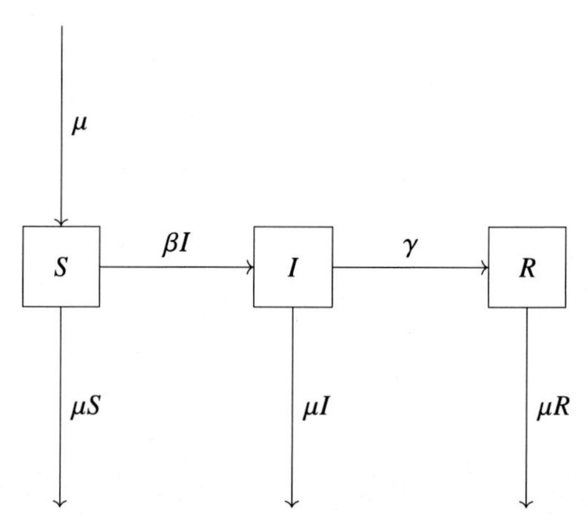

Figure 2.11 In the naive birth-isotropic mortality model, all births accrue to S, whereas deaths flow proportionally from each compartment.

The beauty of this formalism is that it can be expanded to an arbitrary number of compartments without much change, with the caveat that if there are other sources of mortality, and these are treated as outflows without a separate compartment, the constant population constraint would require those to be offset by adding it to the birth term μ in $\frac{dS}{dt}$.

Practice Note 2.5 When to account for vital dynamics

Vital dynamics do not play a significant role in every infectious disease. Especially in human populations, the rate of vital dynamics, when compared to the time dynamics of infectious diseases is tremendously slow. The human gestation period, at 270 days on average, is far from the longest in the mammalian world, but it is certainly very significantly longer than that of many other animals. It is over ten times longer than that of small rodents, such as hamsters and gerbils, and nine times as long as the gestation period of rabbits. Humans are also relatively long-lived. For this reason, in most human disease, vital dynamics have a negligible effect in the short term.

The practitioner must assess in devising the infectious disease model whether accounting for vital dynamics would have a meaningful impact on the results.

Cases where population dynamics ought to be typically modeled are

- infections in animals with relatively short gestation periods (a good rule of thumb is a maximum of $8 - 10\tau$ as a cutoff point),
- infections of long or indeterminate duration (see Subsection 2.3.2), and
- significant heterogeneities in subpopulations with regard to population dynamics.

2.2.6 $\mathfrak{R}_0$ in combined demographic models

Given the constant population naive birth isotropic mortality formalism introduced in the previous section, we obtain for the SIR model the form

$$\begin{aligned}
\frac{dS}{dt} &= \underbrace{\mu}_{\text{births}} - \beta S I - \underbrace{\mu S}_{\text{deaths from } S}, \\
\frac{dI}{dt} &= \beta S I - \gamma I - \underbrace{\mu I}_{\text{deaths from } I}, \\
\frac{dR}{dt} &= \gamma I - \underbrace{\mu R}_{\text{deaths from } R}.
\end{aligned} \tag{2.35}$$

From this, we can derive $\mathfrak{R}_0$ for a model with births and deaths. If we denote the mean time spent in the infectious compartment as τ, we can formulate it as the effect of two processes: for every unit of time spent in the infectious phase (every unit of τ), an infectious individual has γ^{-1} chance to leave the compartment by way of recovery and μ^{-1} chance of leaving due to mortality. Together, $\tau = (\gamma + \mu)^{-1}$.

We know from our derivation of $\mathfrak{R}_0$ in Subsection 2.1.5 that $\mathfrak{R}_0$ is the total number of secondary cases that can be generated by a case during their infectious state. This is, of course, the product of their propensity to generate cases (the transmission coefficient β) and the time they spend in the infectious state (τ). Consequently,

$$\mathfrak{R}_0 = \beta\tau = \frac{\beta}{\mu + \gamma}. \tag{2.36}$$

The consequence of this is that in short-lived host populations, a pathogen must be quite well-transmissible to persist. For humans, with a mean lifespan of seven decades, mortality is not going to be a particular concern when discussing an illness with the mean infectious period of 7–10 days. On the other hand, many vectors of infectious diseases, especially arthropods, are rather short-lived. Where μ is not very significantly smaller than γ, the infection must either be lifelong ($\gamma = 0$ therefore the infection will sustain itself if $\beta > \mu$) or significantly more contagious to offset it. As we shall see in Section 4.1, the former is quite often the case among short-lived vectors.

2.2.7 Funerary transmission

In recent years, increasing attention has been directed to funerary transmission, the phenomenon of infections generated by a deceased individual after death but before burial, especially in the context of Ebola haemorrhagic fever [63]. Tiffany et al. [64] estimated that on average, each unsafe burial during the 2013–2016 West African ZEBOV outbreak generated approximately 2.58 secondary cases.

Definition 2.25 (Funerary transmission). Funerary transmission denotes transmission of a pathogen from recently deceased individuals to susceptible individuals, typically during procedures that involve physical contact in funerary rituals.

Legrand et al. [65] first proposed a model that took account of funerary transmission. This model effectively considered the time pending burial a separate compartment within a SIRD model. Deceased individuals first move to the intermediate compartment (typically denoted F) of the unburied dead, from which they then transition, at a rate typically specified by the time to burial, to the ultimate compartment (D). Fig. 2.12 illustrates the flow diagram of the Legrand et al. model's operation.

SIRFD models allow us to understand not only funerary transmission risk but also the impact that certain public health measures have, such as Safe and Dignified Burials (see Practice Note 2.6).

The effect of reducing the risk of funerary transmission is illustrated by Fig. 2.13. Assuming a time-dependent rate of infections, earlier funeral results in a significantly less severe infectious dynamic. Of course, in the same vein, we may use time as merely a proxy for overall encounters. Reducing the number of encounters during the burial process, e.g., by not embalming the body and by the use of barrier precautions in preparing the body for burial, has a similar effect.

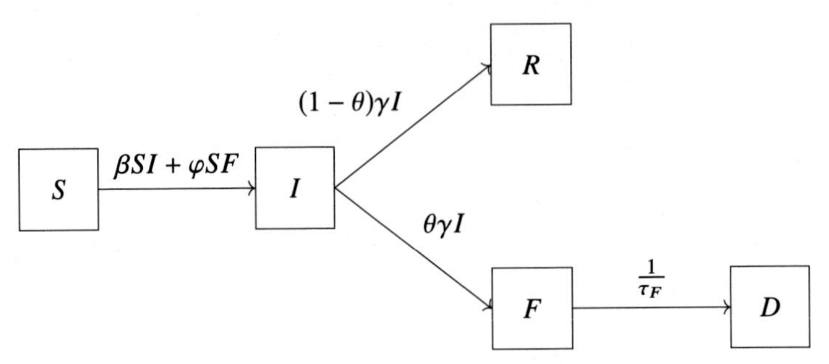

Figure 2.12 SIRFD model of funerary transmission. θ denotes the mortality fraction of removal. Deceased but unburied individuals (F) remain infectious, at the rate of φ, for the time until burial, τ_F.

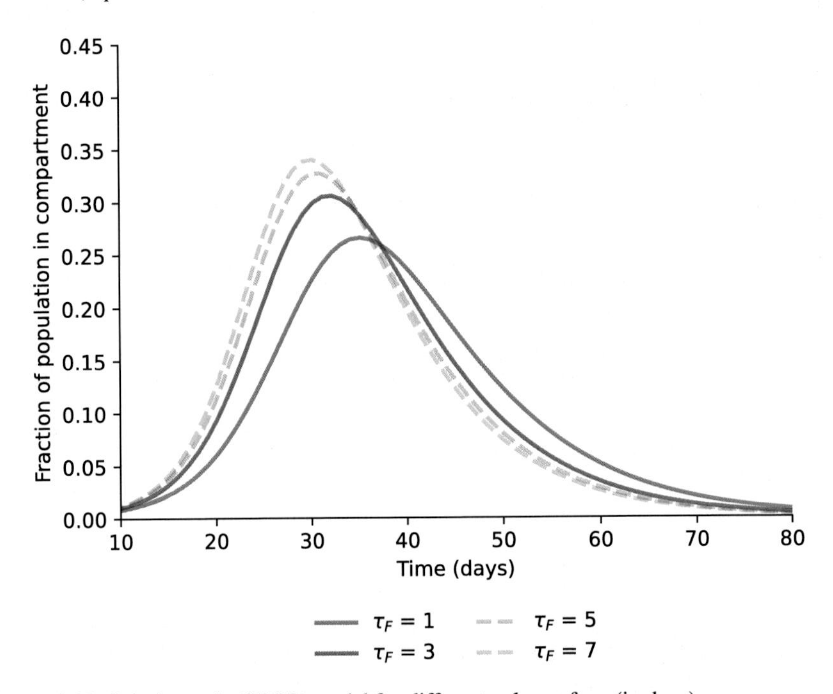

Figure 2.13 Solutions of a SIRFD model for different values of τ_F (in days).

Practice Note 2.6 Safe and dignified burials

Ceremonies and rites that mark the end of life and concern the disposition of the body are almost universal in all cultures and across all of human social evolution. Such ritual activity is often religiously and culturally sensitive, as well as highly emotive for the relatives and the wider community of the decedent. For this reason, funerary transmission raises important ethical and cultural dilemmas. Many funerary practices involve steps, such as ritual cleansing and ablution of the body, that expose the participants to contracting certain pathogens from the decedent, as has been documented, e.g., for haemorrhagic fever outbreaks in West Africa [66]. Funerals are also common social events, often involving geographically disparate family members, and may thus act as superspreader events (see Subsection 3.1.4) [67].

When addressing the contribution of funerary practices to pathogenic spread, it is important to act with understanding and sensitivity for the specific cultural context. During the 2013–2016 West African ebolavirus outbreak, the WHO developed the practice of Safe and Dignified Burials (SDB) [68]. which amended traditional funerary practices with steps intended to reduce the risk of funerary transmission [69]. The SDB guidelines had the benefit of being informed by a cross-cultural perspective and a deep anthropological understanding of the ways in which traditional rituals, especially the preparation of the body, could be altered, while retaining its ritual meaningfulness to the participants. For instance, the Islamic practice of *ghusl*, an ablution of the deceased, is replaced with *tayammum* or "dry ablution," which reduces the risk of pathogenic spread from the decedent.

Though by no means perfect, the development of SDBs is an example of exercising cultural sensitivity and discarding both pure utilitarianism and Eurocentric biases in creating a new ritual that the population will accept. SDBs are also a model of shared decision-making in an epidemic context, allowing the decedent's family to be active participants in decision-making. Even though—in contrast with common ritual—they will not be preparing the decedent's body, the guidelines provide for a family-appointed observer, and adopt a participatory approach that integrates the grieving family [68], Thus although they may not be fulfilling their traditional roles in preparing the body, the SDB remains something that happens "with," rather than merely "to," the family and the community.

SDBs have been credited with avoiding up to 10,000 secondary cases in that outbreak through preventing community spread [64], and it is beyond doubt that this would have been hardly attainable had the practices that we now collectively call SDB not been formulated in a manner that fosters shared decision-making, respect for cultural and religious traditions, and open communication.

2.3 Models of immunity

Immunity is the permanent or temporary resistance of an organism to an infection. Our models until now assumed perfect and lifelong immunity in all survivors. In this section, we are examining more differentiated models of immunity.

In general, when it comes to immunity, we do not quite care as to the origins of the immunity. In the rare cases where there are clinical differences, e.g., the rate of waning, there might be a good case to model, say, post-infectious and vaccine-induced immunity separately. On the other hand, we do care about the likelihood of immunity. Not everyone who is vaccinated or recovers from an illness develops immunity, and often, immunity has a shelf life, as Subsection 2.3.1 discusses in detail. In compartmental models, this is typically accommodated by discounting the rate of immune recovery, or providing for a "flowback" to reflect waning. This section will explore techniques to model immunity in detail, whereas Section 6.1 discusses vaccine-induced immunity specifically in more depth.

2.3.1 Case study: periodicity in syphilis incidence

Syphilis is a bacterial infection caused by *Treponema pallidum spp. pallidum*, which is most often spread through sexual activity, though vertical transmission (see Subsubsection 2.2.3.2) has also been documented (congenital syphilis). In the United States, Grassly, Fraser, and Garnett observed that the peak spectral density of syphilis incidence was around 10 years [70]. Spectral density determines the contribution of individual frequencies to a time series, and its peak determines the frequency of the dominant periodic process. Grassly, Fraser, and Garnett also observed that unlike syphilis, gonorrhoea did not exhibit a similar periodicity.

Syphilis exhibits an oscillating pattern because of waning immunity over time. As we have seen in Subsection 2.1.5, epidemics need a suitably large pool of susceptible individuals to be viable. An infectious disease that engenders immunity will eventually exhaust this pool of susceptibles, and "burn out." However, if immunity gradually wanes, the pool of susceptibles is gradually refilled, and when S once again reaches the critical mass, the epidemic will flare up.

Definition 2.26 (Waning immunity). **Waning immunity** denotes the reduction of immune response against the same pathogen as a function of time.

Biologically, waning immunity refers to the reduction in the effectiveness of the immune response over time.

In epidemiology, we primarily consider waning immunity as a binary-distributed variable over a population. As such, waning immunity denotes the reversion of immune individuals to susceptibility.

Waning immunity thus explains why certain epidemics, absent any external intervention, may recur with astonishing regularity, following multiyear cycles. In this section, we will examine the effect of immunity on our models. Until now, we generally assumed full immunity, i.e., everyone who survived infection was indefinitely immune. This section looks at what happens in the absence of that assumption.

2.3.2 No-immunity models (SI)

Previously examined models have generally assumed that survival bestowed indefinite immunity. We will depart from this assumption in the present section. In fact, we will do so by examining the polar opposite of indefinite immunity, namely pathogens that induce no immunity at all, but rather result in lifelong infection. The classic example of such pathogens is herpes simplex (HSV). HSV infection is widely prevalent and lasts a lifetime. SI models (which some texts, such as Keeling and Rohani [39], refer to as SIS models) allow us to model such infections.

2.3.2.1 SI models with no vital dynamics and no recovery

First, let us consider SI models without vital dynamics and without recovery, i.e., where an infected person remains infected for the duration of the model. We may describe such a model by the system of differential equations:

$$\begin{aligned}\frac{dS}{dt} &= -\underbrace{\beta SI}_{\text{mass action}}, \\ \frac{dI}{dt} &= \underbrace{\beta SI}_{\text{mass action}}.\end{aligned} \tag{2.37}$$

It follows from the above that under the assumption of homogeneous mixing, and in the absence of vital dynamics, the entire population will eventually be infected:

$$\begin{aligned}&\lim_{t\to\infty} S(t) = 0, \\ &\lim_{t\to\infty} I(t) = 1, \\ &\lim_{t\to\infty} \mathcal{I}(t) = N \lim_{t\to\infty} I(t) = N.\end{aligned} \tag{2.38}$$

2.3.2.2 SI models with clearance (SIS models)

SI models with clearance are models where the population alternates between an infectious state and susceptibility. These models are common, where surviving an infection does not result in an appreciable length or degree of immunity, but also where antigenic drift is so significant that immunity developed at one time is unlikely to have a protective effect for all but a trivial time period. This is often observed with rapidly mutating viruses, such as influenzaviruses, which necessitates annual re-immunization due to significant antigenic drift.

Definition 2.27 (Clearance). **Clearance** refers to the reversion from infectiousness into a nonimmune susceptible state. Unlike in the case of recovery, no immunity is built up.

In a SIS model with clearance, the γ parameter, which ordinarily reflects transition from infectious to recovered state, will refer to clearance, that is, return to the susceptible pool. (See Figs. 2.14 and 2.15.)

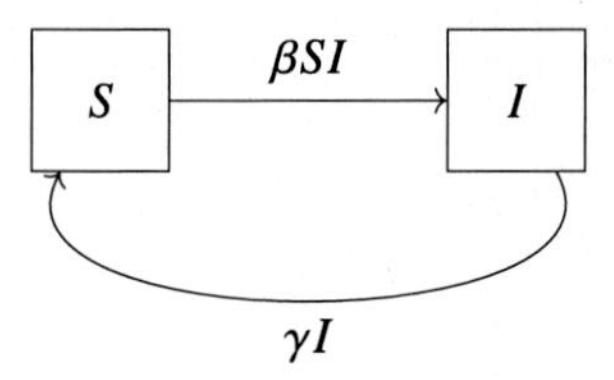

Figure 2.14 In the SI model with clearance, the population moves between the infectious and susceptible state.

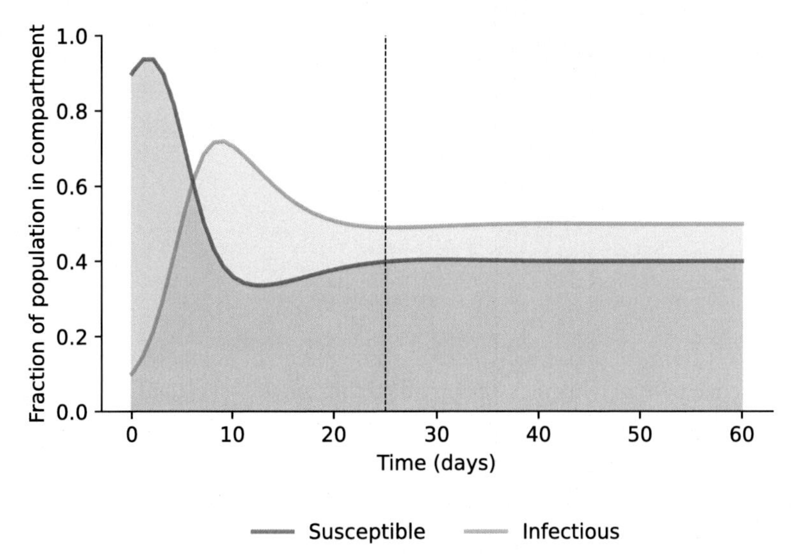

Figure 2.15 Solutions of a SI model with clearance for $\mathfrak{R}_0 = 4$ and $\tau = 6$. The dashed line represents the separation between the oscillatory phase (left) and equilibrium phase (right).

We may characterize this model by the following differential equations:

$$\begin{aligned} \frac{dS}{dt} &= -\underbrace{\beta SI}_{\text{new infections}} + \underbrace{\gamma I}_{\text{clearance}}, \\ \frac{dI}{dt} &= \underbrace{\beta SI}_{\text{new infections}} - \underbrace{\gamma I}_{\text{clearance}}. \end{aligned} \tag{2.39}$$

2.3.3 Modeling loss of immunity

At the heart of immunity is an organism's ability to create an enduring immunological memory against a pathogen. The mechanisms of immunological memory are quite complex, and include both cellular and humoral immunity. Humoral immunological memory results from the persistence of antibodies against the pathogen, even after the effector T cells that have responded to the initial infection have died off. This is

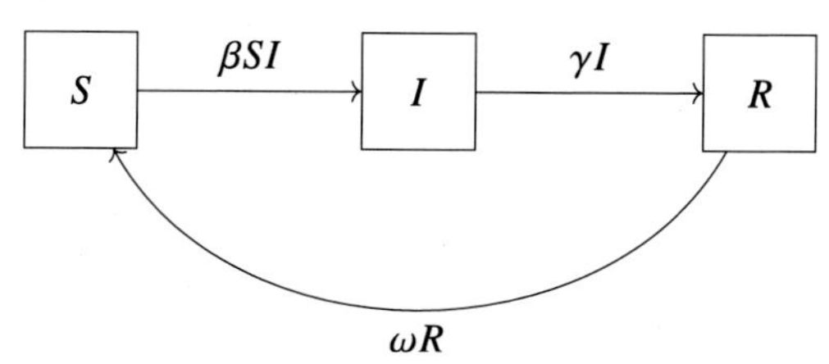

Figure 2.16 Waning immunity is the gradual loss of immune competence against a pathogen. ω denotes the rate of waning immunity, which is inversely related to the duration of immunity τ_R. The dotted lines denote peaks of epidemic waves.

relatively short-lived but highly effective. Cellular immunological memory relies on memory T cells and memory B cells that are long-lived and indeed in some cases confer lifelong immunity.

Waning immunity denotes the process of a loss of immunity over time (see Definition 2.26). In other words, recovered individuals "trickle back" to the susceptible compartment at a certain rate. Our first approximation is to consider this a fixed rate, ω, which is the inverse of the mean duration of immunity engendered by infection (sometimes denoted as τ_R).

Definition 2.28 (Waning rate). The **waning rate** ω is the time-denominated rate at which immunity is lost. It corresponds to the inverse of the duration of immunity.

We may articulate this term by adding it to S and removing it from R, throughout. Thus the governing system of equations becomes

$$\begin{aligned}
\frac{dS}{dt} &= \mu - \underbrace{\beta SI}_{\text{new infections}} + \underbrace{\omega R}_{\text{reversal of waned}} - \mu S,\\
\frac{dI}{dt} &= \underbrace{\beta SI}_{\text{new infections}} - \underbrace{\gamma I}_{\text{recovery}} - \mu I,\\
\frac{dR}{dt} &= \underbrace{\gamma I}_{\text{recovery}} - \underbrace{\omega R}_{\text{waning}} - \mu R.
\end{aligned} \tag{2.40}$$

The formula can, of course, be adapted to take account of vital dynamics similar to (2.35).

A consequence of waning immunity is that the pool of susceptibles is gradually replenished time and time again. We have seen in Subsection 2.1.5 that an epidemic requires a proportion of susceptible individuals to persist in a population. As immunity waxes and wanes, a periodic pattern of "seesaw" infections or "waves" may emerge, with the period of infectious waves coinciding with the time for sufficient replenishment of the susceptible pool. The case study at the beginning of this section (see Subsection 2.3.1) illustrates such a periodic recurrent epidemic. (See Figs. 2.16 and 2.17.)

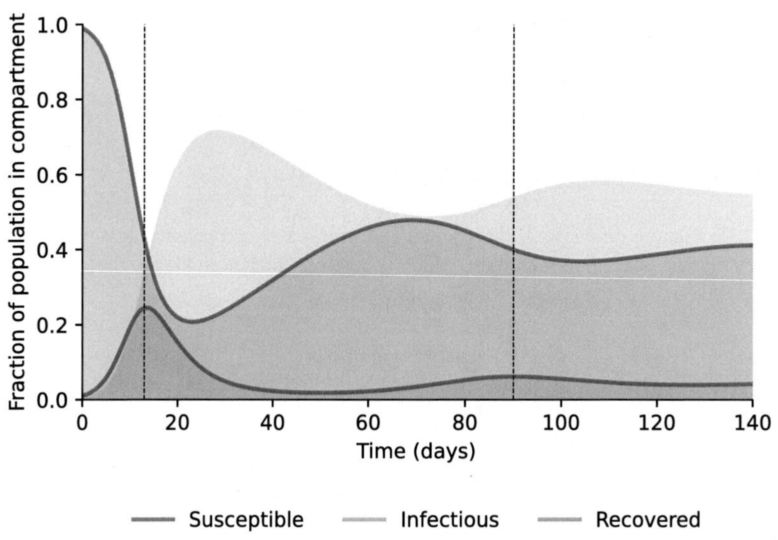

Figure 2.17 Solutions of a SIRS model for $\mathfrak{R}_0 = 2.5$, $\tau = 4$ and τ_R (duration of immunity) of 50 days.

Computational Note 2.5 Waning immunity

By far the most commonly modeled aspect of immunity is periodicity due to waning immunity (as described in Subsection 2.3.3). Fundamentally, a SIRS model is no different in its computational execution than the previous compartmental models. We begin by defining the key initial characteristics, including ω, the rate of waning immunity. It is generally more convenient to represent ω as $\overline{\tau_R}^{-1}$, the inverse of the duration of immunity.

```
I_0 = 1e-2
S_0 = 1 - I_0
R_0 = 0

y_0 = (S_0, I_0, R_0)
R0 = 2.5
tau = 4
gamma = 1/tau
tau_R = 50
omega = 1/tau_R
beta = R0 * gamma
```

We can then use the approach used for prior ODE solutions to arrive at a numerical result.

> *A notebook implementing the contents of this Computational Note is available on the book's companion Github repository in the folder* `/ch02/sir_models`.

In practice, however, waning is not a homogeneous rate process. Often, a better approximation is to assume that immunity lasts a certain time ω^{-1}. Instead of defraying a fraction of the immune population at every step, we can use a delay differential equation (DDE) that simply returns the quantity of individuals who have become immune at $t - \omega^{-1}$ to susceptibility:

$$\begin{aligned}
\frac{dS_t}{dt} &= \mu - \beta S(t)I(t) + \gamma I(t - \omega^{-1}) - \mu S(t),\\
\frac{dI_t}{dt} &= \beta S(t)I(t) - \gamma I(t) - \mu I(t),\\
\frac{dR_t}{dt} &= \gamma I(t) - \gamma I(t - \omega^{-1}) - \mu R(t).
\end{aligned} \tag{2.41}$$

Ordinary ODE solvers cannot solve delay differential equations, which makes them computationally somewhat more demanding and less convenient than simple ODE models. However, they are computationally solvable using DDE solvers. Computational Note 6.2 lays out the use of a DDE solver for waning of vaccination-induced immunity, which can be trivially adapted to cover waning of post-infectious immunity, as detailed in this subsection, too.

2.3.4 Maternal immunity

Maternal immunity, sometimes called vertical immunity, is the immunity counterpart to vertical transmission of disease (on which see Subsection 2.2.3.2). Unlike the forms of immunity we have considered until now, maternal immunity is a form of passive immunity. There are multiple mechanisms for developing maternal immunity. In primates and humans, as well as other mammals that have a haemochorial placenta, the main method of acquiring maternal immunity is through placental transfer, where maternal IgG passes to the foetus through the close contact between the fetal capillary endothelium and the maternal chorionic endothelium. In a number of other animals important for infectious disease, especially canines and chiropterans (bats), this is not the case. These animals have an endotheliochorial placenta, in which the contact between maternal and foetal circulation is less proximate, and consequently, does not offer an avenue for the efficient transfer of IgG. For this reason, offspring of these species typically acquire IgG through the gastrointestinal tract from colostrum [71].

Regardless of how it was obtained, maternal immunity shares two principal features:

1. It is present *ab initio*, i.e., the child is not born into the susceptible compartment.

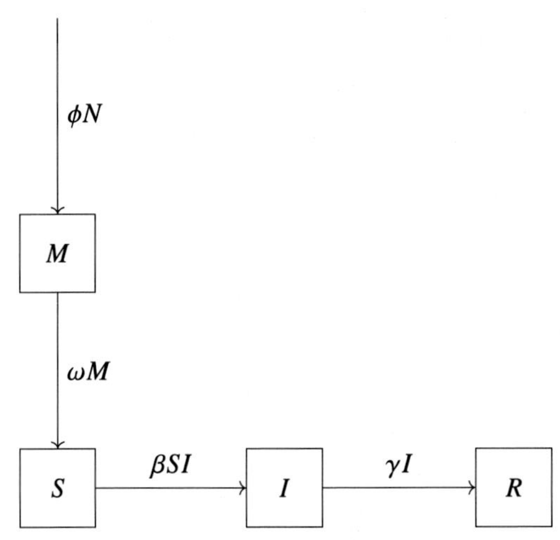

Figure 2.18 The SIRM model with waning immunity allocates births to an intermediate compartment M of maternally immune individuals, whose immunity wanes at the rate of ω.

2. It is relatively short-lived, offering protection for the first months to a year of the child's life [72].

Where maternal immunity is relevant (including by way of maternal vaccination, as is the case for maternal TDaP vaccination), and where infants in their first years of their lives are a significant part of the population, we must therefore reflect this in our models. (See Fig. 2.18.)

A SIRM model does so by reserving a compartment for newborns, who are presumed to be born with some degree of immunity that wanes after a while. The flow rate from M to S is defined by ω, which in turn is the inverse of the mean duration of maternal immunity $\overline{\tau_{Mm}}$. We may thus describe this model by the system of differential equations in Eq. (2.42).

$$
\begin{aligned}
\frac{dM}{dt} &= \underbrace{\phi N}_{\text{vertically immune births}}, \\
\frac{dS}{dt} &= -\underbrace{\beta SI}_{\text{mass action}} + \underbrace{\omega M}_{\text{waning of vertical immunity}}, \\
\frac{dI}{dt} &= \underbrace{\beta SI}_{\text{mass action}} - \underbrace{\gamma I}_{\text{recovery}}, \\
\frac{dR}{dt} &= \underbrace{\gamma I}_{\text{recovery}}.
\end{aligned}
\tag{2.42}
$$

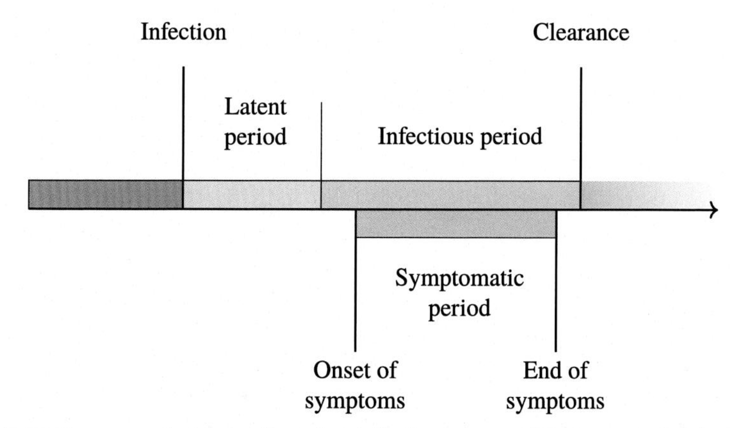

Figure 2.19 Key events in the timeline of an infectious process. The latent period (yellow) denotes the period between infection and infectiousness. Note that the symptomatic period does not exactly coincide with the infectious period. In many, but not all, infectious diseases, the abatement of symptoms tends to coincide with the end of infectiousness. Equally, some diseases may not be accompanied by a noticeable symptomatic period.

2.4 Models with latent periods, asymptomatic infection, and carrier states

So far, we have assumed that an infected individual becomes immediately infectious. (See Fig. 2.19.) We break with this assumption in this section, and look at ways to account for three different scenarios:

- latent periods (the time between infection and onset of infectiousness),
- asymptomatic infections (where a part of the infected population does not experience symptoms), and
- carrier states, where the person recovers from clinical infection but remains infectious.

Definition 2.29 (Latent period). The **latent period** (also known as the **latency period**) denotes the timespan between infection and onset of infectiousness.

Practice Note 2.7 Asymptomatic infection

The notion of asymptomatic infection is sometimes a little muddled, due to confusion with the latent period (Definition 2.29). True asymptomatic infection occurs when a patient transits into the infectious compartment and subsequently leaves it, without showing clinically appreciable symptoms. Strictly speaking, this may include biomarkers (such as elevated leukocyte counts), but in practice, we consider a patient asymptomatic if the condition resolves (into susceptibility, immunity, or some other post-infectious status) without a noticeable clinical picture.

Asymptomatic patients pose two main concerns:

- Because illness often limits a person's interactions with society, asymptomatic patients are more likely to go about their ordinary business and infect others than symptomatic infectious individuals, who are both aware of their status and potentially prevented from pursuing their ordinary activities thereby.
- On the other hand, where symptoms of the illness are intrinsically connected to the mode of transmission, as is the case, e.g., for droplet infections, asymptomatic patients may have a lower infectiousness.

It is, in addition, not easy to identify asymptomatic carriers. Seroprevalence studies are costly, and often yield little information. Antibody seroprevalence studies do not indicate when the patient had the disease (although the ratio of IgG to IgM may differentiate acute phase infection from a resolved infection). Antigen seroprevalence studies, on the other hand, only return positive results for a relatively short period of time. Where asymptomatic transmission is a significant part of a pathogen's life in a population, antigen seroprevalence studies should be carried out and correlated with clinical symptoms to elicit the ratio of asymptomatic patients.

2.4.1 Case study: the true story of "typhoid Mary"

Mary Mallon was born in 1869 in Cookstown, County Tyrone, then part of British-ruled Ireland. Like many of her countrymen, she immigrated to the United States at a young age, where she eventually found employment as a cook. During her lifetime, it was suspected that she has unintentionally (albeit perhaps negligently) infected over fifty people with typhoid [73].

Typhoid fever is a bacterial disease caused by gastrointestinal infection by *Salmonella enterica serovar Typhi*. In most patients, it causes an unpleasant but manageable disease that resolves fully. However, as many as one in twenty patients become chronic carriers, who continue to be infectious for their lifetimes. Mary Mallon was one of the unfortunate few who fell into that category. It is hypothesized today that she contracted typhoid at birth.

Her case, which involved prolonged quarantine on North Brother Island for almost half her life, raises complex moral and ethical questions about reconciling the interests of public health with the moral imperative to respect individual liberties and treating the sick (even if asymptomatic) with compassion [74]. However, it also highlights the difficulties of modeling infectious diseases, where some patients recover into an asymptomatic carrier state. This is the subject of the present section.

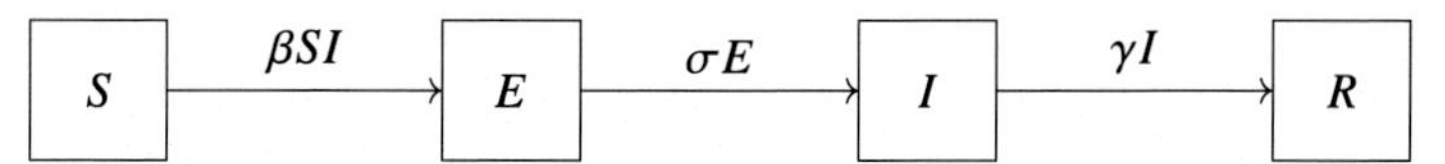

Figure 2.20 In a SEIR model, patients go through an exposed (E) phase, during which they are considered to be infected but not yet infectious.

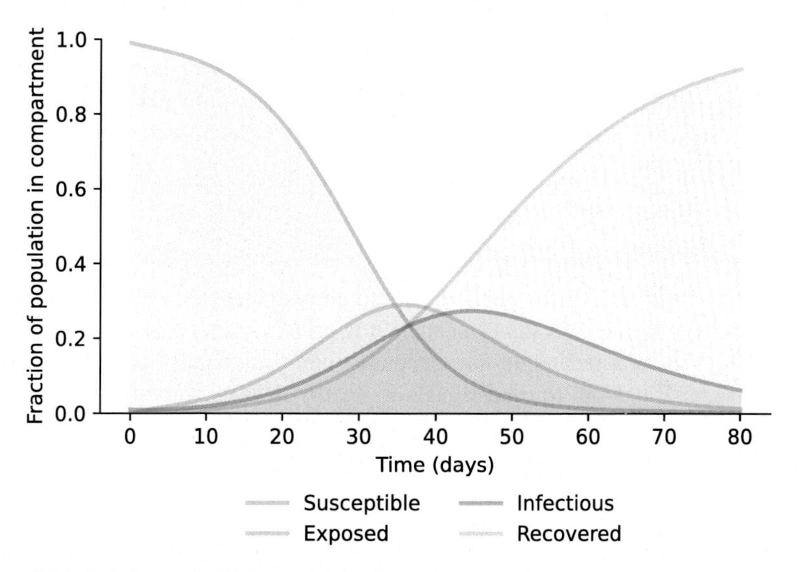

Figure 2.21 Solutions of a SEIR model for $\mathfrak{R}_0 = 6.0$, $\tau = 12$ and $\sigma = 0.1$.

2.4.2 Modeling the latent period: SEIR models

In a SEIR model, we split the hitherto monolithic compartment of infected individuals into exposed (E), who are infected but not infectious, and infectious (I), who are infectious as well. Consequently, patients will flow from S, initially, into E. The rate of transition σ from E to I is the inverse of $\overline{\tau_l}$, the average duration of the latency period. (See Figs. 2.20 and 2.21.)

It is important to remember that the governing mass action is still between the susceptible and the infectious population (see Subsection 2.1.3). Consequently, the flow from S to E is determined by S and I, and largely independent from E.

The system of governing differential equations is thus

$$\begin{aligned}\frac{dS}{dt} &= -\underbrace{\beta SI}_{\text{mass action}},\\ \frac{dE}{dt} &= \underbrace{\beta SI}_{\text{mass action}} - \underbrace{\sigma E}_{\text{exposure}},\\ \frac{dI}{dt} &= -\underbrace{\sigma E}_{\text{exposure}} - \underbrace{\gamma I}_{\text{recovery}},\end{aligned} \tag{2.43}$$

$$\frac{dR}{dt} = \underbrace{\gamma I}_{\text{recovery}} .$$

Because the E compartment acts as an "anteroom" to infectiousness, SEIR models tend to result in a slower development of the pathogenic dynamics. The reverse of the model is that during the latent period, the infection might be much harder to detect. Since by definition the latent period precedes the onset of symptoms, individuals are typically asymptomatic and unaware that they have been infected. For this reason, in spatial models of dissemination and diffusion, longer latent periods may result in wider spatial spread.

Practice Note 2.8 Privacy-preserving exposure tracing

One of the new tools in the public health toolbox that came to be tested for the first time during the COVID-19 pandemic is privacy-preserving exposure tracing [75]. This is a technique that depends upon the ubiquity of smartphones that can use Bluetooth to send signals to nearby devices without the need for an organized network. Devices send anonymous random identifiers to all nearby devices. The low power and thus spatial limitation of Bluetooth is, in this context, a feature, rather than a bug; the limited range corresponds to a fairly good estimate of exposure for a droplet infection. Upon a positive test result, users can then "report" their status to the public health authorities anonymously using their random identifier. The authorities can then trigger exposure notifications on all devices that have been in contact with the infected user's device recently.

This technique is not foolproof, and abounds in false negatives and false positives alike. For instance, since radio waves can propagate in different ways from infectious droplets, a person working on a different floor of the same building could get an exposure notification when in fact actual exposure would not have occurred. At the same time, exposure notifications rely greatly on a cooperative userbase, who report positive tests and maintain tracking throughout. Nevertheless, it is a cost-effective solution that is technologically easy to implement and imposes minimal social burdens in return for the benefits of relatively reliable exposure notifications.

Computational Note 2.6 Solving SEIR models

The functional definition for SEIR models is quite analogous to previous models, namely:

```
def deriv(t, y, beta, gamma, sigma):
    S, E, I, R = y

    dSdt = - beta * S * I
    dEdt = beta * S * I - sigma * E
    dIdt = sigma * E - gamma * I
    dRdt = gamma * I

    return dSdt, dEdt, dIdt, dRdt
```

The versatility of compartmental epidemic models and their representation—and solution—as systems of ODEs is that these models can be expanded with ease, by simply changing the derivative function's definition. Other than ensuring that the `sigma` parameter is taken, solving a SEIR model's IVP computationally is no different from a SIR model.

A notebook implementing the contents of this Computational Note is available on the book's companion Github repository in the folder `/ch02/sir_models`.

2.4.3 Models of recovery into carrier state

In the preceding models, recovery might have conferred immunity, but it was in any case the end of infectiousness. In carrier state models, a proportion of infectious patients recover into asymptomatic carriers, who remain infectious potentially for life, even though their symptoms have subsided. SIRC models allow us to characterize such situations.

Definition 2.30 (Carrier). A **carrier** is an individual who is asymptomatic but capable of transmitting the disease, typically, where this period of time lasts relatively long compared to acute symptomatic infection. Depending on the model, the carrier compartment may include the primary asymptomatic infected, who never develop symptoms, or be limited only to recoveries into a carrier state.

2.4.3.1 Recovery into carrier state with equal infectiousness

The SIRC model assumes that carriers eventually clear the infection and become immune, in $\overline{\tau_c}$ time. We will denote the rate of carriers who become immune as $\omega = \overline{\tau_c}^{-1}$, in line with the way we denoted waning immunity. (See Figs. 2.22 and 2.23.)

From the perspective of mass action, a susceptible individual may contract the infection from a carrier or an infectious individual. Therefore we have to treat I and C as a single compartment for the mass action term, assuming for this iteration of the model that infectiousness is identical for symptomatic and asymptomatic carriers. We denote the fraction of individuals who recover into a carrier state (C), rather than immune recovery (R) by θ. Finally, we use the transfer term ωC to denote the rate at which carriers recover. This gives us the system of differential equations

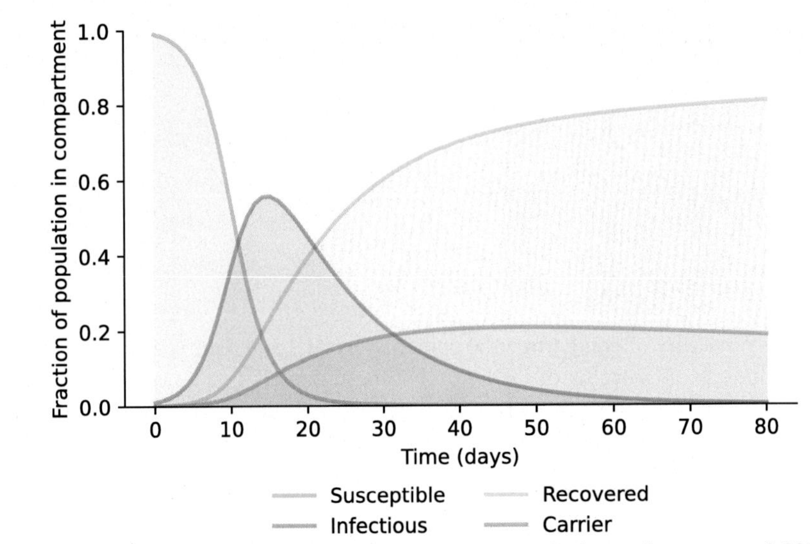

Figure 2.22 Solutions of a SIRC model for $\mathfrak{R}_0 = 6.0$, $\tau = 12$, the waning rate $\omega = 0.005$ and the fraction of carriers $\theta = 0.25$. Note the elongation of the infectious curve due to carriers.

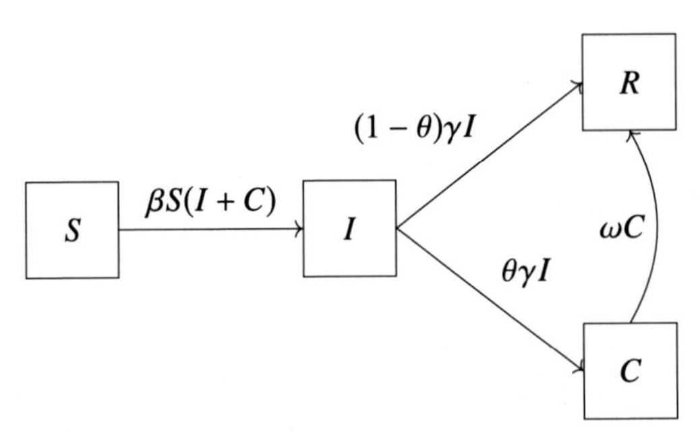

Figure 2.23 The SIRC model results in a carrier state C and a recovered state R. θ denotes the proportion of infected who go on to be carriers. ω is the rate at which carriers convert to noninfectious, immune recovery.

$$
\begin{aligned}
\frac{dS}{dt} &= \underbrace{-\underbrace{\beta SI}_{\text{from infected}} - \underbrace{\beta SC}_{\text{from carriers}}}_{\text{mass action}}, \\
\frac{dI}{dt} &= \underbrace{\beta S(I+C)}_{\text{mass action}} - \underbrace{\gamma I}_{\text{recovery}}, \\
\frac{dR}{dt} &= \underbrace{(1-\theta)\gamma I}_{\text{recovery into } R} + \underbrace{\omega C}_{\text{recovery from } C},
\end{aligned}
\tag{2.44}
$$

$$\frac{dC}{dt} = \underbrace{\theta\gamma I}_{\text{recovery into carrier state}} - \underbrace{\omega C}_{\text{recovery to } R}.$$

Computational Note 2.7 Solving SIRC models

Numerical solutions for SIRC models are arrived at much the same way as previous ODEs. The functional definition of the ODE in (2.44) is

```
def deriv(t, y, beta, gamma, omega, theta):
    S, I, R, C = y

    dSdt = - beta * S * (I + C)
    dIdt = beta * S * (I + C) - gamma * I
    dRdt = (1 - theta) * gamma * I + omega * C
    dCdt = theta * gamma * I - omega * C

    return dSdt, dIdt, dRdt, dCdt
```

A notebook implementing the contents of this Computational Note is available on the book's companion Github repository in the folder `/ch02/sir_models`.

2.4.3.2 Accounting for reduced infectiousness

In the model above, we have assumed that symptomatic and asymptomatic carriers are equally infectious. This is not always warranted; symptoms of infectious disease are not mere epiphenomena, they are often evolutionary adaptations by the pathogen to accelerate its spread. A pathogen does not intrinsically benefit from causing a violent cough or nasal discharge, for instance, except that such activities enable it to be transmitted more easily. The consequence is that asymptomatic carriers, by reason of their asymptomatic state, might have a lower infectious propensity. We will use a discount factor α to reflect this.

The overall structure of a model that accounts for reduced infectiousness is the same as that of the equiinfectious model discussed in Subsection 2.4.3.1, with the exception that the infection term attributable to C is multiplied by the correction factor α. We implement this compensation by amending the mass action term to take account of the reduced rate of transmission, so that it becomes $\beta S(I + \alpha C)$. Including that into Eq. (2.44) yields

$$\frac{dS}{dt} = \underbrace{-\underbrace{\beta S I}_{\text{from } I} - \underbrace{\beta S \alpha C}_{\text{from } C \text{ (reduced)}}}_{\text{mass action}},$$

$$\frac{dI}{dt} = \underbrace{\beta S(I + \alpha C)}_{\text{mass action}} - \underbrace{\gamma I}_{\text{recovery}}, \tag{2.45}$$
$$\frac{dR}{dt} = \underbrace{(1-\theta)\gamma I}_{\text{recovery into } R} + \underbrace{\omega C}_{\text{recoveries from } C},$$
$$\frac{dC}{dt} = \underbrace{\theta\gamma I}_{\text{recovery into } C} - \underbrace{\omega C}_{\text{recoveries from } C}.$$

Computational Note 2.8 Solving SIRC models with reduced infectiousness

To adapt the solution for Eq. (2.44) for Eq. (2.45), we simply add the factor `alpha` to represent the discount factor α:

```
def deriv(t, y, beta, gamma, omega, theta, alpha):
    S, I, R, C = y

    dSdt = - beta * S * (I + (C * alpha))
    dIdt = beta * S * (I + (C * alpha)) - gamma * I
    dRdt = (1 - theta) * gamma * I + omega * C
    dCdt = theta * gamma * I - omega * C

    return dSdt, dIdt, dRdt, dCdt
```

A notebook implementing the contents of this Computational Note is available on the book's companion Github repository in the folder `/ch02/sir_ models`.

2.4.3.3 $\mathfrak{R}_0$ in models with a carrier state

Where a pathogen has a clinically significant infectious asymptomatic carrier state, the $\mathfrak{R}_0$ will have to be adjusted to reflect this. We do so by adding a term to $\frac{\beta}{\gamma}$ to reflect the infected who become carriers. We know that carriers are less efficient at producing secondary infections, therefore the numerator term will have to be multiplied by the adjustment factor α. We multiply this by θ to reflect the relative proportion of carriers and divide it by ω to reflect the loss over time of carriers to the waning of the carrier state. Putting it all together, we may calculate $\mathfrak{R}_0$ as

$$\mathfrak{R}_0 = \underbrace{\frac{\beta}{\gamma}}_{\mathfrak{R}_0 \text{ without carrier state}} + \underbrace{\frac{\theta \overbrace{\alpha}^{\text{adjustment factor}} \beta}{\omega}}_{\text{carrier state}} = \beta\left(\frac{1}{\gamma} + \frac{\theta\alpha}{\omega}\right). \tag{2.46}$$

2.5 Empirical parameter estimation

Many parameters of an infectious disease model lend themselves to easy, or at least feasible, measurement or inference:

- the length of the infectious period τ can be ascertained from studies of the serial interval,
- the recovery rate γ can be calculated from τ,
- the case-fatality ratio (CFR) can be derived from case reports and mortality reports, and
- the duration of the latency interval can be ascertained from contact tracing studies.

$\mathfrak{R}_0$, however, is a derived quantity, and it is not amenable to direct measurement. For this reason, we need ways to estimate it in practice, and ways to identify it in relation to the model's conditioning variables in complex models.

2.5.1 *Case study: early estimation of the $\mathfrak{R}_0$ of Covid-19*

In the waning years of 2019, a new respiratory infection surfaced in Wuhan, People's Republic of China. Within a few short weeks, it attained global spread, and by the end of January 2021, twenty-seven countries were affected. The WHO declared this new respiratory disease a Public Health Emergency of International Concern on 30 January 2020 and a pandemic six weeks later, on 11 March. The rest, of course, is history.

The early days of the COVID-19 pandemic highlight the fast-moving, dynamic nature of infectious disease outbreaks. To adequately support public health responses to rapidly developing outbreaks, it is important to have methods that allow us to ascertain key parameters from real-world data that we can then fit models onto. First and foremost among these is, of course, $\mathfrak{R}_0$.

A systematic review by Alimohamadi, Taghdir, and Sepandi [76] has surveyed statistical and probabilistic models of COVID-19 in China. The 29 models examined by the authors found values ranging between 1.90 and 6.49. Exponential growth-based models alone ranged between 1.90 and 6.30 [76]. The authors arrived at a pooled estimate of 3.32 (95% CI: 2.81–3.82), but what is more revealing is the wide variance. Even based on the same data, there were significant disparities in $\mathfrak{R}_0$ based on the particular implementations and algorithms.

The attempts to quantify the $\mathfrak{R}_0$ of COVID-19 are an instructive example of the sensitivity of $\mathfrak{R}_0$ to methods used to estimate it. $\mathfrak{R}_0$ is often seen as a convenient encapsulation of a pathogen's behavior in a population, but it is rarely recognized how malleable it is as a guiding figure for interventions. It is an instructive tool to illustrate pathogenic dynamics and a useful value to characterize some parts of the infectious process, but it should not become the subject of exaggerated focus. Nor should it be presented uncritically, without clearly conveying that its value greatly depends on the method used to ascertain it.

In a 2020 article, Pandit pointed out the paradox of $\mathfrak{R}_0$: it is "relatively easy to explain, more complicated to understand (even graphically), and very difficult to calculate" [77]. We will concentrate on $\mathfrak{R}_0$, because unlike other variables in simple compartmental models, it is not directly ascertainable or trivially related to a variable that is (the way, e.g., γ is trivially related to the mean infectious period τ by way of being its inverse).

Practice Note 2.9 Communicating $\mathfrak{R}_0$

$\mathfrak{R}_0$ is one of the more descriptive parameters of an infectious disease. It is surprisingly easy to explain relatively accurately what it accomplishes. However, the $\mathfrak{R}_0$ is not a trivially measurable quantity.

In fact, as we shall see throughout this chapter, how we measure $\mathfrak{R}_0$ affects its value greatly. For this reason, excessive attention devoted to $\mathfrak{R}_0$ might mislead the public. The transmission of infections and the dynamics of infectious diseases are complex and multifactorial. They depend on more than a single figure, and using $\mathfrak{R}_0$ alone to illustrate pertinent facts about an infectious disease might not be sufficient. It is up to disease modelers not only to know how to calculate $\mathfrak{R}_0$, but also to know which calculation to use when, and in their communications guard against undue attention directed at $\mathfrak{R}_0$ (and, by extension, $\mathfrak{R}_t$).

2.5.2 Next-generation matrices

The next-generation matrix is by far the most popular method of determining the $\mathfrak{R}_0$ of a complex model, as we shall encounter increasingly often in later steps. We proceed in three steps to calculate $\mathfrak{R}_0$ using the next-generation method:

1. We divide the model into infected and noninfected subsystems. We denote these as the vectors x and y, which are vectors comprising infected and noninfected compartments, respectively.
2. We construct the vector $F(x, y)$, which comprises flows into x, and the vector V, which comprises flows out of x. Note the nomenclature here: F is a vector-valued function, whereas $\mathbf{F}$ is a matrix (some texts use $\mathcal{F}$ for the vector-valued functions and F for the matrix, but this may altogether muddy the waters). These, together, make up $\frac{dx}{dt}$.

$$\frac{dx}{dt} = F(x, y) - V(x, y) \tag{2.47}$$

We can represent this change as the Jacobian matrices $\mathbf{F}$ and $\mathbf{V}$:

$$\begin{aligned} \mathbf{F} &= \left(\frac{\partial F}{\partial x}\right), \\ \mathbf{V} &= \left(\frac{\partial V}{\partial x}\right). \end{aligned} \tag{2.48}$$

3. The $\mathfrak{R}_0$ of the system is obtained by taking the spectral radius of $\mathbf{F}\mathbf{V}^{-1}$.

Definition 2.31 (Next-generation matrix). The matrix $\mathbf{FV}^{-1}$ is called the **next generation matrix**. Its spectral radius (see Definition 2.32), as evaluated at the disease-free equilibrium, equals the $\mathfrak{R}_0$.

Consider, for instance, the SIR model described in (2.6). Our x and y vectors are, of course,

$$x = (I), \quad y = \begin{pmatrix} S \\ R \end{pmatrix}. \tag{2.49}$$

The vector-valued functions can then be evaluated as

$$\begin{aligned} \vec{F}(I) &= \beta S I, \\ \vec{V}(I, S, R) &= \gamma I. \end{aligned} \tag{2.50}$$

Taking the partial differentials and evaluating it at the disease-free equilibrium, we get

$$\begin{aligned} \mathbf{F} &= \left(\frac{\partial \beta S I}{\partial I}\right)\Bigg|_{(1,0,0)} &= \beta, \\ \mathbf{V} &= \left(\frac{\partial \gamma I}{\partial I}\right)\Bigg|_{(1,0,0)} &= \gamma. \end{aligned} \tag{2.51}$$

And thus, we obtain $\mathfrak{R}_0$ as the spectral radius $\varrho(\mathbf{FV}^{-1})$, which in this case is of course $\frac{\beta}{\gamma}$.

Next-generation matrices can be helpful in understanding the dynamics of complex systems. In practice, the next-generation matrix is most useful in estimating the $\mathfrak{R}_0$ of multicompartmental models, as we will encounter them in Chapter 4, in which there are separate compartments for hosts and vectors, or in Chapter 3, where we separate individuals by their behavioral risk into high- and low-risk groups. Symbolic evaluation is often helpful in making sense of such models. Computational Note 2.9 discusses the use of symbolic computation for such a scenario.

Definition 2.32 (Spectral radius and dominant eigenvalue). Let $\Lambda(\mathbf{A})$ be the set of all eigenvalues $\lambda_{1 \ldots i}$ of $\mathbf{A}$.

The spectral radius of $\mathbf{A}$ is the eigenvalue of $\mathbf{A}$ with the largest absolute value. We denote this as $\varrho(\mathbf{A})$. Every matrix has a defined spectral radius.

If there exists an eigenvalue λ_d in $\Lambda(\mathbf{A})$ so that its absolute value is larger than any other eigenvalue in $\Lambda(\mathbf{A})$, it is a **dominant eigenvalue** of $\mathbf{A}$. Not every matrix has a dominant eigenvalue. Every dominant eigenvalue, where it exists, is equal to the spectral radius of the matrix.

Computational Note 2.9 Symbolic computation of $\mathfrak{R}_0$ in a complex model

So far, we have used computational tools to give us numerical solutions to problems. However, we can use computational tools to solve analytical problems, too. Symbolic computation deals with arriving at analytical solutions using the manipulation of symbols (which in this case is just a fancy term for functions, variables, operators, and anything else mathematics is "made of"). Computer algebra systems (CAS), from the simple solvers on a graphing calculator to complex automated theorem-provers, do exactly that.

In Python, we can make use of the `SymPy` package to perform symbolic manipulations. Consider a SEIR model with births and mortality:

$$\begin{aligned}
\frac{dS}{dt} &= -\beta SI - \mu S + \phi(S + E + I + R), \\
\frac{dE}{dt} &= \beta SI - (\sigma + \mu)E, \\
\frac{dI}{dt} &= \sigma E - \gamma I - \mu I, \\
\frac{dR}{dt} &= \gamma I - \mu R.
\end{aligned} \tag{2.52}$$

We can obtain a neat symbolic solution for this using `SymPy`. First and foremost, we need to import `SymPy`, and define our symbols:

```
import numpy as np
from sympy.interactive.printing import init_printing
init_printing(use_unicode=True, wrap_line=True)
from sympy.matrices import Matrix
from sympy import symbols
from sympy import factor

S, E, I, R, beta, mu, phi, sigma, gamma
 = symbols("S E I R beta mu phi sigma gamma")
```

Next, we define x:

```
X = Matrix(np.array([E, I]).T)
```

We can now define the vector-valued functions F and V:

```
Fvec = Matrix(np.array([beta * S * I, 0]).T)
```

```
Vvec = Matrix(np.array([(mu + sigma) * E, -sigma * E
       + (gamma + mu) * I]).T)
```

We evaluate `Fvec` at $S = 1$:

```
Fvec = Fvec.subs({S: 1})
Vvec = Vvec.subs({S: 1})
```

`SymPy` has a convenient function to calculate the Jacobian of a vector-valued function. Recall that the next-generation matrix $\mathbf{FV}^{-1}$ is the product of the Jacobian of F with respect to x multiplied by the inverse of the Jacobian of V in respect of x. We may write this as

```
ngm = Fvec.jacobian(X) * Vvec.jacobian(X).inv()
```

The function `.eigenvals()` returns the spectrum (the list of all eigenvalues) of the matrix. Unfortunately, SymPy cannot automatically determine the dominant eigenvalue, but this should not be difficult manually.

Factoring the dominant eigenvalue gives us a nice-looking output:

```
factor(list(ngm.eigenvals().keys())[0])
```

$$\frac{\beta\sigma}{(\gamma+\mu)(\mu+\sigma)}$$

Symbolic application of the next-generation matrix may often be your best choice to obtain $\mathfrak{R}_0$ from the description of a compartmental model as a differential equation. This is so in particular if the system has many compartments and a large number of variables, as we will indeed encounter later on in models of more complex disease.

A notebook implementing the contents of this Computational Note is available on the book's companion Github repository in the folder `/ch02/symbolic_r0`.

2.5.3 Determining $\mathfrak{R}_0$ from epidemiological data

$\mathfrak{R}_0$ is by far one of the most useful parameters to obtain from epidemiological data, as it holds within itself the values of both β and γ, but unlike both of those variables, it is more amenable to measurement. Together with information about the length of the infectious period, which is also measurable, the $\mathfrak{R}_0$ can help fit a SIR model, or any of the variations on that theme, onto a real-world epidemic.

2.5.3.1 Contact tracing

In practice, contact tracing is one of the most widespread methods of estimating $\mathfrak{R}_0$. The idea of contact tracing is to identify secondary cases by asking confirmed cases

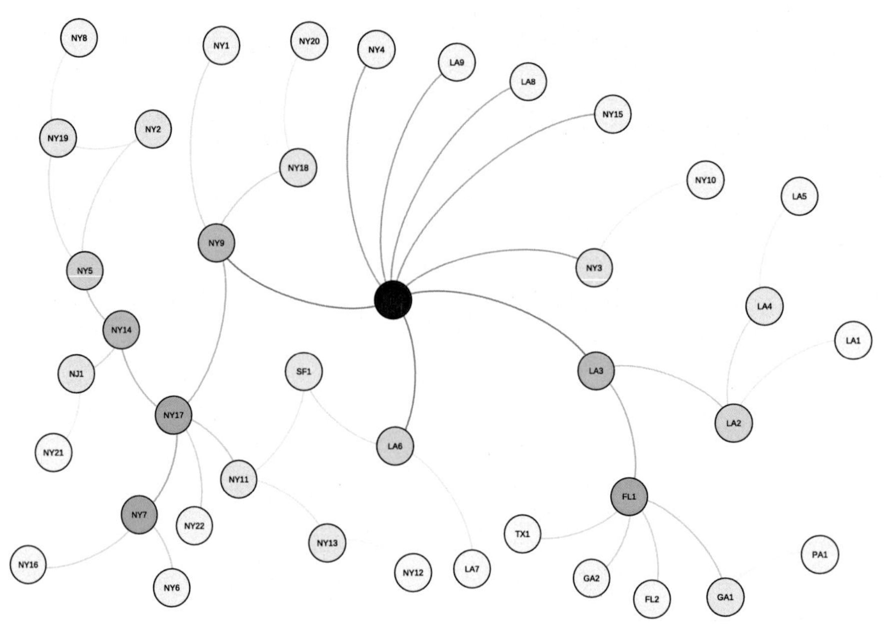

Figure 2.24 Network of sexual interactions among 40 homosexual men, based on Auerbach et al. [78]. Nodes are colored by their degree, i.e., the number of nodes connected to them.

to list potential contacts, then surveilling those contacts for symptoms and/or seroconversion. Though contact tracing is time-consuming, resource-intensive, and depends greatly on patient cooperation for its success, it is nonetheless the most effective ways of ascertaining $\mathfrak{R}_0$.

Consider the study by Auerbach et al. , one of the most famous examples of network analysis used in epidemiology [78]. This study was one of the early examinations of HIV/AIDS, proceeding by identifying the sexual contacts of 40 homosexual men. It is also the source of the colloquialism "patient zero" to mean what ought more appropriately be called the index case of an infection. If we conceive of the results of contact tracing as the directed graph $\mathcal{G}(V, E)$, then $\mathfrak{R}_0$ will be the average outdegree of $\mathcal{G}$.

This network is visualized in Fig. 2.24. The mean outdegree is approximately 1.026, which almost definitely underestimates the actual $\mathfrak{R}_0$ figure by as much as a factor of two. This highlights a fundamental fact about using contact tracing data to get to an $\mathfrak{R}_0$ estimate: the result is only going to be as good as the contact tracing exercise itself. It is difficult to remember everyone one has been in contact with on a given day, never mind over the span of years (could you list everyone you were within 6ft of in the last 48 hours, everyone you touched in the last 30 days, and everyone you were intimate with over the last five years?). In this case, the matter was complicated by the fact that many of the persons were, sadly, deceased by the time contact tracing was carried out, and the researchers had to rely on information from friends and partners. In other contexts, a person might not be aware of all the other individuals they have interacted with, or at least not be able to provide their details. This latter case has been a particular issue during the early AIDS epidemic due to anonymity of homosexual encounters, where such encounters were legally or socially unacceptable at the time, and it continues to be an issue for airborne infections, where contact tracing might have to extend to a surprisingly large number of people, many of whom we never have a profound enough social interaction with to even know their names. Such studies therefore are bound to underestimate $\mathfrak{R}_0$ by some degree. Nevertheless, contact tracing is the most concrete and direct approach to identifying $\mathfrak{R}_0$, and for that reason, it remains an important pillar of epidemic surveillance.

Computational Note 2.10 Contact tracing data with NetworkX

In this example, we will be using NetworkX to estimate $\mathfrak{R}_0$ from contact tracing data.

We begin by instantiating a directed graph (`DiGraph`) object:

```
G = nx.DiGraph()
```

There are many ways to build up a graph object in NetworkX, but where we do not intend to attach complex properties, the easiest is by far to provide edges, and let NetworkX create the nodes for us.

To make data entry easier, it is often a good idea to build a dictionary object, where the keys are the source and the values are a list of destinations of the directed edges.

```
EDGES = {"0": ["LA3", "LA8", "LA6", "NY15", "NY9", "NY4", "NY3",
               "LA9"],
         "LA3": ["LA2", "FL1"],
         "FL1": ["TX1", "GA2", "GA1", "FL2"],
         "GA1": ["PA1"],
         "LA2": ["LA1", "LA4"],
         "LA4": ["LA5"],
         "LA6": ["LA7", "SF1"],
         "NY9": ["NY18", "NY1", "NY17"],
         "NY18": ["NY20"],
         "NY17": ["NY22", "NY21", "NY14", "NY7"],
         "NY7": ["NY6", "NY16"],
         "NY11": ["SF1", "NY13"],
         "NY13": ["NY12"],
         "NY14": ["NY5", "NJ1"],
         "NJ1": ["NY21"],
         "NY5": ["NY2", "NY19"],
```

```
        "NY2": ["NY19"],
        "NY19": ["NY8"]}
```

We can then merge this into the graph using a simple nested iterator:

```
for o in EDGES:
    for d in EDGES[o]:
        G.add_edge(o, d)
```

Finally, we can obtain the mean outdegree, which is our estimator for $\mathfrak{R}_0$, by obtaining the outdegree of each node (which is what `G.out_degree` returns), slice the array to get an array only of the outdegree values, then obtain the average:

```
np.mean(np.array(list(G.out_degree))[:,1].astype("uint8"))

>>> 1.0256410256410255
```

2.5.3.2 *Wallinga-Lipsitch method*

The most popular way of estimating $\mathfrak{R}_0$ in early infection is the method outlined by Wallinga and Lipsitch [79]. This method is most useful in the early phases of an outbreak, when S largely approximates 1, i.e., almost everyone is susceptible. Under such conditions, the disease spreads exponentially, so that after n generations, $\sum_{i=0}^{n} R_0^i$ have been infected at some point. Note that this assumption does not hold for long, or, indeed, mathematically strictly speaking, it does not hold at all, but is in the very beginning of an epidemic an acceptable approximation.

The basis of the Wallinga–Lipsitch approach is that early enough in an epidemic, growth is exponential. This exponential growth rate r governs initial spread so that at least in the beginning, $I(t) = ce^{rt}$. We can, of course, obtain c and r by fitting an exponential model to the data. In practice, it is often simpler to apply a log transformation to the data, then apply a simple linear regression.

Next, we must determine the moment generating function M of the generation time distribution $g(t)$. We can obtain this from epidemiological field studies that compared the symptom onset time of the index case and their secondaries.

Definition 2.33 (Generation time). The generation time or generation interval of a case is the time between the index case's infection and the secondary case's infection.

For instance, if Alice (the secondary case) is infected on Friday by Bob (the index case), who was infected on Monday, the generation time with respect to Alice is 4 days.

Some texts define the generation interval as the time between symptom onsets. Under the assumption that the latency period is largely the same for everyone, the two definitions are equivalent.

Table 2.2 Moment generating functions of distributions frequently used in modeling infectious diseases.

	Distribution	Moment-generating function $M(t)$
Normal	Normal(μ, σ)	$e^{t\mu+\frac{1}{2}\sigma^2 t^2}$
Chi-squared	Q_k^2	$(1-2t)^{-\frac{1}{2}k}$
Gamma	$\Gamma(k, \theta)$	$(1-t\theta)^{-k}$ $(t < \theta^{-1})$
Exponential	Exp(λ)	$(1-t\lambda^{-1})^{-1}$ $(t < \lambda)$
Poisson	Poisson(λ)	$e^{\lambda(e^t-1)}$
Laplace	Laplace(μ, b)	$\frac{e^{t\mu}}{1-b^2t^2}$ $(\lvert t\rvert < b^{-1})$

$\mathfrak{R}_0$ is then calculated as the inverse of the moment-generating function evaluated at $-r$, i.e.,

$$\mathfrak{R}_0 = \frac{1}{M(-r)}. \tag{2.53}$$

Table 2.2 lays out the moment-generating functions of commonly used distributions.

Computational Note 2.11 Symbolic determination of the moment-generating function

SciPy's `optimize` subpackage offers a very useful function, `curve_fit`, which can fit an arbitrary function to data. Given a time series, we can fit the model $I(t) = ce^{rt}$ to our actual data. First, we define the function that represents this model:

```
def exp_model(t, c, r):
    return c * np.exp(r * t)
```

Note that this is equivalent to applying a logarithm transform and fit a simple linear model. We will, however, for the sake of clarity, use the exponential formulation.

Next, we fit this model to a time series. In this example, we will be using a COVID-19 data set for the United States during the first wave (March–June 2020), using `pd.read_csv()`:

```
data = pd.read_csv("https://raw.githubusercontent.com/nytimes/\
            covid-19-data/master/rolling-averages/us.csv",
            usecols=["date", "cases"],
            dtype={"date": str, "cases": int})
```

Next, to control for weekly variations in reporting, we take the seven-day moving average:

```
data["cases"] = data.cases.rolling(7).mean()
```

We filter the time series for our desired dates:

```
data = data[(data.date >= "2020-02-01")
        & (data.date <= "2020-04-11")]
```

Now, we can fit our model against the time series to obtain the growth rate:

```
popt, pcov = optimize.curve_fit(exp_model,
                                data.index,
                                data.cases,
                                p0=(1e-2, 1e-6),
                                maxfev=10000)
```

The resulting `popt` object is a 2-length array that contains the values of `c` and `r` obtained by fitting the model against the data. Fig. 2.25 shows actual case counts during the period under examination and the exponential fit using the parameters we obtained through the above.

We now have r (as `popt[1]`), which allows us to calculate the moment-generating function at $-r$. We know from metaanalyses (e.g., Griffin et al. [80]) that Normal(5, 1.15) a good estimate of the parameters for the generation time of COVID-19. We can symbolically evaluate the moment-generating function

$$M(t)_{\mu,\sigma} = e^{t\mu + \frac{1}{2}\sigma^2 t^2}$$

using SymPy. First, we initialize the variables:

```
mu, sigma, t, blamda = sympy.symbols("mu sigma t lambda")
```

Next, we express the moment-generating function in those terms:

```
mgf_normal = E ** (t * mu + (sigma ** 2 * t ** 2)/2)
```

In the above, `E` is the constant e, which is imported from SymPy. We can now estimate the $\mathfrak{R}_0$ by evaluating the moment-generating function at $\mu = 5$ and $\sigma = 1.15$, at $-t$. $-t$, of course, is `-1 * popt[1]`.

```
1/mgf_normal.subs({"t": -1 * popt[1],
                   "mu": 5,
                   "sigma": 1.15})
```

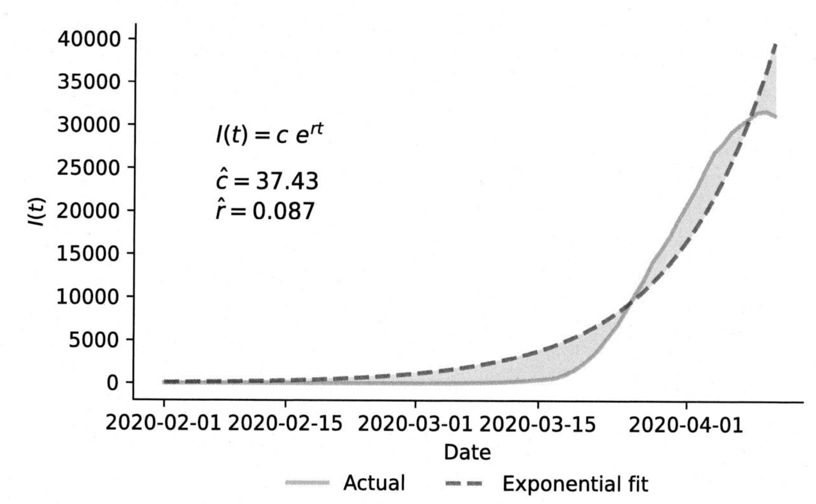

Figure 2.25 Exponential fit on a time series of COVID-19 in the United States in early 2020. Data from The New York Times, based on reports from state and local health agencies.

The `.subs()` method provided by SymPy applies to all kinds of symbolic objects. `.subs()` takes a dictionary of symbols and replaces them with a specified value, essentially parametrically evaluating the expression.

Notably, our result of an $\mathfrak{R}_0$ of 1.49 is rather lower than what is commonly understood to be the $\mathfrak{R}_0$ of COVID-19. This is for a number of reasons, which highlight some of the shortcomings of the Wallinga–Lipsitch method. Our estimate is only as good as our fit for the growth rate and the distribution of the generation time. Obtaining the generation time is often nontrivial in practice, and often depends on the availability of monitoring and tracing tools, which might not be available at the very beginning of the epidemic, just when this model is at its most powerful.

A notebook implementing the contents of this Computational Note is available on the book's companion Github repository in the folder `/ch02/symbolic_mgf`.

The Wallinga–Lipsitch solution is a convenient method in the very beginning to obtain the value of $\mathfrak{R}_0$ as long as there is some information about the generation time or serial interval, but its accuracy hinges greatly on the adequacy of this information. In other words, the wrong distribution fit over the data will yield erroneous results. Regardless of distribution used, it is crucial to document the assumptions employed when describing a Wallinga–Lipsitch estimate of $\mathfrak{R}_0$.

Practice Note 2.10 Using the Wallinga-Lipsitch method

The Wallinga–Lipsitch method ultimately provides us a very convenient way of easily estimating $\mathfrak{R}_0$ in the earliest days of an outbreak. As it relies greatly on the exponential nature of early epidemics, it is not suitable for the modeling of endemic disease. Keeling and Rohani [39] note two additional difficulties with the Wallinga–Lipsitch approximation:

1. The model assumes that at time 0, there will be a negligibly small number of infected individuals amidst a susceptible population, with no part of the population immune. This may hold true for a truly novel pathogen or a hitherto isolated population (as is the case, e.g., when a new disease is introduced to a geographically and epidemiologically isolated group), but in general, even naive populations have some small number of immune individuals owing to cross-immunity with similar pathogens.
2. In the early days of an epidemic, owing to the low number of cases, the epidemic dynamics may be quite turbulent due to stochastic effects, and by the time such stochastic effects subside, the epidemic might no longer be following an exponential growth pattern.

For this reason, $\mathfrak{R}_0$ figures estimated from the exponential model are best treated with a degree of caution.

2.5.3.3 Hesterbeek-Dietz (final epidemic size) method

A relatively simple method of estimating $\mathfrak{R}_0$ from empirical data relies on the difference of the initial state and the state at the end of an epidemic. We have noted earlier that most epidemics reach an equilibrium, where some of the population remains susceptible and some of the population is immune and recovered. We denote this as the final state $S(\infty)$, $I(\infty)$, $R(\infty)$.

In the absence of demographic change, and with some initial immunity (which is almost always a warranted assumption, even for a novel pathogen),

$$\mathfrak{R}_0 = \frac{\log S(0) - \log S(\infty)}{S(0) - S(\infty)}. \tag{2.54}$$

The main drawback of this approach is that it assumes constant behavior throughout the epidemic, from beginning to the end. This is, however, manifestly not the case in most outbreaks. As time moves on, public health authorities as well as individuals respond to an outbreak, which means that the dynamics are not homogeneous throughout, and consequently this method typically overestimates $\mathfrak{R}_0$. In addition, this method is strictly retrospective; it only works where a final state has been established.

2.5.3.4 Mean age at infection

Another method that commends itself by easily ascertainable inputs is discussed by Mollison [81] and Dietz and Schenzle [82]. This model assumes a population of naive births and recovery to immunity, along with perfectly homogeneous mixing, in particular with regard to age.

Definition 2.34 (Mean age of infection). The mean age of infection A is the population mean of the age at which individuals become infected.

Then, $\mathfrak{R}_0$ is proportional to the fraction of the mean lifetime L divided by the mean age of acquiring the disease, A.

$$\mathfrak{R}_0 = 1 + \frac{L}{A}, \tag{2.55}$$

and conversely,

$$A = \frac{L}{\mathfrak{R}_0 - 1}. \tag{2.56}$$

The mean age at infection method is widely used where the mean age of an individual is easy to measure, but other methods, especially contact tracing, are prohibitive. In the wildlife veterinary setting, for instance, information about the mean age of animals and the age of any individual (especially in a "tagged" or otherwise observed cohort) is much easier to come by than any other form of information.

2.5.4 Estimating $\mathfrak{R}_t$

Similarly to $\mathfrak{R}_0$, the time-dependent reproduction number $\mathfrak{R}_t$ also permits a number of methods of estimation. Harvey and Kattuman [83] proposed a particularly convenient estimator for $\mathfrak{R}_t$, namely

$$\hat{R}_{k,\tau,t} = \frac{\overbrace{\sum_{j=0}^{k-1} y_{t-j}}^{k-\text{length moving average}}}{\underbrace{\sum_{j=0}^{k-1} y_{t-\tau-j}}_{\tau\text{-shifted } k\text{-length moving average}}}, \tag{2.57}$$

where τ is the generation interval, and k is a smoothing constant. Harvey and Kattuman notes that setting k to 7 eliminates intra-week variations, whereas a value used for τ depends on the infectious process. Conveniently, the fraction in Eq. (2.57) can be expressed as the fraction of a k-day rolling sum and a second k-day rolling sum shifted by τ days. We shall use this to our advantage in Computational Note 2.12.

Computational Note 2.12 Estimating $\mathfrak{R}_t$

As noted above, Eq. (2.57) for $\mathfrak{R}_t$ can be expressed as the fraction of the k-day rolling sum and the k-day rolling sum at $t - \tau$. Following Harvey and Kattuman [83], we will be using a k-value of 7 to smooth out intra-week variations in reporting. In addition, we will be using data on COVID-19 incidence in the United States from Ritchie et al. [84], and in accordance with common modeling practice (as mentioned, e.g., in Harvey and Kattuman [83]), we will use a value of 4 for τ.

We begin, after ingesting our data into a Pandas data frame, by creating a column for the 7-day rolling sum:

```
df["7ms"] = df.new_cases.rolling(7).sum()
```

We assign another column to represent the denominator, which is essentially the `7ms` column shifted back by τ:

```
df["7ms_tau"] = df["7ms"].shift(4)
```

Finally, we obtain our $\mathfrak{R}_t$ by dividing the two columns:

```
df["Rt"] = df["7ms"]/df["7ms_tau"]
```

The result of this calculation is shown in Fig. 2.26. Note that because of the rolling sum and the shift, as well as because of rapid initial dynamics, there is often an unrealistically high spike in the earliest days of an outbreak. It may be worthwhile to limit $\mathfrak{R}_t$ plots to the period once the epidemic is no longer in its earliest phases.

A notebook implementing the contents of this Computational Note is available on the book's companion Github repository in the folder `/ch02/rt_ estimation`.

2.5.5 Estimating recovery rate

The recovery rate γ is the inverse of the mean duration of illness $\overline{\tau}$, i.e., $\gamma = \overline{\tau}^{-1}$. For simple SIR models, the best estimate in practical settings is from empirical studies. As a first line of departure, we may consider the mean duration of illness to be the time between onset and resolution of symptoms. However, this is not always quite accurate. Recall that γ describes the rate at which individuals transit out of the infectious compartment. Therefore what is actually at concern is the time between onset and end of infectiousness.

A problem with duration of illness data is that it is often difficult to come by in the early stages of an outbreak. Where the pathogen is relatively well-characterized,

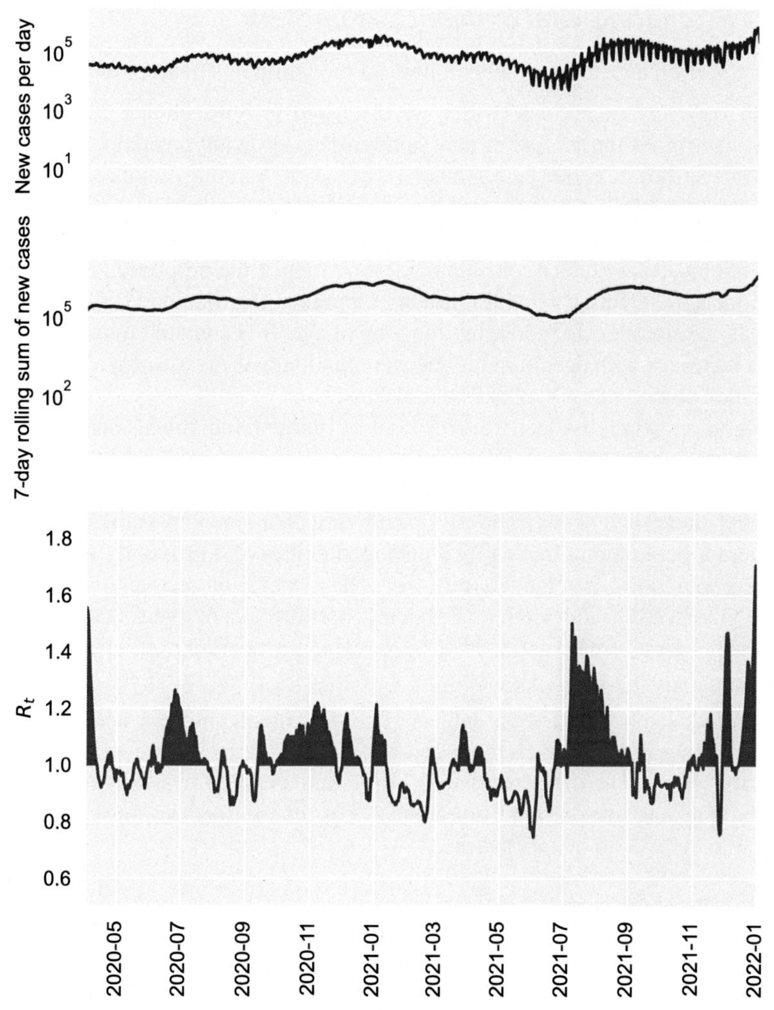

Figure 2.26 Cases, 7-day rolling sum of cases and $\mathfrak{R}_t$ of COVID-19 in the United States between April 2021 and January 2022 estimated using the estimator in Eq. (2.57). Note the prominent increases in $\mathfrak{R}_t$ in mid-2021 (Delta) and late 2021/early 2022 (Omicron). Incidence data obtained from Ritchie et al. [84].

typically from previous outbreaks, information from those cases is a good starting point. Ascertaining the mean duration of illness is helped by contact tracing, which can illuminate the length, if any, of a latent or asymptomatic but infectious period. It is worth noting however that such data typically excludes individuals who do not seek medical care, either because it is not accessible to them, because they do not exhibit sufficiently severe symptoms (or indeed any symptoms at all) or, in rare cases, because they succumb to the disease before they find medical attention.

2.5.6 Estimating vital dynamic parameters

Demographic change affects epidemiological modeling in a handful of ways:

- With very few exceptions (which we discussed in Subsection 2.2.3.2), the birth rate determines the accrual of new susceptibles. Growing populations can support pathogens, because they face a much lower risk of "running out of room to spread," i.e., depleting the pool of susceptibles necessary to sustain chains of transmission.
- Mortality (by which we generally mean mortality from causes other than the infectious disease under consideration) may reduce the proportion of immune individuals. If infection provides lifelong immunity, the likelihood of immunity is a strictly monotonously increasing function of age. Since natural mortality in adults also increases with age, immune-recovered individuals are disproportionately represented among those dying of natural causes.
- Emigration generally decreases the pool of immune individuals and immigration increases the pool of susceptible individuals. We are more likely to be immune to the pathogens that are endemic where we live, because we have spent our lives around those pathogens. The reverse of this phenomenon, pathogenic invasion (where a population affected by a pathogen comes into proximity with a susceptible population), is relatively rare (for a description of such an event, see Huag and Nilssen [85]'s description of phocine morbillivirus outbreaks due to an invading population of harp seals, *Phoca groenlandica*, sparking off an episodic in the immunologically naive population of harbor seals, *Phoca vitulina*).

Migration is generally not the subject of compartmental models; spatial and agent-based modeling is often much more appropriate to examine the effect of migrational population flows. On the other hand, births and deaths do affect the balance of pathogenic spread at nontrivial timeframes. For this reason, we must now turn our attention to the way to quantify these.

2.5.6.1 Mortality

The normal mortality rate is typically considered to be the inverse of the mean life expectancy. Actuarial tables, also known as life tables, contain much additional information in addition to life expectancy. Reading life tables is something of a practiced skill, but as epidemiologists, we are principally concerned only with two numbers:

- $q(x)$ is, by convention, the first (leftmost) value column of a life table. It shows an x-aged person's probability of dying within the index period, typically a year. These probabilities can be averaged, or they can be used to construct a statistical distribution.
- $e(x)$ is typically on the rightmost side of the table, and indicates the average number of "remaining" years expected for a person of age x. Thus $e(18)$ is the time remaining to reach the mean life expectancy for a person aged 18 today.

It is often important to use the right life tables. Life tables should be reflective of the population under consideration. This requires obtaining life tables for the right population.

Practice Note 2.11 Avoiding bias by weighting demographic parameters

Often, researchers use national figures for $e(x)$ and $q(x)$, commonly referred to as "total populations." This approach loses its validity when considering sub-populations, whose distribution is significantly different from that of the general population. The consequence is that the inherent differences in life expectancies between ethnicities and genders—often mediating complex socio-economic determinants of health—are ignored. For this reason, if the population being modeled differs from the national average, it may need to be re-weighted.

Re-weighting consists of calculating a proportion-weighted average of values. Thus for a population consisting of K segments (e.g., "Caucasian males," "Caucasian females," "African-American males," and so on), where $p(k)$ is the proportion of a segment k in the entire population,

$$\overline{e(x)} = \frac{\sum_{i=1}^{K} p(i) e_k(x)}{\sum_{i=1}^{K} p(i)} \tag{2.58}$$

and

$$\overline{q(x)} = \frac{\sum_{i=1}^{K} p(i) q_k(x)}{\sum_{i=1}^{K} p(i)}. \tag{2.59}$$

2.5.6.2 Birth rate

The birth rate, also known as natality, is the person-time-denominated number of births per unit population per unit time. Typically, the crude birth rate (i.e., the birth rate that is not adjusted to reflect differences in age groups) is reported in births per 1000 persons per year.

Ordinarily, the birth rate is by far the preferred metric from which to infer demographic change in a compartmental model with vital dynamics. In some cases, where the population is significantly imbalanced with regard to gender, it may be helpful to use an adjustment. For instance, the birth rate in the US in 2020 was approximately 11 births per year per 1000 persons, whereas the fertility rate was 56.0 births per 1000 women aged 15–44 years. If there are 70% women aged 15–44 years in a population and 30% men, the effective birth rate will be 39.2 births per year per 1000. Typically, the world population is approximately 3–6% weighted towards males. Where the difference between genders in a population is over 10%, rebalancing using the weighted average method might be indicated.

2.5.7 Estimating waning rate

The waning rate is the inverse of the duration of immunity. In general, our metric of concern is the duration of effective immunity. This is not necessarily the same as the decrease in biomarkers evidencing immunity, such as antibody levels, and it is crucial to look at real-world data, i.e., studies of effective duration of protection, alongside biomarkers.

The mean waning rate is, of course, the inverse of the mean duration of immunity. This is generally quite simple to apply. However, in practice, one should not forget to take into account the variance, which can be quite significant. Human and nonhuman animals' immune systems are equally unique and complex. For instance, we know that certain MHC haplotypes are associated with inferior immune response to the Hepatitis B vaccine [86]. Similarly, it is relatively well understood that individuals suffering from immune suppression (e.g., long term users of steroids or immunosuppressive medications) and older individuals typically develop weaker immune responses (quantified as the number of antibodies per unit volume of serum), and their immune responses are often short-lived in comparison to the general population [87].

Practice Note 2.12 Acquired vs. induced immunity: effects on duration and waning rate

In general, there tends to be no significant difference between acquired immunity (through infection and recovery) and vaccine-induced immunity (through vaccination) in terms of duration and strength of immune response. That being said, there exist well-documented outliers, with *Vibrio cholerae* infection being the most significant from a public health perspective. As Wrammert et al. demonstrated, vaccine-induced immunity to *V. cholerae* wanes quite rapidly, whereas acquired immunity is much broader and endures for up to a decade [88]. Though equal waning rates for vaccinated and recovered individuals are typically a good starting assumption, infectious disease modelers must be ready to adapt their models if real-world data indicates a meaningful divergence.

2.5.8 Multiparameter estimation by nonlinear curve fitting

Curve fitting is a general method that covers a multitude of things. In general, curve fitting methods converge in their overall aim: given a series y and a function $\hat{y} = f_t(\theta)$, find θ so as to minimize a distance metric or "loss" between $\hat{y}$ and y. By far, the most frequent method for this is least-squares estimation, i.e.,

$$\arg\min_{\theta} \sum_{i=1}^{k} \left(y(i) - f_i(\theta)\right)^2. \tag{2.60}$$

This is mathematically convenient, because the minimum of the expression above occurs where the gradient is zero. This obtains where for all $j \in \theta$,

$$\sum y(i) - f_i(\theta) \frac{\partial(y(i) - f_i(\theta))}{\partial \theta_j} = 0. \tag{2.61}$$

For nonlinear systems, this is rarely an easily ascertainable figure, but can be numerically approximated with relative ease.

The caveat of least-squares solvers is that they are the sledgehammers of parameter estimation: highly effective but rather crude. Among the most significant shortcomings of curve fitting algorithms is the fact that outbreaks do not neatly conform to the simple compartmental model's neat sinusoidal curves. It may be somewhat unfair to lay blame for that at the door of curve fitting algorithms, since the root of the issue lies not with the curve fitting algorithm itself but with compartmental models in general, and their inability to accommodate more complex nonlinearities, such as successive waves of infection with accuracy. Curve fitting algorithms are, however, very effective where there is relatively little data and speed is prioritized over accuracy. Thus, a curve-fitting approach can yield useful estimates of multiple parameters in a single outbreak, but would typically struggle with aperiodic waves from emergent strains, as we witnessed with, e.g., the Delta and later the Omicron VOCs of COVID-19.

Computational Note 2.13 Multiparameter estimation with lmfit and Emcee

The advantage of multiparameter estimation is that—as the name suggests—it estimates multiple parameters contemporaneously and in relation to each other. This is particularly useful if we have information about more than one compartment (e.g., number of infectious cases and number of decedents). In this computational example, we are using a SEIRD model to estimate a whole host of parameters all at once: β, γ, σ, and CFR. We do so in three steps:

1. First, we define the model, our starting conditions, and our best guesses as to the parameters.
2. Next, we obtain a best estimate for β, γ, σ, and CFR by performing a Nelder–Mead minimization with `lmfit`. This is often better than the more commonly used Levenberg–Marquardt algorithm, which tends to have a somewhat more difficult time with exponential processes, which is of course the kind of problem that describes epidemics.
3. Finally, we use a Markov chain Monte Carlo (MCMC) simulation to get an idea of how good our estimates are.

Once we have obtained the data (in this case, from Ritchie et al. [84], for the United States, between 06 March 2020 (the first day when cases exceeded a hundred new diagnoses per day) and the end of July 2021) and cut it down to the columns on cases and deaths, we calculate a ten-day rolling sum of cases as a rough estimate of active cases at any one time. From this, we create our X and Y vectors (in keeping with the convention in scientific Python as to what the dependent and independent variable vectors ought to be named):

```
Y = df[["total_cases", "total_deaths"]].dropna().to_numpy()
```

```
X = np.array(list(range(0, len(Y))), dtype=float)
```

Next, we define our SEIRD model:

```
def SEIRD_model(t, y, beta, gamma, sigma, CFR):
    S, E, I, R, D = y
    dSdt = - beta * S * I / N
    dEdt = beta * S * I / N - sigma * E
    dIdt = sigma * E - gamma * I
    dRdt = (1 - CFR) * gamma * I
    dDdt = CFR * gamma * I

    return dSdt, dEdt, dIdt, dRdt, dDdt
```

We also need to define some sensible starting values. We have data on $I(0)$ and $D(0)$. We assume (somewhat naively) that $E(0) = 2I(0)$. Since it is early in the epidemic, we assume there are no recovered individuals yet. This gives us the initial conditions:

```
N = 329.5e6
E0 = 2 * Y[0, 0]
I0 = Y[0, 0]
R0 = 0
D0 = Y[0, 1]
S0 = N - E0 - I0 - D0
starting_values = (S0, E0, I0, R0, D0)
```

Next, we need to describe our parameters, i.e., the vector θ. The "values" we provide here actually matter rather little; their primary purpose is to give the fitting algorithm somewhere to start.

```
theta = lmfit.Parameters()
theta.add("beta", value=0.5, min=0, max=5)
theta.add("gamma", value=1/10, min=0, max=1)
theta.add("sigma", value=1/3, min=0, max=1)
theta.add("CFR", value=0.05, min=0, max=1)
```

Now, we need to provide the solver with a quantification of how good a fit is, given X, Y, the starting conditions, and θ. `lmfit` expects a function that, given X, θ, the starting conditions, and Y, returns $\hat{y}_{i,j} - Y_{i,j}$ as a one-dimensional array. We obtain this by taking any given value of θ and performing the integration, then taking the third and fifth columns (corresponding to the I and D

compartments, for which we have observed data) and subtracting the predicted from the observed value. This is our residual function, which allows `lmfit` to evaluate each proposed value for θ:

```
def residual(theta, X, Y, starting_values=starting_values):
    parvals = theta.valuesdict()
    beta, gamma, sigma, CFR = parvals["beta"], parvals["gamma"], \
                              parvals["sigma"], parvals["CFR"]
    res = solve_ivp(fun=SEIRD_model,
                    t_eval=X,
                    t_span=(X[0], X[-1]),
                    y0=starting_values,
                    args=(beta, gamma, sigma, CFR))

    return (Y.T - res.y[[2, 4], :]).ravel()
```

Note that we need to deconstruct the `Parameters` object `theta`. Ordinarily, iterating over a `Parameters` object yields the keys, i.e., the names of the parameters as strings. Since that is not all that helpful to us, we obtain a dictionary of values with the `.valuesdict()` method, and obtain each of the values by their key.

Now, we can call `lmfit.minimize` on the residual, and the parameter set `theta`:

```
fit = lmfit.minimize(residual,
                     theta,
                     method="nelder",
                     args=(X, Y, starting_values))
```

Calling the `fit` object will provide a quite helpful short summary of the values, including the estimates with their standard error. Our model seems to indicate, for this period, a best fit for $\beta = 0.492$, $\gamma = 0.408$, and a CFR of 0.006 (i.e., 0.6%). These suggest a $\mathfrak{R}_0$ of around 1.2, which is a quite reasonable approximation if we consider that case recognition and reporting in the early days of the pandemic was rather imperfect.

We are finally interested in how good our fit is. Out of the box, calling a fit object or the `report_fit()` function in `lmfit.printfuncs`, provides a range of standard metrics (Q^2, Akaike's information criterion and Bayesian information criterion), but a much better and more illuminating way is to obtain the posterior probability distribution of the parameters, given observed data. For this, we create a fit with Emcee, a package for Markov chain Monte Carlo sampling:

```
res = lmfit.minimize(lambda x: residual(x, X, Y, starting_values),
                     method="emcee",
```

```
                          burn=300,
                          steps=1_500,
                          thin=20,
                          params=theta,
                          is_weighted=False,
                          progress=True)
```

Note that this is not really a fit; we are not fitting anything, and this only works if the results have already been fit once, as we did above. Rather, this uses the fitting API to estimate the posterior distributions. From this, we can construct a corner plot as seen in Fig. 2.27. The corner plot shows the posterior distributions of the parameters with respect to each other. This allows us to reason about the degree to which our estimates for one value are contingent on the value of another, and their relationship.

Since we did not weight our data (we assume that measurement uncertainty is equal for all measurements of the residual), Emcee will try to encapsulate it as the parameter $\ln(\varsigma)$ (which is not the same as σ, the inverse duration of the latent period; unfortunately, we are limited in our choices of symbols far beyond our needs). This appears as the last row of the corner plot, serving as the estimate of uncertainty in the data.

Reading corner plots is nontrivial and requires some experience, but it is a helpful tool to understand how the parameters were estimated and what the effect of an error into a particular direction would be on the other parameters.

A notebook implementing the contents of this Computational Note is available on the book's companion Github repository in the folder `/ch02/multiparameter`.

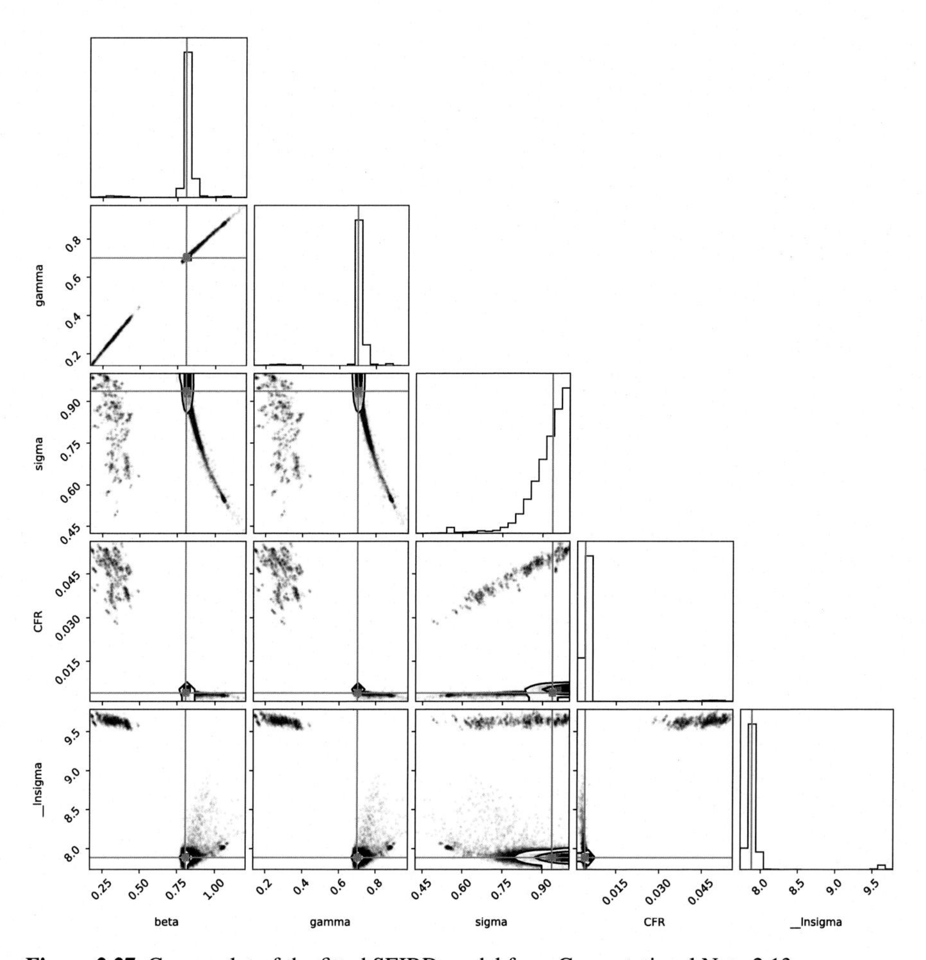

Figure 2.27 Corner plot of the fitted SEIRD model from Computational Note 2.13.

Host factors
Who gets sick and why

[A]ll good things (e.g., stability) are more fragile than bad things. It seems that in good situations a number of requirements must hold simultaneously, while to call a situation bad even one failure suffices.

Arnol'd, Catastrophe theory, 1986 [89]

The term "host factors" denotes those aspects of infectious disease transmission that are inherent in the potential host. Simply put, host factors are what sets an individual apart from others in the context of contracting an infectious disease. Because we are modeling populations, we need to represent host factors often as groups of individuals who share meaningfully similar values of a particular host factor (e.g., age groups, binned numbers of sexual partners, binned values of an exposure variable). We refer to these groupings as strata.

Definition 3.1 (Strata). A **stratum** (*pl.* **strata**) is a modifier that sets individual types of compartment apart by a population characteristic.

For instance, in a model that has separate strata for high- and low-risk behaviors, the high-risk stratum would comprise high-risk susceptible and infectious compartments (S_H and I_H, respectively), whereas the low-risk stratum would comprise low-risk susceptible and infectious compartments (S_L and I_L, respectively).

This chapter deals with situations of heterogeneous populations. In computational epidemiology, strata are our way of representing sets of compartments that "belong together" as members of groups with groupwise heterogeneity, such as higher transmission risk. It is sometimes common to refer to such models as "structured," e.g., "age-structured" for an age stratified model, but in what follow, we will use the term stratification, where risks are represented as categories and strata.

3.1 Heterogeneity of transmission risk

In models where transmission risk is heterogeneous, we are no longer assuming that the transmission likelihood between every individual is the same. Recall that the mass action term βSI assumed homogeneous mixing and equal risk of transmission from any member of I to any member of S (see Subsection 2.1.3). In this section, we will be breaking the first half of that symmetry. In reality, when it comes to passing on or sustaining infections, all are not created equal. For reasons that are not as well understood as their practical importance would warrant, a small number of infected individuals are often responsible for a disproportionately large number of infectious

Computational Modeling of Infectious Disease. https://doi.org/10.1016/B978-0-32-395389-4.00012-8

events (see Subsection 3.1.4). On the other hand, some factors of the clinical course may reduce the likelihood of infecting others, such as isolation and hospitalization (see Subsection 3.1.6). This section describes the mathematical models we can use to make sense of these heterogeneities.

Practice Note 3.1 Meaningful heterogeneity

To an extent, all populations are heterogeneous when it comes to the transmission of infectious disease. When creating models, it is up to you to know whether these differences are relevant. For instance, though women are somewhat more likely to acquire influenzaviral infections, as noted by e.g. Klein, Hodgson, and Robinson [90], the difference is rarely large enough to justify modeling.

As a guideline, heterogeneity is meaningful if

- there is a significant difference in terms of infection between strata (RR $\geq$ 1.5), or
- there is a significant difference in terms of outcomes between strata (RR $\geq$ 1.5), or
- there is a significant issue of health equity involved, e.g., an underserved stratum with a much higher risk.

3.1.1 Case study: determinants of hepatitis C transmission

Hepatitis C virus (HCV) affects approximately 2% of the global population. It is one of the leading causes of liver failure worldwide, and curative treatment is expensive [91].

In the United States and most developing nations, the majority of HCV infection are attributable to intravenous drug use (IVDU), where the pathogen is spread through the reuse and sharing of hypodermic needles. Shiffman [92] notes that over half of intravenous drug users have been exposed to HCV, driving a new epidemic of hepatitis C, especially among younger people.

The additional risk from a certain behavior (in this case, IVDU) may put a subpopulation at a significantly higher risk for exposure. Where such differences in exposure or transmission risk are meaningful, it may be useful to consider accommodating this risk difference in our models. In such cases, we denote the population as risk-heterogeneous.

Definition 3.2 (Risk heterogeneity). A population is **risk-heterogeneous** if a part of the population has a significantly higher or lower risk of acquiring, transmitting, or carrying an infectious disease (i.e., a higher or lower risk of transitioning out of the susceptible, and into an infected, compartment).

Heterogeneities may come in one of three forms:

- behavioral heterogeneities (e.g., high-risk sexual behaviors),
- exposure heterogeneities (e.g., working in a profession where exposure to blood-borne pathogens is more likely, such as EMTs and emergency physicians), and

- susceptibility heterogeneities (e.g., iatrogenic immunosuppression or old age).

Practice Note 3.2 Avoiding undue stigma

Discussing higher and lower risk behaviors can easily move into the territory of stigmatizing people who engage in what we epidemiologically regard as "higher risk." Not only is this analytically unhelpful, it is also counterproductive from a public health perspective. When we speak of high-risk behaviors, we must understand that some—indeed, many!—of these behaviors are not voluntary choices. Although we use the terminology of behaviors, these are often the consequence of givens and circumstances, such as certain socioeconomic determinants of health.

When writing about risk heterogeneity, remember to be sensitive to the subject and avoid stigmatising language wherever possible.

Heterogeneities of multiple sources may often coexist. For HCV, the main heterogeneity is, of course, behavioral (i.e., IVDU). For other pathogens, however, there may be multiple risk factors. For HIV, for instance, high-risk sexual behaviors constitute a behavioral heterogeneity in the population. On the other hand, some alleles of the CCR5 protein (specifically, CCR5-Δ32) have a protective effect against infection by macrophagotropic strains of HIV-1, because they interfere with viral entry. The latter is a susceptibility heterogeneity, present in the protective homozygous form in about 0.1% of the population. When encountering a large number of heterogeneities, it is often useful to assess whether they are meaningful enough to be modeled (see Practice Note 3.1).

3.1.2 Modeling risk-heterogeneous populations

A risk-heterogeneous population has two or more distinct subgroups that exhibit risk heterogeneity (see Definition 3.2).

Let us consider a simple case of a population made up of two groups, high-risk and low-risk, denoted by the subscripts H and L, respectively (see Fig. 3.1). We know from our examination of the mass action term (in Subsection 2.1.3) that the infectious effect is the product of the transmission coefficient β, the susceptible population, and the infected population. There are four distinct processes going on in our model of two separate risk groups: transmissions inside the groups (high-risk to other high-risk individuals and low-risk to low-risk individuals) and transmissions between groups (high-risk to low-risk individuals and low-risk to high-risk individuals). We may conveniently represent this as a matrix $\mathbf{b}$ of transmission probabilities, where by convention $\mathbf{b}_{i,j}$ means the likelihood of transmission from j to i

$$\mathbf{b} = \begin{pmatrix} \beta_{H,H} & \beta_{H,L} \\ \beta_{L,H} & \beta_{L,L} \end{pmatrix}. \tag{3.1}$$

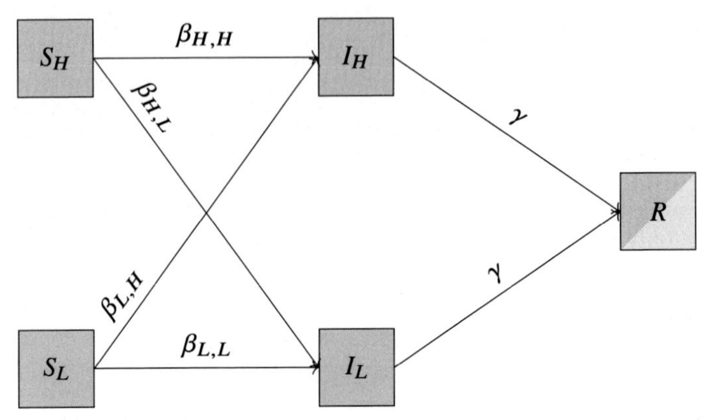

Figure 3.1 A SIR model with two differential risk classes. $\beta_{i,j}$ denotes the transmissibility from j to i.

This matrix is sometimes referred to as a WAIFW (who acquires infection from whom) matrix.

Definition 3.3 (WAIFW matrix). A WAIFW (who acquires infection from whom) matrix $\mathbf{b}$ for a model with m strata is an $m \times m$ matrix, where $\mathbf{b}_{i,j} = \beta_{j \to i}$.

The WAIFW matrix is a convenient mathematical shorthand for the values of β within the separate differential equations for the infectious subsystem $\{I_H, I_L\}$ in a $SI_{H,L}R$ model:

$$\begin{aligned} \frac{dI_H}{dt} &= \overbrace{\beta_{H,H} S_H I_H}^{H \to H} + \overbrace{\beta_{H,L} S_H I_L}^{H \to L} - \overbrace{\gamma I_H}^{\text{recovery}}, \\ \frac{dI_L}{dt} &= \underbrace{\beta_{L,H} S_L I_L}_{L \to H} + \underbrace{\beta_{L,L} S_L I_L}_{L \to L} - \underbrace{\gamma I_L}_{\text{recovery}}. \end{aligned} \tag{3.2}$$

Eq. (3.2) corresponds, given the matrix $\mathbf{b}$ from Eq. (3.1) and a vector $\vec{\gamma} = \begin{pmatrix} \gamma_H \\ \gamma_L \end{pmatrix}$, to

$$\frac{dI_i}{dt} = \underbrace{S_i \sum_{j=1}^{n} \overbrace{\mathbf{b}_{i,j}}^{\text{WAIFW matrix}} I_j}_{\text{mass action}} - \overbrace{\gamma_i}^{\text{vector of recovery rates by stratum}} I_i. \tag{3.3}$$

Where there are multiple subpopulations of each of the compartments, it may often be a sensible choice to represent different transitions between those populations as WAIFW matrices and nondirectional subpopulation-dependent factors (such as γ or μ, which depend solely on the subpopulation to which they relate) as vectors.

3.1.2.1 Analysis of the WAIFW matrix

In our previous models, we assumed homogeneous mixing (see Definition 2.11), and thus we had a single compartment S and a single compartment I, with a single coefficient β. The WAIFW matrix breaks down β into components. Thus where the WAIFW matrix's elements have been empirically ascertained, we can obtain some insights about the underlying process.

- The symmetry of $\mathbf{b}$ indicates the equality of response. If $\mathbf{b}_{i,j} = \mathbf{b}_{j,i}$, then the response is not directionally dependent. This is the case where belonging to a risk group affects the likelihood of transmission and exposure, but not the likelihood of becoming infected upon exposure. We refer to this as equiresponse (see Definition 3.4). Conversely, where being in a particular risk group is associated with risk factors that also make infection upon exposure more likely (e.g., the ratio of persons suffering from immune deficiencies is higher among intravenous drug users than the general population), the matrix will not be symmetrical, because the risk groups are not equirespondent.
- The dominance of the diagonal (see Definition 3.5) indicates the degree of assortative mixing. In general, members of risk groups tend to mix among each other. In fact, for behaviorally defined risk groups, interacting with a high-risk individual is part of the way high-risk behaviors are defined (e.g., needle sharing in the context of IVDU). The degree to which the diagonal is dominant indicates the degree of assortative mixing.
- The relative values of $\mathbf{b}$'s columns indicate the overall "spreading potential" of the pathogen. The out-of-group spreading of a risk group indicates the exposure hazard to which a compartment's members of that group expose the other groups.

Definition 3.4 (Equiresponse). The risk groups P and Q are said to be **equirespondent** to the infection in question if $\beta_{P,Q} = \beta_{Q,P}$ (i.e., if the likelihood of acquiring infection upon interaction is the same, regardless of which of the equirespondent risk groups an individual belongs to).

Definition 3.5 (Diagonal dominance). A matrix $\mathbf{b}$ is **diagonally dominant** if for all i, j, $\sum_{i \neq j} |\mathbf{b}_{i,i}| \geq |\mathbf{b}_{i,j}|$.

3.1.2.2 Coupled dynamics in risk-heterogeneous models

Infectious processes in risk-heterogeneous models go through three distinct phases, as shown in Fig. 3.2:

1. Initially (1), the proportions of the model are driven by the initial distributions of high- and low-risk individuals within the initially infected population.
2. In the next period (2), the number of infected individuals rises exponentially.
3. In the asymptotic phase (3), the model converges on the endemic equilibrium $\lim_{t \to \infty} I(t)$.

During the second phase, both risk groups follow the same dynamics, determined entirely by $\mathfrak{R}_0$. The risk groups in this state are said to be coupled (or, in some texts, such as Keeling and Rohani [39], *slaved*).

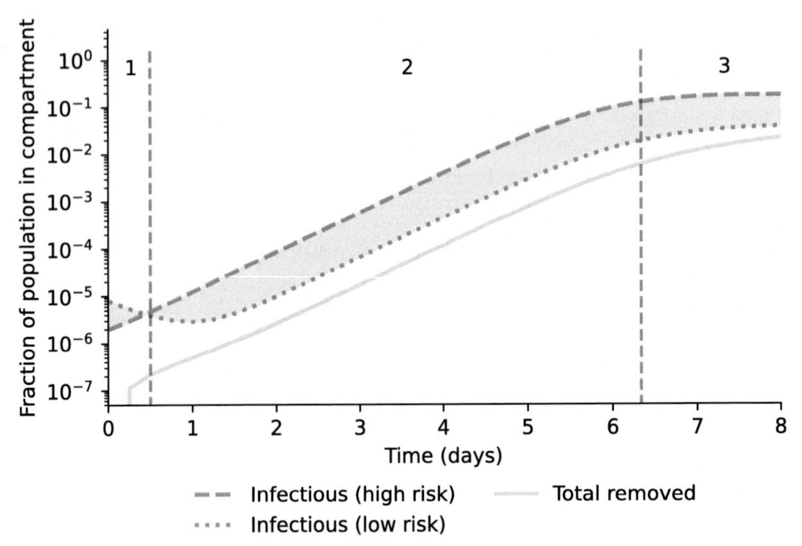

Figure 3.2 Solutions for the infectious compartment in a SIR model with two risk classes. The stages of the model's evolution are noted above the plots.

3.1.3 $\mathfrak{R}_0$ *for risk-heterogeneous populations*

As we have seen, $\mathfrak{R}_0$ determines the dynamics in the coupled phase of the infection's evolution. However, unlike in previous models, there is not a single value of β, and hence, of $\mathfrak{R}_0$ (recall that $\mathfrak{R}_0 = \frac{\beta}{\gamma}$). Consequently, we will have to adjust our calculation of $\mathfrak{R}_0$.

During the coupled phase, $\mathfrak{R}_0$ is driving the dynamics of both I_H and I_L:

$$\begin{aligned} I_H(t) &\propto e^{\mathfrak{R}_0 t}, \\ I_L(t) &\propto e^{\mathfrak{R}_0 t}. \end{aligned} \tag{3.4}$$

Note that because of the coupling, $\mathfrak{R}_0$ is the same for both risk groups.

Given the WAIFW matrix and population proportions of each risk group, as well as reasonable approximations for γ, we can use symbolic computation followed by numerical evaluation to characterize the dynamics of the infectious process during the coupled phase (see Computational Note 3.1).

Computational Note 3.1 Calculating the $\mathfrak{R}_0$ of complex stratified models

Calculating the $\mathfrak{R}_0$ of a stratified model, even one apparently quite simple, can lead to extremely complex (and mathematically ugly) results. Symbolic computation with later evaluation can be used in this situation with great effect.

Consider the system described in (3.2). We will primarily consider the infected subsystem, i.e., I_H and I_L.

Let us first define our variables:

```
import numpy as np
from sympy.interactive.printing import init_printing
init_printing(use_unicode=True, wrap_line=True)
from sympy.matrices import Matrix
from sympy import symbols

I_H, I_L, beta_HH, beta_HL, beta_LH, beta_LL, gamma_H, gamma_L,
    n_H, n_L =
    symbols("I_H I_L beta_HH beta_HL beta_LH beta_LL \
              gamma_H gamma_L n_H n_L")
```

Next, we need to define the ODE of the infectious system as two column vectors: one containing the compartments and one containing their definitions.

```
infectious_system = Matrix(np.array([I_H, I_L]).T)

d_Is = Matrix(np.array([(beta_HH * n_H - gamma_H) * I_H \
                        + (beta_HL * n_H) * I_L,
                        beta_LH * n_L * I_H \
                        + (beta_LL * n_L - gamma_L) * I_L]).T)
```

Now, we can take the Jacobian of the compartments to get the matrix of coefficients:

```
coeffs = d_Is.jacobian(infectious_system)
```

$$\begin{bmatrix} \beta_{HH}n_H - \gamma_H & \beta_{HL}n_H \\ \beta_{LH}n_L & \beta_{LL}n_L - \gamma_L \end{bmatrix}$$

We could, technically, proceed entirely symbolically from here onwards. However, the number of variables means that the resultant expression will be inevitably rather messy. Fortunately, `SymPy` allows us to replace symbols with actual values using the `.subs` method, which takes a dictionary of symbol keys paired to the values that are to be substituted for them. We will be using a rather reasonable WAIFW matrix, namely

$$\mathbf{b} = \begin{pmatrix} 10 & 0.5 \\ 0.5 & 2 \end{pmatrix}.$$

Alongside that, we will assume that 20% of the population belonging to the high risk group, and we will also assume that both groups have the same removal rate of 0.05. We create the matrix J through substitution using `.subs` on the Jacobian created in the previous step:

```
J = coeffs.subs({beta_HH: 10,
                 beta_HL: 0.5,
                 beta_LH: 0.5,
                 beta_LL: 2,
                 n_H: 0.2,
                 n_L: 0.8,
                 gamma_H: 0.05,
                 gamma_L: 0.05})
```

$$\begin{bmatrix} 1.95 & 0.1 \\ 0.4 & 1.55 \end{bmatrix}$$

Now, we can calculate an approximation of $\mathfrak{R}_0$ for the coupled period. This is best accomplished by creating a list of the absolute values of the spectrum (returned by `J.eigenvals()`) and taking the maximum:

```
R_0 = max([abs(i) for i in J.eigenvals()])
```

```
>>> 2.03284271247462
```

Thus we can now characterize the infectious dynamics during the coupled phase as proportional to the term $e^{\mathfrak{R}_0 t}$.

A notebook implementing the contents of this Computational Note is available on the book's companion Github repository in the folder `/ch03/stratified_r0`.

The computational approach outlined above is particularly useful, because it gives us insights about the eventual fate of the pathogen in the population. $\mathfrak{R}_0$ alone does not determine the dynamics of the whole disease process, especially not the transient interval in the beginning, which is largely governed by initial population ratios and, in real-world use cases, stochasticity, but it does determine the long-term fate of the infectious process. As such, if the overall $\mathfrak{R}_0$ as calculated using the method in Computational Note 3.1 is below zero, the infection will eventually die out. This can be accelerated through targeted interventions, in particular vaccinating the strata that have a higher stratum-wise $\mathfrak{R}_0$.

3.1.4 Superspreading and supershedding

During the SARS outbreak, it was observed that a relatively small number of infected persons were responsible for a relatively large number of secondary cases. In epidemiology, groups of individuals with a significantly higher likelihood to infect others are called "supershedders," although the term does not really have a settled definition. For instance, the WAIFW matrix

$$\mathbf{b}_1 = \begin{pmatrix} 10 & 3 & 1 \\ 9 & 3 & 2 \\ 9 & 2 & 3 \end{pmatrix} \tag{3.5}$$

suggests supershedding by the first compartment (first column). Note that supershedding merely indicates higher likelihood to transmit the disease, not to acquire it. Supershedding is primarily a biological phenomenon and only relatively weakly behavioral.

On the other hand, there also exist individuals who are both more likely to be infected and to pass on infection. Consider the WAIFW matrix

$$\mathbf{b}_1 = \begin{pmatrix} 10 & 7 & 8 \\ 9 & 3 & 2 \\ 9 & 2 & 3 \end{pmatrix}, \tag{3.6}$$

which indicates that people in the first compartment are more likely not only to pass on the infection but also to be infected in the first place. We call these individuals superspreaders.

Lloyd-Smith et al. [93] define superspreading in a statistically rigorous way as instances drawn from the right-hand tail of a continuous probability distribution centered around $\mathfrak{R}_0$. In other words, there will be individuals who are significantly more likely to spread the infection, just as there will be individuals who are significantly less so. The latter, of course, generally escape epidemiological notice. An often used metric of superspreading likelihood is the t_{20} proportion, which plots the part of transmission that is attributable to the most infections 20% of cases. As Lloyd-Smith et al. [93] have shown, the often-cited Pareto "20/80" relationship, where 20% of cases cause 80% of transmission [94] does not universally hold. In other words, the dispersion parameter of the probability distribution from which an individual's propensity to cause secondary cases is drawn varies from disease to disease and outbreak to outbreak. Table 3.1 shows the t_{20} values of a number of outbreaks, illustrating the wide disparity that exists between diseases in their potential to cause superspreading events. Interpretation and discussion of "superspreading" must therefore be in light of the limitations, both definitional and practical, of the concept.

Table 3.1 t_{20} of some infectious diseases at vaccination rates, based on Lloyd-Smith et al. [93].

Pathogen	% vaccination level	t_{20}
SARS	0%	0.88
Measles	95%	0.82
Smallpox	20–70%	0.74
Smallpox	80%	0.71
Monkeypox	70%	0.62
Pneumonic plague	0%	0.47
Hantavirus	0%	0.45
ZEBOV	0%	0.34

Practice Note 3.3 Superspreaders and superspreader events

There is an unhelpful tendency to regard superspreaders and events where superspreading has occurred as anomalies out of the ordinary. This contributes relatively little to our understanding of infectious dynamics and is bound to exacerbate the stigmatization of individuals, as it has, e.g., during the early years of AIDS, when much sensationalistic and unjustified blame was laid at the feet of early HIV patient Gaëtan Dugas (see McKay [95]). Rather, superspreading is one "tail" of a distribution prominent mainly because it is noticeable: statistical models predict that there are generally an equal number of "greatly inferior spreaders" who are particularly ineffective in spreading the illness. The significance of "superspreading" therefore is peculiar to the pathogen, the population and a wide range of other factors.

It should not be assumed, absent evidence, to apply to any pathogen. Nor is superspreading particularly predictable. While all things being equal, events with close physical proximity and no barrier precautions are more likely to become superspreader events, there exists no scientifically sound methodology to extrapolate the likelihood of this occurring purely from the number of attendees or the manner an event is organized. The heavy-tailed distribution of superspreading indicates that the right (or, rather, wrong) person at the right time can turn an intimate family gathering into a superspreader event. It is therefore crucial not to equate superspreading with large-scale events and be aware of the superspreading risk at small events. Superspreaders can create vast numbers of secondaries, even without meeting larger groups of people at any point. Thus avoiding mass gatherings, though a sensible precaution amidst an epidemic, is no guarantee against falling victim to superspreading.

When discussing superspreading, infectious disease modelers should refrain from assigning blame or engaging in moral discourse. Superspreading is frequently not within the patient's control, and especially amidst an epidemic, branding individuals as particularly infectious or even reckless may elicit hostility, exclusion, and even violence. Many superspreader events continue to elude a clear causal explanation, and approaching the issue from a clear scientific perspective is paramount, as is avoiding stigma and steering clear of moral discourse.

3.1.5 Treatment effect

For many infectious diseases, treatment is largely supportive. This is the case in particular for those viral diseases that do not have a specific antiviral therapy. For other diseases, treatment may affect transmission rates. In those cases, it may be important to account for the effect of treatment as a separate compartment.

We consider a hypothetical treatment that, upon initiation, reduces transmission by a given percentage. We reflect this in the value β_T. Furthermore, treatment lasts τ_T days, after which patients are cured (recovered). For the sake of simplicity, we will assume furthermore that this treatment has a uniform effect and perfect effectiveness. Fig. 3.3 illustrates this model structure.

This gives us the system of ordinary differential equations outlined in Eq. (3.7).

$$\begin{aligned}
\frac{dI}{dt} &= \underbrace{\underbrace{\beta S I}_{\text{from untreated}} + \underbrace{\beta_T S T}_{\text{from treated}}}_{\text{mass action}} - \underbrace{\underbrace{\gamma I}_{\text{untreated}} - \underbrace{\theta I}_{\text{treated}}}_{\text{removals}}, \\
\frac{dT}{dt} &= \underbrace{\theta I}_{\text{treatment influx}} - \underbrace{\frac{1}{\tau_T} T}_{\text{treated recoveries}}, \\
\frac{dR}{dt} &= \underbrace{\gamma I}_{\text{recovery}} + \underbrace{\frac{1}{\tau_T} T}_{\text{treated recoveries}}.
\end{aligned} \tag{3.7}$$

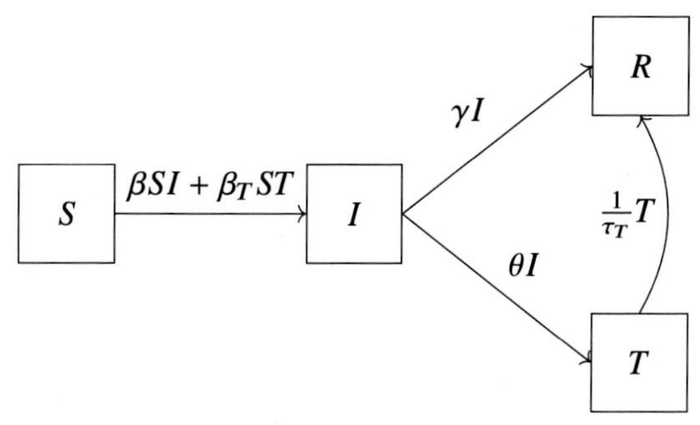

Figure 3.3 A SIRT model of differential treatment. θ is the rate at which patients are treated. β_T is the value of β specific to the treated.

	G	H
G	10	5
H	0	6

Figure 3.4 WAIFW matrix with hospitalized and community transmission. G denotes the general population, H denotes the hospitalized population. Dark red indicates community transmission, dark blue denotes nosocomial transmission. Since only infected persons are in the hospitalized compartment, there is no $G \to H$ transmission.

It emerges from this model that a higher rate of treatment, as well as a higher treatment efficacy θ, can suppress the overall intensity of an outbreak.

A variant of this model is the delayed-effect treatment model. This model accounts for the fact that initiation of therapy does not immediately stop infectiousness. How long this period is, and consequently how meaningful it is to model such a period, depends both on the disease and the treatment. Bactericidal antibiotics typically result in a faster suppression of infectiousness than bacteriostatic agents, and some diseases have significantly longer post-treatment infectious periods than others.

3.1.6 Hospitalization

Where a part of infectious cases are hospitalized, the transmission dynamics change. Hospitalized cases have relatively few interactions with the general population, and as a consequence, the two main routes of transmission are within the general population and within the hospitalized population (nosocomial transmission). Fig. 3.4 shows a typical WAIFW matrix for such a scenario.

- The general population is usually relatively more likely to transmit (community transmission) than in-hospital transmission (nosocomial transmission). There are some exceptions to this assumption, however: some pathogens are relatively more frequent in healthcare settings than "in the wild." The classic example is, of course, methicillin-resistant *Staphylococcus aureus* (MRSA).
- Nosocomial spread tends to be lower than spread in a general population: patients are often confined spatially, to wards or treatment units. Indeed, this is largely the governing principle in the organization of SITTUs (severe infectious temporary treatment units) and Ebola treatment units (ETUs) [96].
- $H \to G$ spread involves transmission from hospitalized individuals to the general population who come into contact with them. By far, the largest category of $H \to G$ spread is infection of healthcare workers. The COVID-19 pandemic has highlighted the risk healthcare workers are exposed to in the course of their work, and their contacts with the community in turn can spark off tertiary chains of transmission.

Hospitalization models can, of course, be combined with treatment models. In recent years, especially in the context of the COVID-19 pandemic, multitiered hospitalization models have also been widely used. These distinguish between patients by level of care, e.g., ICU, medical wards, and ER/urgent care. Such models are useful to chart healthcare utilization and remaining capacity. For this reason, models of hospitalization are often useful, even where only a small fraction of individuals suffering from a particular infectious disease ever seek care or are hospitalized.

3.1.7 Vulnerability estimation from WAIFW matrices

A vulnerability score is an individual metric of a particular population's risk relative to others. With other factors, a vulnerability score might form part of a vulnerability index, an overall estimate of a population's risk. For instance, during the COVID-19 pandemic, the pandemic vulnerability index (PVI) was used to aggregate risk from infectious dynamics (number of cases), public health interventions (social distancing and testing), interaction factors (population density), and exogenous variables (air pollution, comorbidities) [97]. In the WAIFW matrix, the n-th row represents the transmission risk of the n-th stratum from all strata (including itself). We will call this the stratal exposure of the n-th stratum given the WAIFW matrix $\mathbf{b}$.

Definition 3.6 (Stratal exposure). The stratal exposure of the n-th stratum, given the $m \times m$ WAIFW matrix $\mathbf{b}$, denoted $\mathcal{E}_n(\mathbf{b})$, is the sum of the n-th row of $\mathbf{b}$, i.e.,

$$\mathcal{E}_n(\mathbf{b}) = \sum_{j=1}^{m} \mathbf{b}_{n,j}.$$

We can now represent each stratum's risk score as the fraction

$$\frac{\mathcal{E}_n(\mathbf{b})}{\sum_{k=1}^{m} \mathcal{E}_k(\mathbf{b})}.$$

This is the proportionate share of the overall risk that the n-th stratum is exposed to.

3.2 Continuous and semicontinuous heterogeneities

Some heterogeneities are discrete; for instance, a person may or may not engage in intravenous drug use. On the other hand, heterogeneities may also be continuous. Consider, for instance, the number of distinct sexual partners per unit time. This may take any nonnegative real value. Clearly a higher number of distinct sexual partners puts one at a higher risk, statistically speaking, of sexually transmitted infections. However, this may not be particularly easy to model. In practice, three approaches have emerged: semidiscrete heterogeneities, discretization, and continuous modeling. Discretization is by far the most widely used method in the computational realm. Many continuous variables are amenable to discretize modeling, and the most frequently used continuous variable, age, is particularly so.

3.2.1 Case study: age-dependent transmission heterogeneities

Age is a crucial factor in epidemiology, because it mediates a range of factors that affect infectious processes in populations.

Age affects mixing affinity. In assortativity of subpopulations, the rule "like attracts like" holds true. Most people spend a considerable part of their time with others who are, roughly, within the same age group. A high school student's time away from the family home will be spent almost entirely with others who are within a few years' of their age. Moreover, age mediates behaviors and social interactions: young children spend relatively little time interacting with others, and reaching the age of mandatory schooling (typically around the 6th year of life) rapidly increases one's number of social encounters. Departure from the active workforce, as is typical towards the 6th to 7th decade of life in Western societies, has the reverse effect.

Age also affects immune response, however, and consequently, the likelihood that an infection will lead to transmission. The very young and the very old are less likely to mount robust immune responses to a pathogen, and consequently more likely to suffer prolonged illness (decreasing γ and increasing the mean infectious period). They are also less likely to have effective immunity; some children may be too young to immunize, and a reduced immune response is common among the elderly and is addressed, e.g., by using vaccines that are more immunogenic, such as adjuvanted influenzavirus vaccines. Meanwhile, young children are often less able to perform the fundamental hygiene behaviors that keep older humans healthy.

3.2.2 Semidiscrete heterogeneities

The power (and beauty) of computational epidemiology is that it can leverage computation to easily solve what would otherwise be quite difficult. Consider the simulated network of sexual contacts created by Rocha, Liljeros, and Holme [98].

This simulation comprises 50,632 sexual encounters between over 16,000 simulated individuals, whose number of distinct sexual partners ranges from one to 615. We could, of course, stratify this into a number of sexual risk classes. However, we may be able to do better if we consider every person's degree (the number of distinct individuals they are connected to) as a risk stratum of its own.

The problem is, of course, that would yield a 149-by-149 sized WAIFW matrix, which is unwieldy to manage at the best of times. Fortunately, we do not need to. We can compute the WAIFW matrix for a vast problem space quite easily based on empirical data.

Practice Note 3.4 Patient zero

"Patient zero" is one of the most widely known terms of epidemiology, which is startling, in view of the fact that epidemiologists do not use it, the correct term being "index case." In fact, its origins lie in a report on a sexual contact network in the early days of the HIV/AIDS pandemic, included in this book as Fig. 2.24 [78]. This study denoted the node in the middle of the transmission network as "O," since he had a number of sexual contacts in California but was out-of-state (hence "O"). Through a misreading, "O" became "zero," and linked the tragic fate of a young man to a perception of being the "source" of HIV in America.

Nothing could be further from the truth. In fact, the Canadian flight attendant Gaëtan Dugas, the person behind the moniker, was neither an extraordinarily prolific spreader of HIV, nor a particularly early case. A phylogenetic study by Worobey et al. conclusively disproves the claim that Dugas was a "first" in any sense [99]. In fact, the numbering is not in temporal order, but rather order of receiving reports from state health agencies. The alleged "zero" was, in fact, the fifty-seventh reported case [99]. And a decade and a half prior to Auerbach et al.'s 1984 paper, a 15-year-old male has succumbed to what today is rather unambiguously considered to be the first case of AIDS in the United States [100].

The lesson of Patient "Zero" is that networks are rarely comprehensive. The networks we examine are often subsets of larger networks, and thus notions of origin and centrality are often limited to this subset of data, rather than being reflective of wider realities.

We assume that $\mathbf{b}_{i,j} = \frac{\mathbf{M}_{i,j}}{n_i n_j}$, where $\mathbf{M}$ is the degree mixing matrix, that is, $M_{i,j}$ is the number of individuals of the ith degree who are connected to an individual of the jth degree. Note that our matrix does not include compartments that have no members. Therefore the ith column or row might not necessarily denote degree i, but rather the ith in the ascending list of degrees. The computational approach to this is outlined in Computational Note 3.2.

Computational Note 3.2 Calculating the mixing matrix from a contact network

In this computational example, we will be using `networkx`, a widely used Python package for network analysis, with the data set from Rocha, Liljeros, and Holme [98]. This data set is available as a CSV file as supporting information (S1) to the article, which can be read into NetworkX by first reading it into a Pandas data frame (using the `pd.read_csv` function), then creating a graph object:

```
import pandas as pd
import numpy as np
import networkx as nx

df = pd.read_csv("pcbi.1001109.s001.csv",
                 skiprows=24,
```

```
                    header=None,
                    sep=";")[[0, 1]]
df = df.rename(columns={0: "F", 1: "M"})

n = nx.convert_matrix.from_pandas_edgelist(df,
                                           source="F",
                                           target="M",
                                           create_using=nx.MultiGraph)
```

Now that we have our graph object `n`, we can determine the two parts of the matrix **b**: the mixing matrix (numerator) and the outer product of the number of individuals in each of the "risk groups," i.e., the number of nodes by the number of their sexual partners.

We calculate the mixing matrix using NetworkX's built-in functionality:

```
mixing_matrix = nx.degree_mixing_matrix(n, normalized=False)
```

And we calculate the sizes of the risk groups by counting distinct values of sexual encounters (i.e., the degree list of the graph `n`):

```
number_by_degree = pd.DataFrame(list(n.degree),
                                columns=["id", "degree"]).\
                                groupby("degree").\
                                count().\
                                rename(columns= {"id": "count"})
```

Now, we can obtain our matrix **b** by dividing the mixing matrix by the outer product:

```
betas = mixing_matrix / np.outer(number_by_degree, number_by_degree)
```

Because this uses NumPy's built-in vector arithmetic, it is remarkably fast; results are in a few seconds.

This example illustrates how we can gain insights from very complex models. Transmission networks are particularly important for STIs and for contact tracing in the initial phases of an outbreak. The **b** matrix we obtained can be used to parameterize any of the compartmental models we have discussed previously, with highly granular and individualized data.

A notebook implementing the contents of this Computational Note is available on the book's companion Github repository in the folder `/ch03/contact_waifw`.

The calculation of a mixing matrix with hundreds of classes illustrates a profound point about computational methods: for large problem sizes that would be difficult to

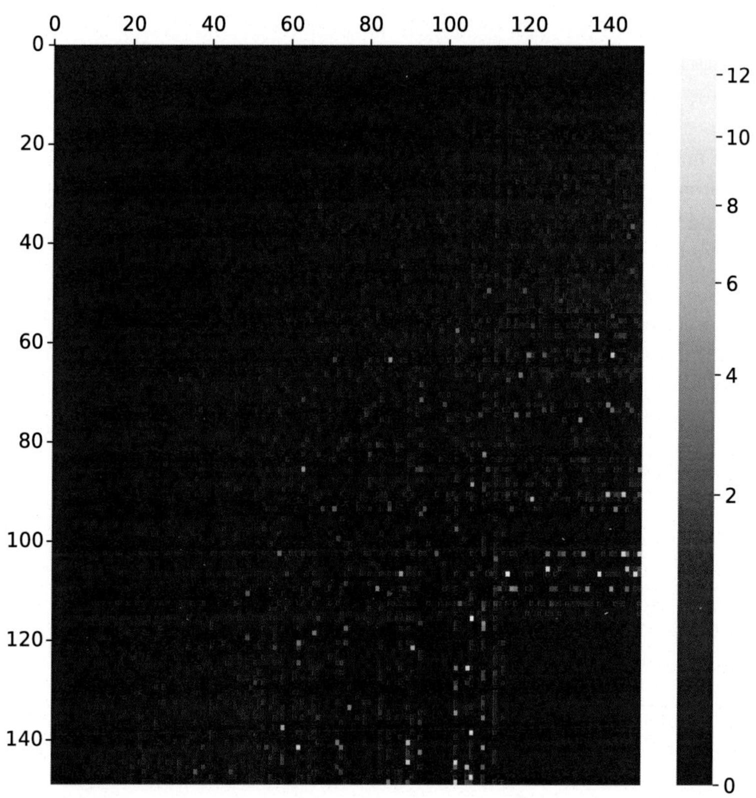

Figure 3.5 WAIFW matrix for the simulated network of sexual contacts in Rocha, Liljeros, and Holme [98] stratified by number of sexual partners.

tackle with analytical methods, computational approaches can provide highly detailed and accurate models. (See Fig. 3.5.)

3.2.3 Discretized continuous heterogeneities

In some cases, there is a meaningful transition in terms of a property to stratify. An example is age: individuals only age in one direction, that being forward. It is possible, especially if age groups typically mix only between themselves as is the case in the context of pediatric infectious disease in school-age children, to create a model that discretizes the continuous heterogeneity of age.

Unlike previous discrete models, a crucial difference is that we have to account for "natural" transitions between age groups. Consider, for instance, a Sixth Form house at a boarding school (Sixth Form comprises Upper Sixth and Lower Sixth, corresponding to years 12 and 13, in the British educational system). Pupils in the Lower Sixth typically associate with others in the Lower Sixth, and vice versa. Nevertheless, there is some interaction between them, being under the same roof, and consequently the mass action term for each group will have to include infectious individuals from both age groups. We account for natural aging-out (transitioning from Lower to Upper Sixth) by the rate ν_L and additions to the population (enrollment) by μ, noting that the two tend to be largely equal.

We can now extend the SIR model with heterogeneous risks (see Subsection 3.1.2) with age progression. (3.8) describes such a model, for the rate of aging ν (the rate of transitioning from Lower to Upper Sixth), a population growth rate μ and a population leaving rate θ (which are typically the class size over a year).

$$
\begin{aligned}
\frac{dS_L}{dt} &= -\overbrace{S_L(\overbrace{\beta_{L,L}I_L}^{L\to L}+\overbrace{\beta_{L,U}I_U}^{L\to U})}^{\text{mass action}} - \nu_L S_L + \mu(S_L + I_L + R_L),\\
\frac{dS_U}{dt} &= -\underbrace{S_U(\underbrace{\beta_{U,U}I_U}_{U\to U}+\underbrace{\beta_{U,L}I_L}_{U\to L})}_{\text{mass action}} + \nu_L S_L,\\
\frac{dI_L}{dt} &= \underbrace{S_L(\beta_{L,L}I_L+\beta_{L,U}I_U)}_{\text{new infections}} - \underbrace{\nu_L I_L}_{\text{aging}} - \underbrace{\gamma I_L}_{\text{recovery}},\\
\frac{dI_U}{dt} &= \underbrace{S_U(\beta_{U,U}I_U+\beta_{U,L}I_L)}_{\text{new cases}} + \underbrace{\nu_L I_L}_{\text{aging}} - \underbrace{\gamma I_U}_{\text{recovery}},\\
\frac{dR_L}{dt} &= \underbrace{\gamma I_L}_{\text{recoveries}} - \underbrace{\nu_L R_L}_{\text{aging}},\\
\frac{dR_U}{dt} &= \underbrace{\gamma I_U}_{\text{recoveries}} - \underbrace{\theta R_U}_{\text{graduation}}.
\end{aligned}
\tag{3.8}
$$

Computational Note 3.3 A simple age-heterogeneity model

In this computational example, we will use the tried and true methods from previous applications to solve the scenario from (3.8). For this, we will use the WAIFW matrix

$$
\mathbf{b} = \begin{pmatrix} 8 & 4 \\ 4 & 8 \end{pmatrix}.
$$

The symmetricity of the WAIFW matrix suggests that there will largely be a coupled behavior, but the differential equations from (3.8) indicate that a lot of the dynamics are highly dependent on population change. Because all population growth initially accrues to the Lower Sixth, and because these are deemed immunologically naive, the relative proportion of recovered will decrease over time as the population grows.

We begin by initializing our parameters $\mu = 0.01$, $\nu = 0.02$ and $\theta = 0.015$, with $\gamma = \frac{1}{20}$. Next, we implement the differential equations above as the differential function. For convenience, this functional definition tracks recovered groups separately for Upper and Lower Sixth, but of course there is no reason why they could not be considered jointly.

```
def deriv(t, y, beta, gamma, mu, nu, theta):
    S_L, I_L, S_U, I_U, R_U, R_L = y

    dSLdt = - S_L * (beta[0] * I_L + beta[1] * I_U) - nu * S_L \
            + mu * (S_L + I_L + R_L)
    dSUdt = - S_U * (beta[2] * I_L + beta[3] * I_U) + nu * S_L \

    dILdt = S_L * (beta[0] * I_L + beta[1] * I_U) - nu * I_L \
            - gamma * I_L
    dIUdt = S_U * (beta[2] * I_L + beta[3] * I_U) + nu * I_L \
            - gamma * I_U

    dRLdt = gamma * I_L - nu * R_L
    dRUdt = gamma * I_U + nu * R_L - theta * R_U

    return dSLdt, dILdt, dSUdt, dIUdt, dRUdt, dRLdt
```

Now, we can solve this as an ordinary initial value problem.

```
res = solve_ivp(fun=deriv,
                t_span = (0, 100),
                y0=y_0,
                args=(beta, gamma, mu, nu, theta),
                max_step=1,
                method="BDF")
```

A notebook implementing the contents of this Computational Note is available on the book's companion Github repository in the folder `/ch03/age_differential_sir`.

As Fig. 3.6 shows, the proportion of individuals who are immune in the Lower Sixth is gradually decreasing, while the proportion of individuals who are immune in the Upper Sixth grows steadily. The reason behind that is that all population growth accrues to S_L, since the only way to access the Upper Sixth compartments is to go through the Lower Sixth. Consequently, the Lower Sixth stratum is fed a constant stream of immunologically naive individuals.

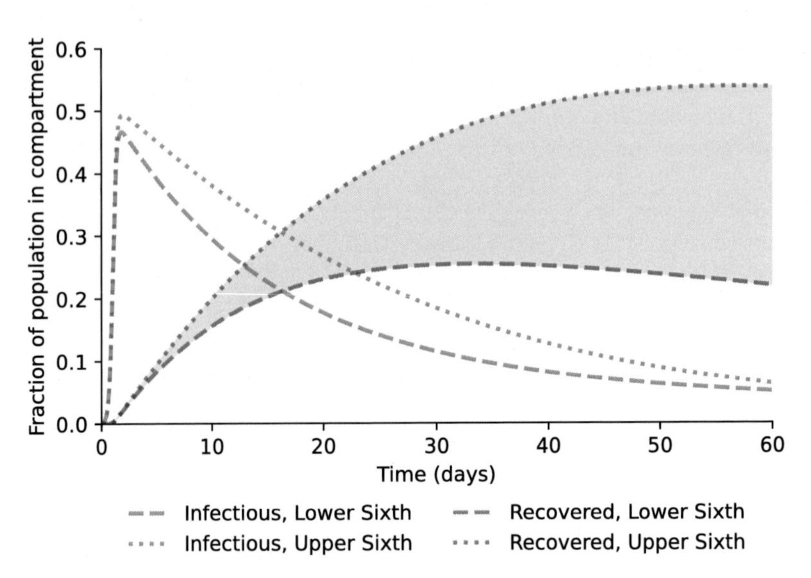

Figure 3.6 Solutions of the SIR model with age strata from Eq. (3.8).

On the other hand, the influx into the Upper Sixth comes, of course, from the Lower Sixth, a proportion of whom have been exposed to the pathogen before, and are thus immune. This highlights a crucial point, which we will discuss in detail in Chapter 6: early acquisition of immunity reduces overall time-integrated cases (i.e., the total number of cases at the end of the epidemic) significantly. This is the reason behind vaccinating as early as it is biologically sensible. Not only will that individual be protected but, as long as the vaccination provides enduring immunity, they will enter later age strata not as susceptibles but individuals with immunity. Since the driving dynamic of most outbreaks is the product of infected and susceptible individuals (the mass action term), they will not participate in the infectious dynamics in later age strata and will not "fuel" the susceptible pool the pathogen requires for its survival. Or, to put it in a somewhat more personifying way, the pathogen will have one less person to infect at its disposal. Thus effects compound over time, and vaccinating a single individual early enough can protect many others from disease over time.

3.2.4 Inference of mixing matrices

Whereas WAIFW matrices tell us about who acquires infections from whom, a mixing matrix tells us about who encounters whom. The reason why mixing matrices are considered separately, especially where it comes to age-stratified models, is that age mediates many of our associative behaviors as humans.

Mixing matrices can be obtained empirically, typically at a rather shocking investment of time, treasure, and manpower. The best known such epidemiological contact survey was POLYMOD, which sampled populations in eight European countries [101]. Since then, the methodology of POLYMOD has become a gold standard, and it touched off a series of POLYMOD-like social contact surveys [102]. Gratifyingly, POLYMOD has proven to be quite amenable to extrapolative projection, as Prem, Cook, and Jit [103] have demonstrated by projecting it to most of the countries on the planet. This is, of course, much less expensive, but also somewhat less accurate, than performing the painstaking social contact research that gave us POLYMOD.

An alternative to empirical studies is the estimation of mixing matrices from synthetic populations. A synthetic population is essentially a population where individuals are drawn from distributions elicited from known data. This is greatly facilitated by the increasing availability of microdata, which gives us a household-level insight into populations. The approach adopted by Mistry et al. [104] is an excellent model of synthetic generation of high-resolution mixing matrices. Mistry et al.'s approach proceeded in three steps:

- Create mixing matrices for different contexts of mixing, e.g., work, school, etc.
- Create simulated assortativities, e.g., simulated households, schools, workplaces.
- Analyze these simulated assortativities to approximate the likelihood that an encounter by an individual of age a_i will be with an individual of age a_j.

This methodology commends itself in two principal ways. First, though identifying distributions in a Bayesian approach then performing adaptive sampling for conditional probabilities is arguably the most elegant way, a good approximation of reality can be obtained by just sampling data as available. Second, such simulated assortativities are easy to build with adequate microdata.

Considering, for instance, family-based assortativities, the approach of Mistry et al. [104] proceeded by obtaining census data on households, and adaptive sampling. Thus we obtain the household size from the unconditional probability distribution of household sizes. The age of the head of household then follows from the conditional distribution given the household size. The age of the head of household's spouse and the number of children, follows from the head of household's age. The age of the children is then drawn from a distribution conditioned on the age of each of their parents. These figures can be obtained quite easily by filtering the data set to within a given tolerance band for ages, then drawing random samples. These amounts, in Bayesian terms, effectively to a uniform prior.

The advantage of more sophisticated Bayesian frameworks is the ability to calibrate the multivariate distributions, from which we draw the various conditional samples (e.g., age of head of household given family size) against empirical data, such as that obtained by POLYMOD. Nevertheless, as Computational Example 3.4 shows, it is possible to get a relatively good approximation of the mixing matrix at a good enough speed (thousands of simulated households per second of simulation time on commodity hardware) to be able to model nation-level processes. Such models can be arbitrarily expanded by introducing multiple different layers that together make up the mesoscale likelihood:

- Account differently for individuals in different settings. The US census microdata, for instance, identifies individuals living in group settings (institutionalization, incarceration). The example of COVID-19 has shown that the spatial-population dynamics and often overcrowded situations in settings of incarceration result in higher incidence and mortality of infectious disease [105]. The age mixing dynamics of a prison, which by definition separates an individual from their family, perturb the ordinary, household/family-based age-assortativity model.

- The overall mixing matrix is typically a linearly weighted combination of the various matrices—representing work, school, home, and so on—by a weighting factor that represents partly the time spent in that setting, and partly the significance of that setting to pathogens. It also highlights the intensity of mixing: a 14-year-old index case may spend his school days among other students in the same range.
- Certain mixing processes are very stratal. We see these as relatively solid squares on the mixing matrix. To infectious disease modelers who have been dealing with age-stratified models, the diagonally successive 4×4 blocks that mark out the four-year segments in which much of the US divides school years (primary, middle, high) must be quite familiar. In England, the segregation is quite different: Sixth Formers (comparable to US 11th and 12th grade) typically live apart from the rest, but especially in boarding schools, lower grades constitute a single assortative mixture.

Computational Note 3.4 Inference of mixing matrices

In this Computational Note, we will be estimating mixing matrices through a population simulation approach derived from Mistry et al. [104]. This approach is generally quite painstaking, and for this reason, only key elements of code will be reproduced here. The full code is, of course, reproduced in the book's companion Github repository (see link at the end of the Computational Note), including the parameters to obtain source data.

The general approach by which we create our synthetic contact matrices is to create synthetic households, schools, and workplaces. From this, we calculate a contact matrix by determining the absolute abundance of contacts. For a single instance of a household, school, or workplace S, which we represent as a vector of individual ages (with each element being one simulated individual), the absolute abundance of contacts is a (symmetric) matrix $\mathbf{C}$ defined as

$$\mathbf{C}_{i,j} = \frac{1}{|S|-1} \sum_{m=1}^{S} [S_m = i] \Big(\sum_{n=1}^{S} [S_n = j] - \delta_{i,j} \Big), \tag{3.9}$$

where $\delta_{i,j}$ is the Kronecker delta function [104]. We obtain the total abundance of contacts for a given number of simulated instances of the setting by adding up their matrices. From this, we obtain the mixing probability matrix

$$\mathbf{M}_{i,j} = \sum \frac{\mathbf{C}_{i,j}}{N_i}, \tag{3.10}$$

where N_i is the number of i-aged individuals in the sample. Thus a value of $\mathbf{M}_{i,j}$ is the *per capita* probability that an individual of age j will encounter an individual of age i. This is, of course, asymmetrical, because it is per-capitalized by reference to the population for each i. The example of schools in Fig. 3.7 shows this quite well: the bright band on the bottom results from the fact that for most adults, the vast majority of contacts will be with those of school age (since students outnumber teachers in most cases). Given various simulated settings, we obtain a composite through linear weighting by a scaling vector η, so that

$$\mathbf{M}_{\text{composite}} = \sum_{p=1}^{\{\mathbf{M}\}} \eta_p \{\mathbf{M}\}_p, \tag{3.11}$$

where $\{\mathbf{M}\}$ is the set of the component matrices corresponding to a setting (e.g., households, schools, and workplaces) ordered correspondingly to their scaling coefficient in η.

η can be ascertained by optimization with respect to a reference matrix $\mathbf{M}^{\text{ref}}$ and a distance metric. The Canberra distance metric, as used by Mistry et al. [104], is a particularly attractive target of optimization. We intend to ascertain η so as to minimize the Canberra distance,

$$\sum_{i=1}^{m} \sum_{j=1}^{n} \frac{|\mathbf{M}^{\eta}_{i,j} - \mathbf{M}^{\text{ref}}_{i,j}|}{|\mathbf{M}^{\eta}_{i,j}| + |\mathbf{M}^{\text{ref}}_{i,j}|}, \tag{3.12}$$

where $\mathbf{M}^{\eta}$ is a composite matrix, as described in Eq. (3.11) with the linear coefficient vector η.

To illustrate the methodology of what Mistry et al. [104] called "adaptive sampling," our method of creating synthetic families proceeds as follows:

1. We use census microdata, obtained from IPUMS [106], to calculate the distribution of household sizes. The American community survey (ACS) product of the United States census uses the `FAMSIZE` attribute for family size, which is a good approximation of the household size. We are interested in a single figure per household, and for this reason, we group by `SERIAL`, the unique identifier assigned to each household, and take the first value (using `groupby("SERIAL").first()`).
2. Next, we determine the age of the head of household given the family size by filtering the data set to families of the desired size, then sampling the `AGE` property.
3. Next, we determine the likelihood of a spouse, given the age of the head of household: we filter all heads of household of the age from the previous step, and join their spouse's age (a spouse would have a `RELATE` property of `2` and share the head of household's `SERIAL`). This yields `NaN` if the individual has no spouse, and the spouse's age otherwise. Thus it acts as both a way of

determining whether the head of household would have a spouse, and what their age would be.

4. We determine the remaining number of individuals to be allocated by subtracting 1 (if there is no spouse) or 2 (if there is) from the household size.
5. We also determine, given the age of the head of household, the size of the household and the age of the spouse, the likelihood that any other person in the family is a child, rather than another relative (e.g., parents, siblings, more distant relatives). We do so by filtering the microdata sample to the age of the head of household, to the age of the spouse, and to families the size of the household from the first step. This gives us p_c, the probability that a person in the family is a child of the head of household.
6. For each remaining individual to be allocated, if any, we generate a random number between [0, 1]. If rand() $< p_c$, the new person will be a child. If not, it will be another relative.
7. If the new person is a child, we filter for the age of the head of household and the age of the spouse, and sample the ages of children (`RELATE` is `3`) who are in households with the corresponding head of household's and spouse's age. We do the same for other relatives, except that we sample the ages of other relatives (`RELATE` is any value other than `1`, `2`, or `3`).

In the same vein, we can sample from various other populations, e.g., to determine the faculty age distribution; we can use occupation-filtered census data and determine an age distribution we can sample. In most cases, for sufficiently large data sets, simple sampling of the empirical data, filtered for age conditionals with a tolerance (± 2 years tend to give a fairly good approximation), is very fast and allows full population simulations of mid- to large US states and mid-size nations on commodity hardware.

An alternative to simple empirical sampling is to use the observed data to fit multivariate probability distributions, and sample these for the conditional probability. The key difference between these approaches is that whereas the latter is more computationally expensive, it gives "smoother" values towards the edges. If the source microdata does not contain at least one 100-year-old individual, then there will be no data at all about them at all. The smaller the available microdata set, the stronger the case for fitting multivariate distributions, rather than simply sampling the empirical observations filtered by conditionals.

A notebook implementing the contents of this Computational Note is available on the book's companion Github repository in the folder `/ch03/contact_ matrix`.

3.2.5 Age-dependent continuous heterogeneities

Anderson and May [110] describe a basic model for age-dependent modeling in a compartmental structure. Let $S(a,t)$, $I(a,t)$, and $R(a,t)$ be the proportion of individ-

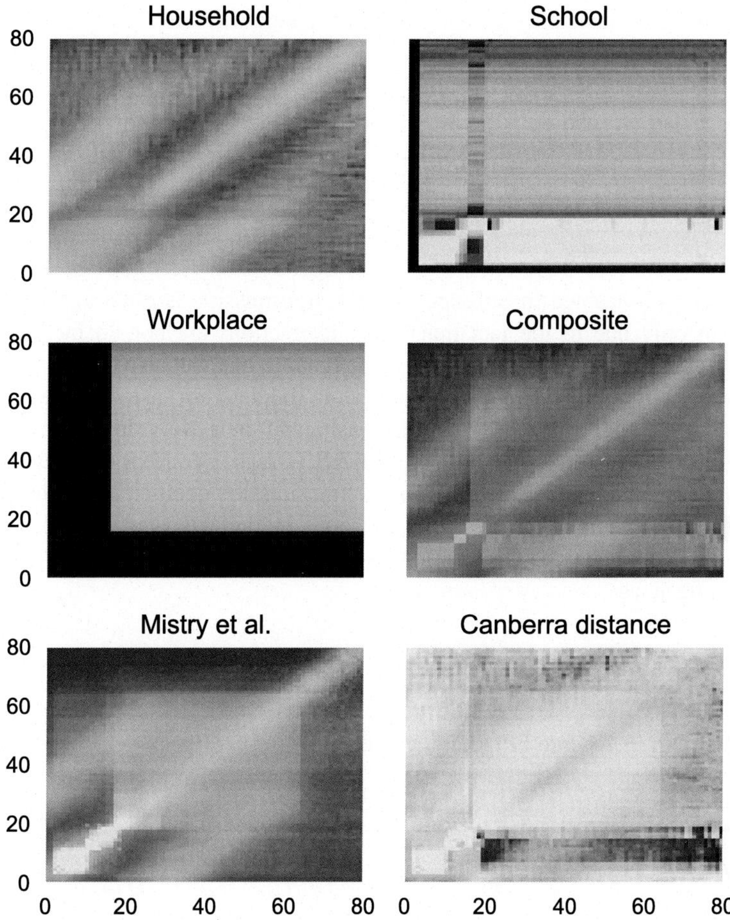

Figure 3.7 Inferred contact matrices for different settings for Virginia. The bottom left figure shows the matrix for the same area estimated by Mistry et al. [104] for comparison. The bottom right figure shows the Canberra distance between the composite from the model described in Computational Note 3.4.
Demographic data on households obtained from IPUMS [106]. Data on schools obtained from the Virginia Department of Education [107]. Data on businesses [108] and number of teachers [109] were obtained from the U.S. Census Bureau. The comparator matrix was obtained from Supplementary Information 1 to Mistry et al. [104].

uals of age a at time t, who are susceptible, infectious, or recovered, respectively. We assume that β is the same for all age groups, i.e., perfect assortative mixing with equal transmission odds, and that the force of infection λ is a pure function of the infectious compartment's size at time t, so that

$$\lambda(t) = \beta \int_0^\infty I(a,t)\, da. \tag{3.13}$$

Then, the individual compartments' rates of change will be

$$\begin{aligned}
\frac{\partial S(a,t)}{\partial t} + \frac{\partial S(a,t)}{\partial a} &= -\lambda(t)S(a,t), \\
\frac{\partial I(a,t)}{\partial t} + \frac{\partial I(a,t)}{\partial a} &= \lambda(t)S(a,t) - \gamma I(a,t), \\
\frac{\partial R(a,t)}{\partial t} + \frac{\partial R(a,t)}{\partial a} &= \gamma I(a,t).
\end{aligned} \tag{3.14}$$

However, as we noted in Subsection 3.2.1, transmission is often age-dependent. This is not only due to the fact that social interactions are not, in fact, perfectly assortative, but rather highly structured with regard to age, but also because of intrinsic differences in both physiology and behavior. Toddlers, for instance, are generally less adept at the application of hygienic measures in their every-day lives than fully grown adults. We therefore generalize the WAIFW matrix concept (see Eq. (3.1)) to a continuous function $\beta(a', a)$, denoting the transmission coefficient for the interaction between susceptibles aged a with infectious individuals aged a' (following the notation by Keeling and Rohani [39]). This allows us to rewrite Eq. (3.13) in an age-dependent form, as

$$\lambda(a,t) = \int_0^\infty \beta(a', a) I(a', t)\, da'. \tag{3.15}$$

Inserting Eq. (3.15) into Eq. (3.14) yields

$$\begin{aligned}
\frac{\partial S(a,t)}{\partial t} + \frac{\partial S(a,t)}{\partial a} &= -S(a,t)\int_0^\infty \beta(a,a')I(a',t)\, da, \\
\frac{\partial I(a,t)}{\partial t} + \frac{\partial I(a,t)}{\partial a} &= S(a,t)\int_0^\infty \beta(a,a')I(a',t)\, da - \gamma I(a,t), \\
\frac{\partial R(a,t)}{\partial t} + \frac{\partial R(a,t)}{\partial a} &= \gamma I(a,t).
\end{aligned} \tag{3.16}$$

It follows from this that

$$\begin{aligned}
\frac{\partial S(a)}{\partial t} &= -S(a)\int_0^\infty \beta(a,a')I(a',t)\, da' - \frac{\partial S}{\partial a}, \\
\frac{\partial I(a)}{\partial t} &= \int_0^\infty \beta(a,a')I(a',t)\, da' - \gamma I(a) - \frac{\partial I}{\partial a}.
\end{aligned} \tag{3.17}$$

In practice, values for $\lambda(a,t)$ can be obtained from contact tracing data. a' and a can be conceptualized as the list of value pairs of each infectious case's age and the age of the newly infected case to whom they have passed on the infection. $\beta(a', a)$ can then be estimated by fitting a probability distribution over the value pairs. $I(a', t)$ is equally ascertainable from contact tracing data, typically by estimating the number of cases whose date of initial symptoms has been fewer than $\frac{1}{\gamma}$ days ago. Thus $\lambda(a,t)$

is often quite easy to numerically estimate if a sufficient volume (and accuracy) of contact tracing data is available.

Practice Note 3.5 To bin or not to bin?

Whereas the method laid out above as to $\lambda(a, t)$ (see Eq. (3.15)) generalizes the WAIFW matrix concept to a continuous function, it does contribute to a degree of additional complexity. In practice, $\beta(a', a)$ is rarely a smooth distribution. Rather, most values are along the diagonal axis, as noted in Subsection 3.2.1; we spend most of our existence with people relatively close to our own age, especially in the institutional educational context. This is true for interpersonal relationships, too: over a third of American couples in 2017 were less than a year apart in age, and only four in ten were more than three years apart.

This makes creating age compartments or "bins" often preferable to dealing with an irregular age-encounter function, whose approximations may be less accurate. Where there is a clear rationale for age grouping due to interactions, such as for primary, middle, high school, and college-aged individuals (most of whose contacts will come from the same age group), a discrete WAIFW matrix by age groups is likely to yield much better results than numerical approximation of $\lambda(a, t)$.

The best test to determine whether a particular binning by age makes sense is to look at the statistical distributions of values in the WAIFW matrix. A good age structure will result in "tight" estimates of each value within the WAIFW matrix, with a low standard deviation. This is reflective of a homogeneous encounter behavior between the two age groups. It is often useful to look at the $\sigma(\mathbf{b})$ matrix, comprising the standard deviation of each WAIFW matrix element, and delineate age groups in a way that minimizes the grand sum (the sum of every element in a matrix, i.e., $\sum_{i,j} \sigma(\mathbf{b})_{i,j}$).

Host-vector and multihost systems

Dynamics of host-vector transmission

The virus now known as Hendra wasn't the first of the scary new bugs. [...] It made its debut near Brisbane, Australia, in 1994. Initially there were two cases, only one of them fatal. No, wait, correction: there were two human cases, one human fatality. Other victims suffered and died too, more than a dozen—equine victims—and their story is part of this story. The subject of animal disease and the subject of human disease are [...] strands of one braided cord.

Quammen, Spillover: animal infections and the next human pandemic, 2012 [111]

Some pathogens live exciting double lives, with entirely separate life cycles and behaviors in different animals. We see this manifest in three different ways:

1. In pure vector-borne diseases, the hosts are incapable of transmitting the pathogen among themselves. Contact with the vector is necessary for infection. Malaria is an example of pure vector-borne diseases.
2. In human-transmissible zoonoses, the human hosts are also capable of transmitting the infection. Contact with the vector is not necessary for infection, as contact with another infected individual can also pass on the pathogen. Examples of human-transmissible zoonoses include most ebolaviruses, pneumonic plague (*Yersinia pestis*), and Lake Victoria marburgvirus.
3. Multispecies pathogens can infect both humans and nonhuman animals. Many, but not all, such pathogens allow for both zoonotic infection (animal to human) and zooanthroponotic infection (human to animal). Examples include Rift Valley fever, foot-and-mouth disease and a recently emerging health concern among domestic animals, the zooanthroponotic transmission of tuberculosis [112].

4.1 Pure vector-borne diseases

A pure vector-borne disease requires every host to come into contact with a vector; there is, in other words, no interhost transmission. Pure vector-borne diseases therefore rely on abundant vectors that have the appropriate mechanisms for transmission—typically, transdermal penetration—to pass on the pathogen. Because of their abundance and presence in almost all biomes, haematophagous insects make up the overwhelming majority of vectors. Table 4.1 outlines some examples of pure vector-borne pathogens with their associated vectors.

Computational Modeling of Infectious Disease. https://doi.org/10.1016/B978-0-32-395389-4.00013-X

Table 4.1 Representative examples of pure vector-borne diseases.

Pathogen	Vector	Disease
Plasmodium spp.	*Anopheles spp.*	malaria
Trypanosoma brucei	*Glossinidae* (tsetse)	trypanosomiasis (sleeping sickness)
Trypanosoma cruzi	*Triatominae*	Chagas disease
genus *Leishmania*	sand flies	leishmaniasis
Onchocerca volvulus	blackflies (*Simuliidae*)	onchocerciasis (river blindness)
Dengue virus (DENV)	*Aedes spp.*	dengue fever
Chikungunya virus (CHIKV)	*Aedes spp.*	chikungunya
Zika virus (ZIKV)	*Aedes spp.*	Zika virus disease
Borrellia spp.	*Ixodes spp.*	Lyme disease
La Crosse virus (LACV)	*Aedes triseriatus*	La Crosse encephalitis

4.1.1 Case study: malaria

In 1895 Sir Ronald Ross, then still unknighted and a mere surgeon in the Indian Medical Service, began an extraordinary journey. It would culminate in an 1897 article that would forever change the way we thought about one of humanity's oldest fellow travelers [113]. The paper, which would eventually earn Ross the Nobel Prize in physiology or medicine five years later, proved that malaria was a parasitic disease transmitted by mosquitoes.

Malaria is an infectious disease caused by eukaryotes of the genus Plasmodium *spp*. Plasmodium is an apicomplexan parasite, with specific cellular forms for effecting life cycle specific tasks. Fig. 4.1 illustrates this life cycle. Upon a mosquito bite, sporozoites (1) enter the human host, and lodge in the liver. This begins the hepatic phase of malaria, which lasts approximately 2–3 weeks. At the end of this stage, the infected hepatocytes burst and release merozoites into the bloodstream (2). This launches the erythrocytic phase of the disease, during which merozoites infect red blood cells and multiply (3). Infected red cells play host to a cycle of asexual reproduction (4). The synchronous waves of merozoite release (and the ensuing immune reaction) are responsible for the hallmark symptom of malaria, namely periodic recurrent fevers (tertian and quartan fever).

Some merozoites differentiate into gametocytes (5), which are ingested when a mosquito bites, or, in the rarefied language of malaria epidemiologists, "takes a blood meal" from an infected human. The gametocytes then multiply in the mosquito (6 and 7), eventually producing sporozoites (8), and as these are transmitted by the next blood meal, the cycle begins anew.

Few diseases have had an impact on human evolution, culture and society on par with malaria [114,115]. It is one of the oldest documented infectious diseases. Indeed, it has been hypothesized by Wellems, Hayton, and Fairhurst [116] that the protective effect bestowed by a heterozygous sickle cell allele explains its survival to the modern

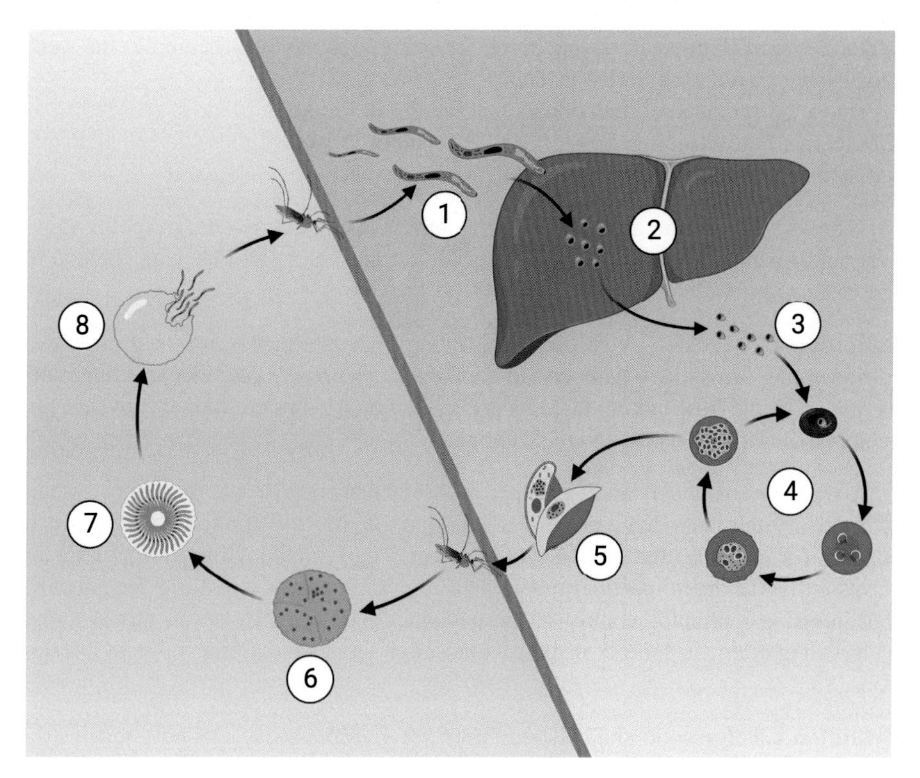

Figure 4.1 Lifecycle of *Plasmodium falciparum*.

day. As such, malaria has left its footprint on human evolution in a profound way few other diseases have.

Yet its true origins were the matter of considerable controversy. The clue is in the name; the prevailing theory until Ross's discovery was that malaria resulted from *mala aria*, that is, "bad air." It took the advent of modern evidence-based medical science to challenge this "miasma theory." Ross's elucidation of the role of mosquitoes in the lifecycle of malaria has opened up a new subject for epidemiological consideration: the vector-borne disease. This chapter deals with the modeling of such diseases.

4.1.2 The basic dynamics of a vector-borne disease

The hallmark feature of vector-borne diseases is the existence, as the name suggests, of at least one species that acts as a vector. Thus we are no longer preoccupied by a single population, but rather by two populations (planes of transmission), each of which is experiencing a disease process of its own. The epidemic among humans mirrors the epizootic, i.e., the disease process among the vectors.

Definition 4.1 (Planes of transmission). A **plane of transmission** refers to a set of unique compartments of individuals that play the same role, such as ultimate hosts, vectors, intermediate vectors, and so on. These individuals may belong to the same species, but do not necessarily have to.

Transmission between individuals that belong to the same plane is referred to as **intraplane** transmission, whereas transmission between individuals that belong to different planes is referred to as **interplane** transmission.

There is some debate as to the definition of a vector [117]. For our purposes, however, and for the sake of simplicity, we shall define vectors and hosts in relation to their role within human disease.

Definition 4.2 (Vector). A **vector** is any living organism that is not itself directly a pathogen, but transmits a pathogen to another organism. We consider that target organism to be the **host**, noting that this is largely a matter of definitional convenience; vectors themselves play host to the pathogen.

A common misunderstanding is that in a vector-borne disease, the vector is "unaffected" by the pathogen's presence. This may be the case for some pathogens. It appears, for instance, that certain bat species may harbor ebolaviruses without themselves falling ill due to peculiarities of their immune systems that allow for effective (but infectious) immune control of the pathogen [118,119]. However, this is by no means a requirement. Indeed, many vectors suffer some loss of fitness due to infestation.

Definition 4.3 (Enzootic and epizootic processes). Most models of host-vector systems have a human host. For this reason, the terms "epidemic" and "endemic" are reserved for the human plane of transmission.

An epizootic is an epidemic process on the vector plane. An enzootic is the equivalent of an endemic disease in the vector plane.

By convention, we will refer to host plane processes as epidemics and endemics, even if they occur in nonhuman animals.

In a pure vector-borne disease, only vectors can transmit the disease to humans. Malaria is an example of a pure vector-borne disease, since it is not generally transmitted between humans. On the other hand, filoviral haemorrhagic fevers, such as Lake Victoria marburgvirus or the Zaire and Sudan ebolaviruses can be passed to humans through zoonotic transfer, but can also pass between humans.

Definition 4.4 (Pure vector-borne disease). In a **pure vector-borne disease**, there is only interplane transmission.

The key differentiation between pure vector-borne disease and host transmissible zoonoses is, of course, in the mass action term:

- For pure vector-borne diseases, the mass action is driven by the number of susceptible individuals and the infected vector population.
- For human-transmissible zoonoses, the mass action is driven by the number of susceptible individuals and both the infected vector population and the population of infected humans.

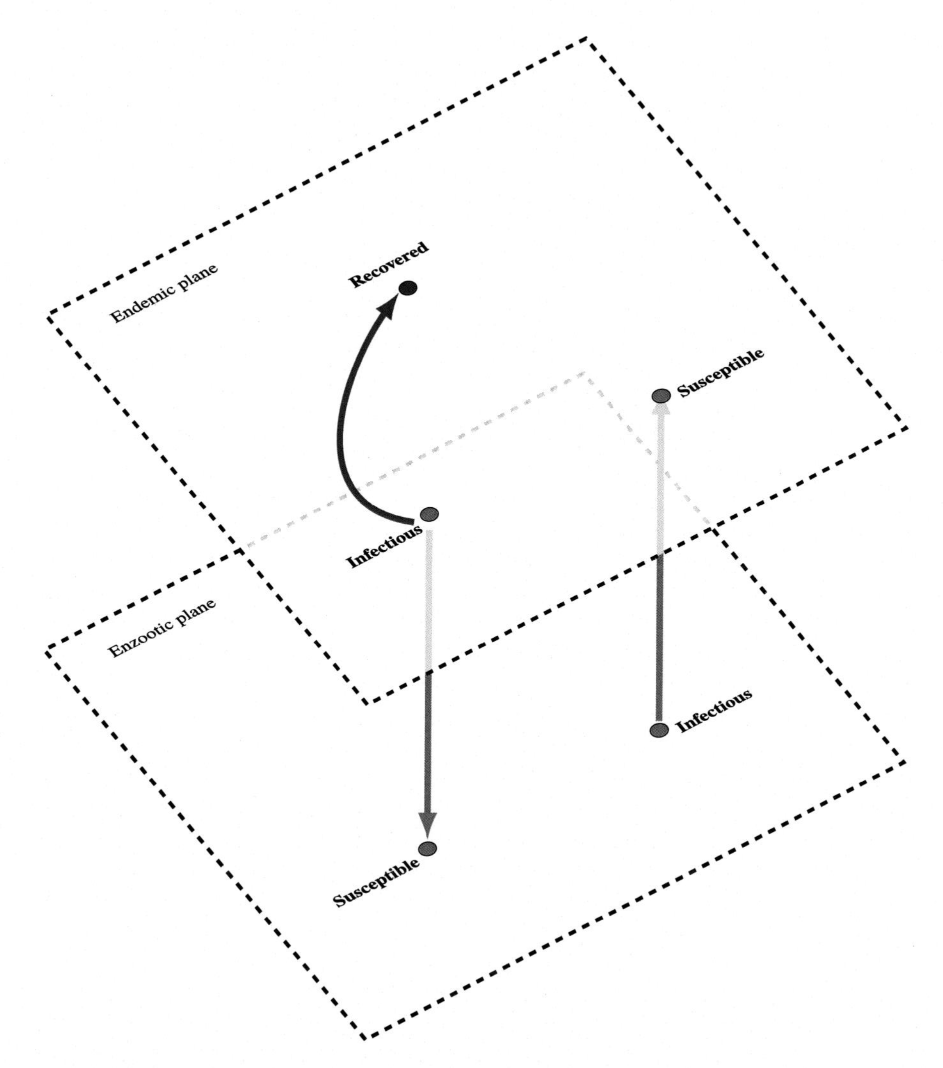

Figure 4.2 Planes of transmission for a pure vector-borne disease.

It is often helpful to map these planes against each other, as in Fig. 4.2. In a pure vector-borne disease, there is no transmission between either vectors or hosts. Vectors can only acquire infection from a host, and hosts can only acquire infection from a vector. Mathematically formulated, the WAIFW matrix **b** of a pure vector-borne pathogen is

$$\begin{pmatrix} 0 & m \\ n & 0 \end{pmatrix}. \tag{4.1}$$

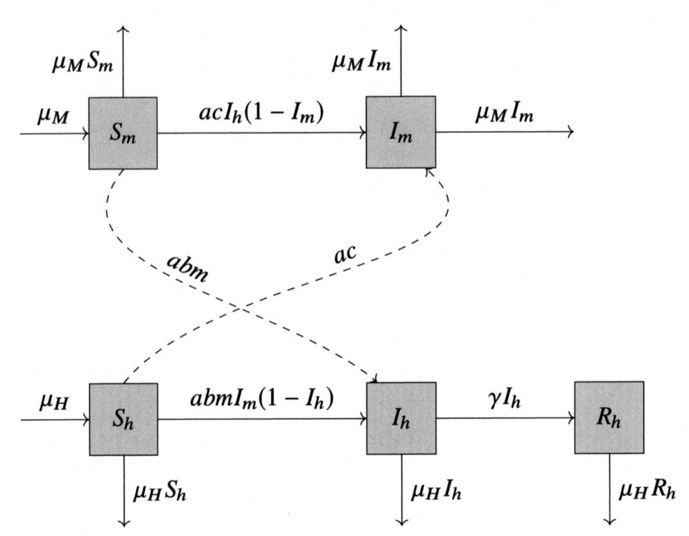

Figure 4.3 Ross's malaria model. The subscripts m and h denote the mosquito and human compartments, respectively. μ_M is the mortality of mosquitoes. a is the bite rate, b and c are the proportion of bites that transmit infection from mosquitoes to humans and humans to mosquitoes, respectively. m is the number of female mosquitoes per human.

4.1.3 The Ross malaria model

There are no limits to the degree of complexity one may, with sufficient time and unfailing enthusiasm, refine any compartmental model. Malaria models attest to this fact: it is far from unusual to find models with dozens of compartments to account for a wide range of processes and states. A gentler introduction into malaria (and more generally, vectored transmission) models is provided by the Ross model, depicted in Fig. 4.3.

The Ross model makes some salient assumptions, one being that mosquitoes, once infected, remain infected for life. This is warranted given the relatively short lifespan of mosquitoes. The dynamics of the infection relies on the interaction between the infectious and susceptible compartments, and is focused around three variables: the bite rate a (the number of bites per unit time), the likelihood that a bite passes an infection from a mosquito to a human b, and the likelihood that a bite passes an infection from a human to a mosquito c.

We may conceive of the infectious process as follows: For unit time, there are a interactions per mosquito. From the perspective of the mosquito, the likelihood of contracting malaria from a human, per interaction, is governed by two factors: the likelihood of biting someone with malaria and the likelihood that the bite will result in transmission. The former is, of course, I_h, the number of humans who are infected, and thus capable of infecting the mosquito. The latter is the rate c. Together, acI_h represent the total volume of infections per unit time. $1 - I_m$ represents the number of susceptible mosquitoes. As such, we see in $acI_h(1 - I_m)$ the mass action term βSI (in the order of βIS).

From the human perspective, the equation is largely the same, except in this case, we must also consider that mosquitoes are rather more abundant. Because S_m, I_m, S_h, I_h, and R_h are all proportions, rather than actual numbers, we must account for this difference through a factor. m thus denotes the number of female mosquitoes per human (since only female mosquitoes bite). Then, the mass action term for humans will be $abmI_m(1 - I_h)$, where I_m corresponds to I and $1 - I_h$ corresponds to S from previous formulations.

Based on this, we can write the WAIFW matrix of Ross's malaria model as

$$\begin{pmatrix} 0 & \overbrace{acI_h(1 - I_m)}^{\text{human-to-mosquito}} \\ \underbrace{abmI_m(1 - I_h)}_{\text{mosquito-to-human}} & 0 \end{pmatrix}. \tag{4.2}$$

As is expected, this is a hollow matrix, since there is no intraplane transmission. Thus the whole system of differential equations works out to

$$\begin{aligned} \frac{dS_m}{dt} &= \mu_m - acI_h(1 - I_m) - \mu_m S_m, \\ \frac{dS_h}{dt} &= \mu_h - abmI_m(1 - I_h) - \mu_h S_h, \\ \frac{dI_m}{dt} &= acI_h(1 - I_m) - \mu_m I_m, \\ \frac{dI_h}{dt} &= abmI_m(1 - I_h - R_h) - \gamma I_h - \mu_h I_h, \\ \frac{dR_h}{dt} &= \gamma I_h - \mu_h R_h. \end{aligned} \tag{4.3}$$

Computational Note 4.1 Implementing the Ross-MacDonald model

By now, we have sufficient practice in solving ODEs numerically to be able to omit a few moving parts. In particular, we can reduce the model and track only the infectious compartments:

```
def deriv(t, y):
    I_m, I_h = y

    dImdt = a * c * I_h * (1 - I_m) - mu * I_m
    dIhdt = a * b * m * I_m * (1 - I_h) - gamma * I_h

    return dImdt, dIhdt
```

We will be using some of the sensible defaults from Table 4.2 for this model:

```
I_m_0 = 0.2
I_h_0 = 0.005

a = 0.1
b = 0.2
c = 0.5
m = 10
gamma = 0.05
mu = 0.05

y_0 = (I_m_0, I_h_0)
```

As we can see from solving this system of differential equations, a stable equilibrium emerges in the absence of human mortality and births both human and mosquito. This is convenient and somewhat accurate in the short term, but including a birth term might be important to accommodate population growth.

A notebook implementing the contents of this Computational Note is available on the book's companion Github repository in the folder `/ch04/ross_ macdonald`.

Fig. 4.4 shows the convergent dynamics of the model. Streamplots are useful tools for understanding the convergent behavior of a system of differential equations, and are for this reason widely used in engineering and mathematics. They do, however, take some practice to read.

The best way of approaching streamplots is to consider their relationship to phase portraits, which we have discussed in Subsection 2.3.3. Recall that a phase portrait shows a single solution of a differential equation by plotting two values against each other. One may consider a streamplot to be a "generalized" version of a phase portrait,

Table 4.2 Some sensible default values for malaria models, based on Mandal, Sarkar, and Sinha [120].

Symbol	Parameter	Value
a	bite rate	0.1–0.05 d^{-1}
b	$m \rightarrow h$ transmission	0.2–0.5
c	$h \rightarrow m$ transmission	0.5
γ	recovery rate, human	0.05 d^{-1}
m	female mosquitoes per human	0.5–40
μ_h	death rate, human	0.017 y^{-1}
μ_m	death rate, mosquito	0.05–0.5 d^{-1}
τ_m	latent period, mosquito	5–15 d
τ_h	latent period, human	10–100 d

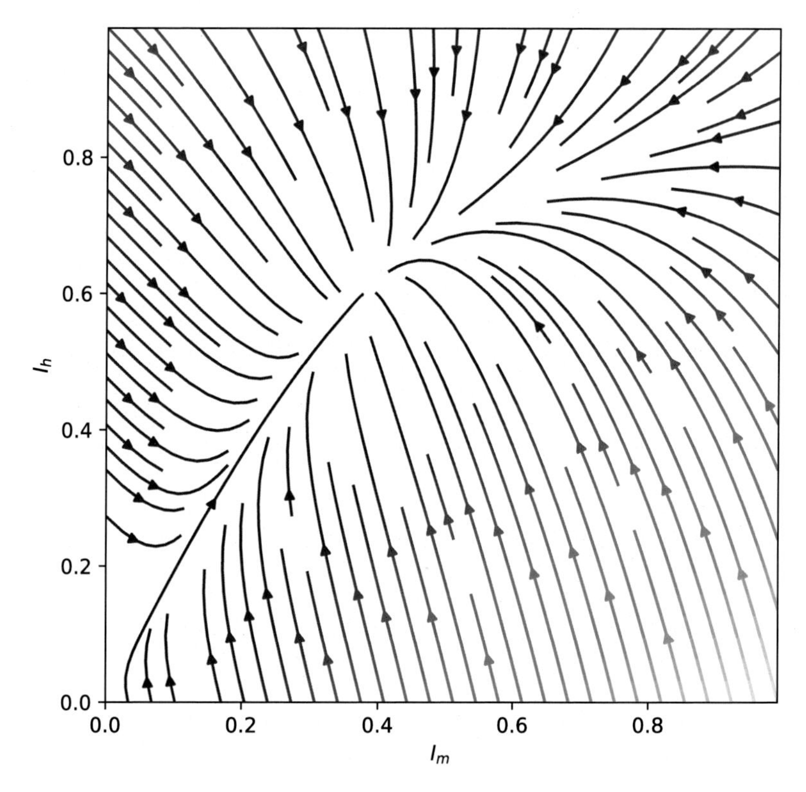

Figure 4.4 Streamplot of the Ross–MacDonald model for the default values in Table 4.2.

in which the starting values are varied. Thus the plot in Fig. 4.4 shows a number of different starting values of I_m and I_h. This gives a visual representation to the more abstract vector field that governs the two interacting quantities. Given a value of I_m and I_h, you can estimate the trajectory of the phase portrait for those values by tracing the nearest line in its direction (noted by the arrows).

Computational Note 4.2 Creating streamplots

Streamplots are a beautiful and convenient way to show more than a single phase profile of a system, and generalize it to show all potential phase profiles, given certain starting values.

Drawing streamplots is not difficult, but it requires some tricks. First and foremost, we need to select two differential equations out of the whole system. We will be able to vary their starting quantities, but not much else. Thus all of that is deemed constant.

In this case, we only have two differential equations. We begin by setting up a grid, which will give us the points at which we will "sample" the values of the vector field. Since this is a numerical solution, we are essentially calculating the values for $\frac{dI_m}{dt}$ and $\frac{dI_h}{dt}$ at every point.

The function `np.mgrid` creates a mesh grid at equally spaced distances; it is essentially a convenience method to give us the product space of two `np.linspaces`.

```
i_h, i_m = np.mgrid[0:1:0.01, 0:1:0.01]
```

Next, we define the two differential equations, in terms of the two axes created in the previous step:

```
u = a * c * i_h * (1 - i_m) - mu * i_m
v = a * b * m * i_m * (1 - i_h) - gamma * i_h
```

Finally, we can move on to plotting:

```
fig = plt.figure(facecolor="w", figsize=(6, 6))
ax = fig.add_subplot(111, axisbelow=True)

ax.streamplot(i_m, i_h, u, v, density=1, color=u + v)

ax.set_xlabel("$ I_m $")
ax.set_ylabel("$ I_h $")
```

A notebook implementing the contents of this Computational Note is available on the book's companion Github repository in the folder `/ch04/ross_macdonald`.

4.1.4 Basic reproductive number of the Ross malaria model

We may be interested in calculating $\mathfrak{R}_0$ for the Ross malaria model. If a human is infected, they will be infectious for $\tau = \gamma^{-1}$, so the total number of bites suffered during the infectious period will be $\frac{a}{\gamma}$. For each of these bites, we assume the mosquito has its full lifespan, μ_m^{-1} left. Thus each of those mosquitoes will go on to bite $\frac{a}{\mu_m}$ times.

The likelihood that biting an infectious human will transfer the infection is, of course, c, so the total number of bites by infectious mosquitoes attributable to the initial human case is $\frac{ac}{\gamma}$. Of these, only a fraction, namely, b, will be transmitted to another human. Consequently, the

$$\mathfrak{R}_0 = \frac{a^2 bcm}{\gamma \mu_m}. \tag{4.4}$$

4.1.5 Stability of the Ross malaria model

From examining the infectious subsystem

$$\begin{aligned} \frac{dI_m}{dt} &= acI_h(1 - I_m) - \mu_m I_m, \\ \frac{dI_h}{dt} &= abmI_m(1 - I_h - R_h) - \gamma I_h - \mu_h I_h, \end{aligned} \tag{4.5}$$

we can obtain the Jacobian of this system as

$$\mathbf{J}_{I_m, I_h} = \begin{pmatrix} -acI_h - \mu_m & ac(1 - I_m) \\ abm(1 - I_h - R_h) & -\gamma - abmI_m - \mu_h \end{pmatrix}. \tag{4.6}$$

Since a, b, and c are by biological necessity positive, and $1 - I_m$ and $1 - I_h - R_h$ are also positive due to the population partition property, whereby $I_m + S_m = 1$ and $I_h + R_h + S_h = 1$, both off-diagonal elements of the Jacobian are positive. We denote such a matrix, where all off-diagonal elements are positive, as a Metzler matrix. Where the Jacobian of a system is a Metzler matrix, the system is described as monotone, and if the matrix is irreducible, the system is strongly monotone. If a matrix is monotone, then, subject to some conditions,

1. the disease-free equilibrium will be globally asymptotically stable for $\mathfrak{R}_0 \leq 1$, and
2. the endemic equilibrium will be globally asymptotically stable for $\mathfrak{R}_0 > 1$.

4.1.6 Temporal dynamics between planes

An important feature of some pure vector-borne diseases is that relative to the host, the vectors may have very short lifespans with pronounced seasonalities. An example is *Aedes aegypti*, an important vector of a number of pathogens, including Zika virus (ZIKV), Chikungunya virus (CHIKV), yellow fever, and the virus responsible for Eastern equine encephalitis. The mean lifespan of *Aedes aegypti* is, according to Meena [121], 19.94 days. Che-Mendoza et al. [122] describe the seasonality of *Aedes aegypti* populations as sinusoidal, peaking during the early rainy season, beginning to decline towards its end, and reaching lowest levels during the dry season. The reason why these temporal dynamics are important in modeling vector-borne infectious diseases is twofold:

1. Differences in the population of vectors affect the mass action term. The mass action term describes the likelihood of contact, and thus transmission of the pathogen. If there are fewer vectors, the number of host-vector interactions will decrease significantly.
2. We generally assume that vectors are born susceptible. The number of infectious vectors can be quite rapidly depleted due to natural mortality.

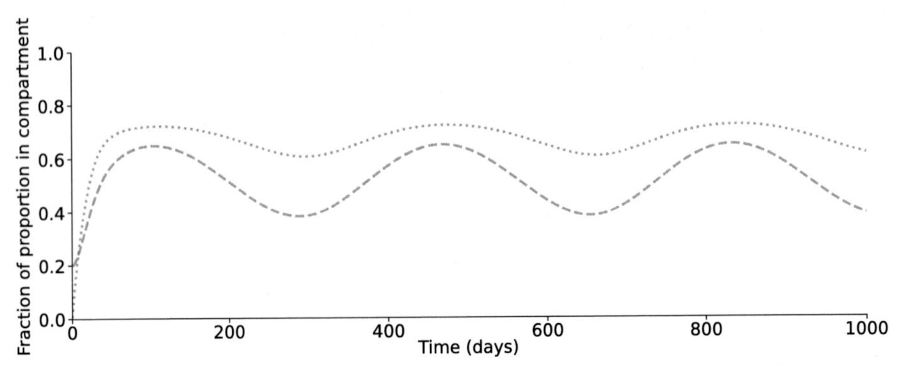

Figure 4.5 The Ross malaria model with time-dependent birth rates of the vector for $\phi_0 = 0.01$ and $\phi_1 = 0.05$. The period of forcing is annual and the phase ψ is -90 days. All other parameters are as described in Computational Note 4.1.

Consider the Ross model, ignoring human births and deaths, in which mosquitoes are born according to a time-dependent forcing function:

$$\frac{dI_m}{dt} = \overbrace{\phi(t)}^{\text{time-dependent birth term}} \quad acI_h(1 - I_m) - \mu I_m,$$
$$\frac{dI_h}{dt} = abmI_m(1 - I_h) - \gamma I_h, \tag{4.7}$$
$$\phi(t) = \overbrace{\phi_0}^{\text{base birth rate}} \left(1 + \overbrace{\phi_1}^{\text{amplitude}} \sin\left(\frac{2\pi}{365}(t - \underbrace{\psi}_{\text{phase}})\right)\right).$$

The seasonality induced by the time dependent birth rate—via $\phi(t)$—does not only induce fluctuations in the vector, but also in the infected host population. This can create complex resonance and interference phenomena when the birth rate of the host population, too, is temporally dependent. (See Fig. 4.5.)

4.1.7 Parameter inference in vector-borne diseases

Vector-borne diseases confront us with the problem that a large number of crucial parameters are quite difficult to ascertain. Humans, for instance, are generally known to their governments and when unwell, seek care. Neither of those could be said for, say, mosquitoes. A task such as estimating the number of mosquitoes in a given area is daunting. Some factors of the Ross–McDonald model can be approximated with reasonable accuracy, such as the lifespan of a mosquito or the length of the viraemic period in humans.

Inferential methods help us where we run out of useful empirical data. Bayesian parametric inference is the currently dominant technique in this field. The Bayesian approach encompasses a diversity of solutions, but they all share a basic reasoning:

given a time series of observations Y, what is the posterior distribution of the parameter θ of the vector-valued function $\hat{f}(\theta)$ so as to minimize $\sum_{i=1}^{|Y|} \delta(Y_i, \hat{f}(\theta)_i)$, where $\delta(v, v')$ is a distance metric?

Computational Note 4.3 Inferring parameters for a vector-borne disease

Dengue fever is a disease caused by the dengue virus (DENV), endemic throughout the tropics. Like malaria, dengue is a mosquito-borne disease, specifically transmitted by *Aedes spp.*. Similarly to malaria, it is not horizontally transmitted among either humans or mosquitoes, but rather require interplane transmission, making it a pure vector-borne disease. In most cases, it is an unpleasant albeit survivable illness that manifests as a 3–4 day long viraemia. In rare cases, it may present as dengue haemorrhagic fever (DHF), which is a rather more severe.

A good approximation of the dynamics of dengue fever is the system

$$\begin{aligned}
\frac{dS_H}{dt} &= \mu_H N_H - b\beta_{V,H} S_H \frac{I_V}{mN_H} - \mu_H S_H, \\
\frac{dI_H}{dt} &= b\beta_{V,H} S_H \frac{I_V}{N_H} - \gamma I_H - \mu_H I_H, \\
\frac{dS_V}{dt} &= \mu_V m N_H - \beta_{H,V} S_V \frac{I_H}{N_H} - \mu_V S_V, \\
\frac{dI_V}{dt} &= b\beta_{H,V} S_V \frac{I_H}{N_H} - \mu_V I_V,
\end{aligned} \tag{4.8}$$

where b is the bite rate, m is the number of female mosquitoes per human, and $\beta_{V,H}$ and $\beta_{H,V}$ are the respective transmission rates upon contact from vector to host and host to vector.

There are many ways of multiparameter inference, but Markov chain Monte Carlo (MCMC) methods are among the most popular. The idea of MCMC to estimate parameters is quite simple: given some priors (i.e., our educated guesses as to the values of model parameters) and evidence (the actual number of cases), what is the most likely vector θ that comprises the model's parameters? In other words, what values of θ maximize the posterior probability density function?

Emcee [123] is a Python package that implements the affine invariant MCMC ensemble sampler described in Goodman and Weare [124]. In its syntax and its approach, it is rather different from other MCMC samplers. First, it defines priors not as probability distributions but as binary ranges. Considering the model above, we are interested in inferring $\beta_{V,H}$, $\beta_{H,V}$, m (the number of female mosquitoes per human), b (the bite rate), and $\frac{I_V(0)}{N_V}$, the fraction of infected mosquitoes at the start (denoted as `pV0` in the model). These together make up our parameter vector θ. Whereas in other tools (such as PyMC), we would ordinarily need to have a good guess as to the distribution from which each element of θ is drawn, Emcee operates on a much simpler principle: the log prior of θ is defined as a binary function, being zero-valued if the value is within the range of acceptable priors, and $-\infty$ otherwise:

```
def log_prior(theta):
    beta_VH, beta_HV, m, b, pV0 = theta

    if 0 < beta_VH < 0.66 and \
        0 < beta_HV < 0.66 and \
        1e2 < m < 1e4 and \
        0 < b < 30 and \
        0 < pV0 < 0.01:
         return 0.0
    else:
         return -np.inf
```

Next, we need to define the log likelihood function. Given a θ and a number of points in time (t), we define this function as

$$\ln \ell(\theta, t, y) = -\frac{1}{2t} \sum_{i=1}^{t} \sum_{j=1}^{|y|} (y_j - \hat{y}_j(\theta))^2, \tag{4.9}$$

where y is the vector of actual infectious cases in humans, and $\hat{y}_j(\theta)$ is the number of infectious cases in humans at time t given θ. In Bayesian terms, this function gives us $p(\theta|y)$.

Finally, we need to build the function for the log probability. Recall that our function for the prior returned a finite value if (and only if) each element of θ was within the range of permissible values. This allows us to express log probability as the result of the `log_likelihood()` function if the log prior is within the permissible range, and $-\infty$ otherwise:

```
def log_probability(theta, t, data):
    lp = log_prior(theta)
    if not np.isfinite(lp):
         return -np.inf
    else:
         return log_likelihood(theta, t, data)
```

We now need to create the initial position matrix. This is an $m \times n$ matrix, where m is the number of "walkers" and n the number of parameters, i.e., $|\theta|$. An efficient way to get our model to work relatively fast is to provide some initial guesses for θ, and perturb them by adding a small random number:

```
pos = np.array([0.3, 0.3, 200, 1/3, 0.001]) \
    + 1e-4 * np.random.randn(32, 5)
```

Because of broadcasting (see Subsubsection A.11.1.3), this will give us a matrix of 32 independent "guesses" for each of the five variables we are estimating.

We are now ready to initialize the sampler:

```
sampler = emcee.EnsembleSampler(nwalkers=pos.shape[0],
                                ndim=pos.shape[1],
                                log_prob_fn=log_probability,
                                args=(np.arange(1, len(data)),
                                      data.to_numpy()))


sampler.run_mcmc(pos,
                 nsteps=25_000,
                 progress=True)


sampler.get_blobs(discard=1_000)
```

The method `sampler.get_chain(flat$=$True)` provides a convenient way to access the results, and taking the 50th percentile of the values gives a good estimate of what the median estimate is for each of the values:

```
beta_VH_hat, beta_HV_hat, m_hat, b_hat, pV0_hat = np.percentile(
    sampler.get_chain(flat=True),
    50,
    axis=0)
```

Fig. 4.6 shows a corner plot fitted for cases of dengue in the Ilocos Region of the Philippines between June 2016 and July 2017. Though the values are somewhat different from what we would expect (in particular, the number of female mosquitoes per human has a fat-tailed posterior distribution, suggesting that the model did not converge with particularly high confidence to a singular value), the example above shows the usefulness of MCMC parameter estimation techniques. Where real-world data is available, parameter estimation can provide fairly good estimates of parameters that are often costly and/or difficult to empirically determine.

A notebook implementing the contents of this Computational Note is available on the book's companion Github repository in the folder `/ch04/host_ vector_ pe`.

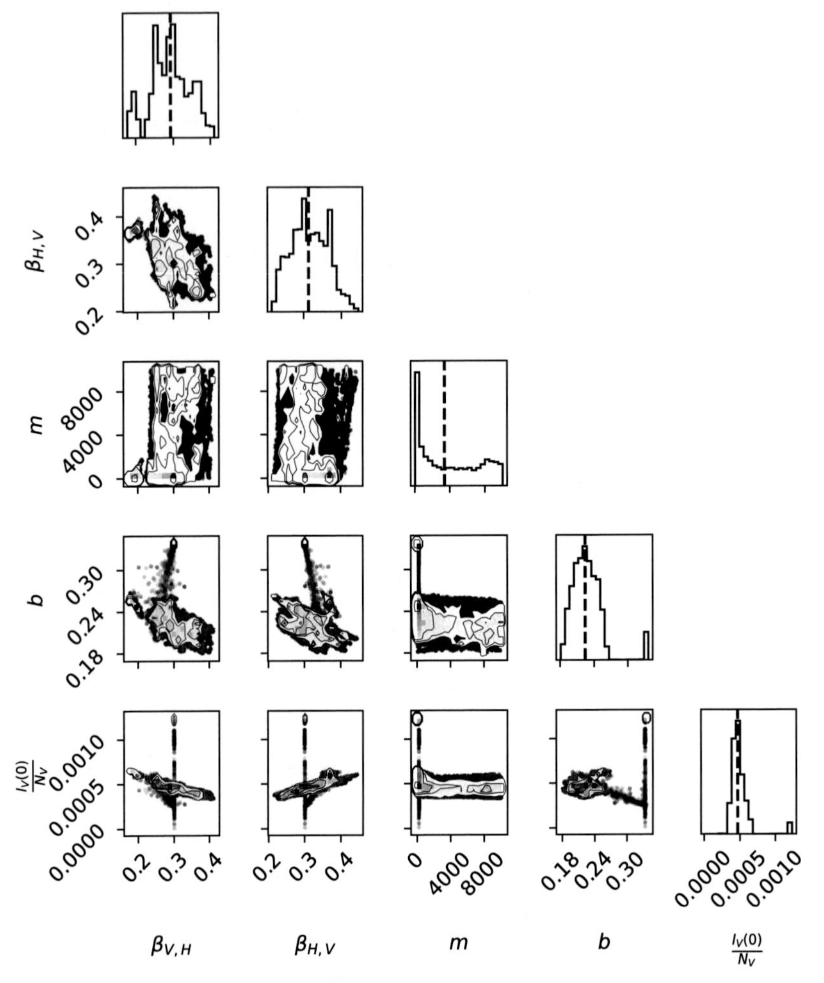

Figure 4.6 Corner plot of parameter estimates for dengue fever in the 2016–17 dengue season in the Ilocos Region, Republic of the Philippines. Case data obtained from the Epidemiology Bureau of the Department of Health, Republic of the Philippines, obtained via the Humanitarian Data Exchange of the United Nations Office for the Coordination of Humanitarian Affairs.

4.2 Zoonotic disease

In the context of malaria, the human and the enzootic planes had no within-plane infectivity. For instance, in the malaria model we examined in Section 4.1, mosquitoes (the epizootic or vector plane) and humans (the epidemic or host plane) each could only "communicate" the pathogen between each other through host-vector interactions. In this section, we will deal with infectious diseases where a pathogen, once it passes through zoonotic transfer (sometimes called a spillover event) into the human (or other host) population, can be transmitted within that population directly. For this reason, a single zoonotic event can touch off a wildfire of infection within the human population. Such events do not necessarily have to be frequent or even probable; these spillover events may in fact be quite rare, but a single spillover event can spark an epidemic that spreads like a wildfire with a death toll in the thousands. The case study of the West African ZEBOV outbreak, in Subsection 4.2.1, is a sad illustration of this effect.

Practice Note 4.1 One Health

In 2005 the British Medical Journal and the Veterinary Record published an influential joint issue, titled *Human and animal health: strengthening the link* (or *vice versa*, for readers of the Veterinary Record). The decades preceding this joint publication saw a number of zoonotic outbreaks that highlighted the interdependence of animal and human populations. As Gibbs's 2014 historical retrospective noted, the outbreaks of bovine spongiform encephalitis (BSE), highly pathogenic avian influenza (HPAI) H5N1, and SARS shone a light on the interdependence of the health of human and animal populations [125]. In 2004 the Manhattan Principles laid the groundwork for what it called One Health: an interdisciplinary framework for public health that considers human and animal populations alike. In the wake of the COVID-19 pandemic, the Berlin Principles updated the Manhattan Principles, calling on world leaders to "recognize and take action to retain the essential health links between humans, wildlife, domesticated animals and plants, and all nature; and ensure the conservation and protection of biodiversity which, interwoven with intact and functional ecosystems, provides the critical foundational infrastructure of life, health, and well-being on our planet" [126].

Climate change, deforestation, and habitat loss (on the consequences of which see also Subsection 4.2.1) have increased the frequency of human-animal encounter events, including with species that are immunologically highly suitable to act as reservoir hosts for a wide range of pathogens [127]. The global commerce in animals and animal products has also greatly increased the potential reach of pathogens. Finally, health disparities and global differences in public health funding have resulted in an uneven ability to respond to a newly emergent pathogen. The One Health perspective has been a significant shift in integrating the domains of human and animal health, along with concerns of conservation ecology and biodiversity. This interdisciplinary approach may well hold the promise of bringing both human and veterinary epidemiological tools to bear on zoonotic disease.

4.2.1 Case study: Zaire ebolavirus and the 2013–16 West African outbreak

Zaire ebolavirus (ZEBOV) is a filovirus that causes a severe and often lethal viral haemorrhagic fever in primates and humans. Since the first documented outbreak in

Yambuku [128], there have been at least eight major outbreaks across Western Africa. The reservoir host of ZEBOV has not yet been definitively identified, but is widely presumed to be a bat species [129].

Emile Ouamouno, a young boy aged two, lived in the village of Meliandou, in Guinea's Nzerekore district. On 2 December 2013, he came down with a high fever, vomiting, and tarry stools suggestive of intestinal haemorrhaging. On Friday, his life was tragically cut short by what was not much later identified as an ebolavirus and presumed to be ZEBOV. It started a rapid chain of propagation, affecting first his close relatives, then spiralling out of control. By mid-May, it was in Sierra Leone, Liberia, and the Ivory Coast. In the waning days of July 2014, it entered Nigeria. Not long after, it spread across the globe, along the economic and commercial arteries that keep the global economy functioning: air routes, shipping, and freight.

By the time the epidemic burned out, over 11,300 were dead, but it all began with a single bite. Research into the zoonotic origins of the 2013–16 West African ZEBOV outbreak has traced the infection back to the index case playing under a hollow tree, where Angolan free-tailed bats (*Mops condylurus*) were roosting [130]. It is assumed that he was bitten or otherwise infected by the bats sometime around December 2013, and succumbed to Ebola virus disease not much later. His family were among the first secondary cases. Traditional funeral practices, which involve close contact with the decedent's body, and thus ample opportunity for secondary transmission, have also played a significant role in the epidemic (see Practice Note 2.6).

The West African ZEBOV outbreak highlights the incredible human cost that circulating zoonoses can exact. For that reason alone, we should turn our attention to zoonotic diseases that are also human-transmissible. Unlike exclusively vector-borne diseases, which only affect persons who have come into direct contact with the vector or reservoir host, human-transmissible zoonoses operate on "parallel planes," capable both of transmission from an animal to a human and from human to human. Indeed, most human-transmissible zoonoses are primarily transmitted by humans. (See Fig. 4.7.)

A spillover event is defined as an event where a pathogen jumps from an affected population, typically but not necessarily a reservoir host, to a different species that until then is largely naive to the pathogen. After the spillover event, such as the unfortunate encounter of the index case with the Angolan free-tailed bats, much of the transmission happens between humans. In this section, we will discuss ways to separately model the reservoir host and the human "planes" of transmission, and quantify interactions between them.

4.2.2 Modeling host-transmissible zoonoses

Our models of pure vector-borne diseases have so far focused on pathogens where transmission was exclusively interplane, i.e., every transmission event occurred between individuals that belonged to different populations. In this subsection, we turn our attention to pathogens that have at least one intraplane mode of transmission (Fig. 4.8).

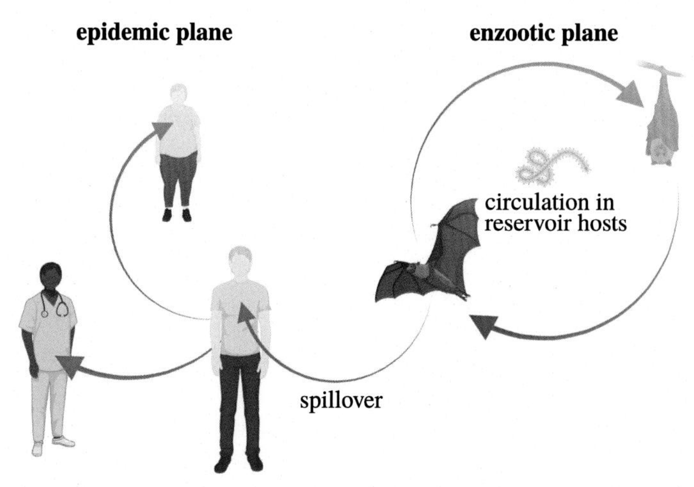

Figure 4.7 Human-transmissible zoonotic cycles: the example of Zaire ebolavirus (ZEBOV). ZEBOV is enzootic in the reservoir host, a bat species, who transmit the pathogen vertically and horizontally, without exhibiting symptoms. Interactions between the enzootically affected population and the human population can lead to spillover events, which can touch off chains of transmission to other humans, including healthcare workers, who are at a particularly high risk of contracting ZEBOV from infected cases.

The rate of new infections in each plane is governed by the interactions between that plane's susceptible population with all infectious populations, multiplied by a given value of β. This is, of course, quite similar to the way we dealt with host heterogeneities (see Subsection 3.1.2). For two planes H and V, with closed populations,

$$\begin{aligned}\frac{dI_h}{dt} &= \overbrace{\overbrace{\beta_{H,H} S_H I_H}^{\text{host-to-host}} + \overbrace{\beta_{H,V} S_H I_V}^{\text{host-to-vector}}}^{\text{mass action, hosts}} - \gamma_H I_H,\\ \frac{dI_V}{dt} &= \underbrace{\underbrace{\beta_{V,V} S_V I_V}_{\text{vector-to-vector}} + \underbrace{\beta_{V,H} S_V I_H}_{\text{vector-to-host}}}_{\text{mass action, vectors}} - \gamma_V I_V.\end{aligned} \tag{4.10}$$

Or, more generally, we may say that given n classes and the $n \times n$ WAIFW matrix $\mathbf{b}$, the mass action term for I_m (for $m \in n$) will be

$$\sum_{i=0}^{n} \mathbf{b}_{m,i} S_m I_i. \tag{4.11}$$

Depending on the structure of the infection, we may model human-transmissible zoonoses either as SI+S(E)IR or SIR+S(E)IR models.

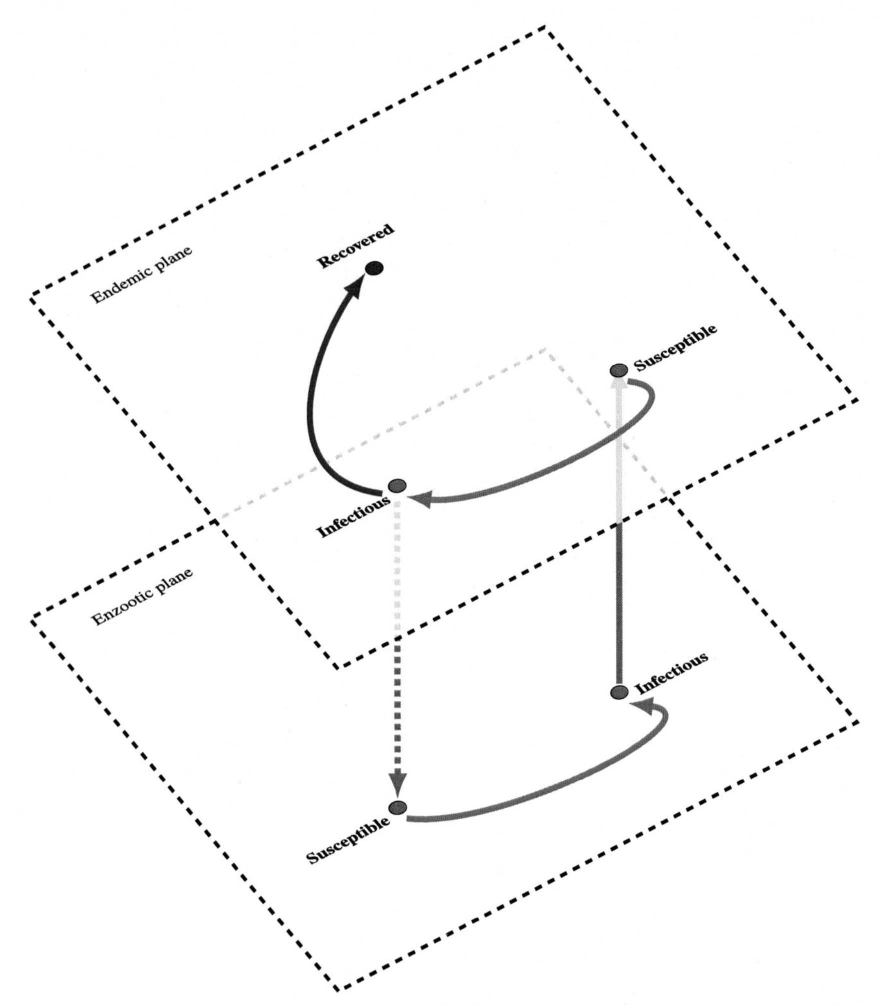

Figure 4.8 Planes of transmission for a zoonotic disease with limited or no transmission from the endemic to the enzootic plane but active transmission both within planes and from the enzootic to the endemic plane.

- In the SI+S(E)IR approach, we assume that infection is lifelong for the vector. This is warranted, in general, in two cases: vectors with relatively short lifespans and infections that significantly reduce the vector's lifespan.
- In the SIR+S(E)IR approach, our assumption is that vectors, too, can recover.

4.2.3 Reservoir hosts and reinfection

A zoonotic infection that has intraclass transmission on the vector plane has a "reservoir host," a host in which it can persist for a long time. Typically, a reservoir host

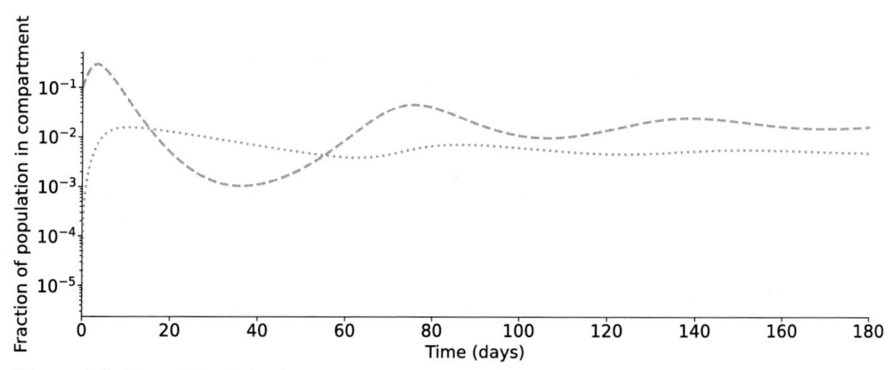

Figure 4.9 Size of the infectious compartments on the host and vector planes for a SIR+SI zoonosis model. The parameters of the model are those given in Computational Note 4.5.

suffers no significant reduction in evolutionary fitness from the infection's presence. Bats, for instance, act as reservoir hosts for certain filoviruses without exhibiting symptoms due to effective immune regulation [118].

The consequence of a reservoir host is that interactions with it can always spark off new chains of human transmission. Consider Lake Victoria marburgvirus (MARV): it is endemic (often maternally transmitted) in the vector population, and may be spread to the human (host) population through bites or the consumption of diseased animals. Among humans, it may be transmitted through bodily fluids. It only takes a single interaction between the human host and the vector to touch off a chain of transmission. In some cases, this will result in a single human case [131], whereas in other instances a longer chain of transmission can establish itself on the host plane. The consequence is that such infections can disappear in the host population altogether, then reappear after a short while.

From an evolutionary standpoint, zoonotic transmission allows a pathogen to afflict a population in which it could not otherwise persist. Consider a SI+SIR model, in which vectors do not recover from their infectious state:

$$\begin{aligned}
\frac{dS_H}{dt} &= \mu_H - \beta_{H,H} S_H I_H - \varrho\beta_{V,H} S_H I_V - \mu_H S_H, \\
\frac{dS_V}{dt} &= \mu_V(1-m_v) - \beta_{V,V} S_V I_V - \varrho\beta_{H,V} S_V I_H - \mu_V S_V, \\
\frac{dI_H}{dt} &= \varrho\beta_{V,H} S_H I_V + \beta_{H,H} S_H I_H - \gamma I_H - \mu_H I_H, \\
\frac{dI_V}{dt} &= \mu_V m_V + \beta_{V,V} S_V I_V + \varrho\beta_{H,V} S_V I_H - \mu_V I_V,
\end{aligned} \tag{4.12}$$

where ϱ is the encounter rate between hosts and vectors.

It follows that in such a model, $\mathfrak{R}_0$ will be composed of two factors: host-to-host transmission and host-to-vector transmission. The first component is, of course,

$$\mathfrak{R}_{0,HH} = \frac{\beta_{H,H}}{\gamma + \mu_H}. \tag{4.13}$$

To this, we add secondary cases in humans that are attributable to each case by way of indirect transmission. We account for three modes of indirect transmission via the enzootic plane:

1. Vectors who obtain infection from humans and pass it on to other humans directly (HVH transmission),
2. Vectors who obtain infection from humans, then have ϱ_V interactions with other vectors, who then pass it on to humans (HVVH transmission),
3. Vectors who are female and obtain infection, then pass it on vertically to their offspring (vertical transmission).

These correspond to each of the terms in

$$\begin{aligned}\mathfrak{R}_{0,HV} = \frac{1}{\gamma + \mu_V} \Big(& \underbrace{\varrho^2 \beta_{H,V} \beta_{V,H}}_{\text{HVH transmission}} \\ & + \underbrace{\varrho^2 \varrho_V \beta_{H,V} \beta_{V,V} \beta_{V,H}}_{\text{HVVH transmission}} \\ & + \underbrace{\frac{1}{2} \varrho^2 \beta_{H,V} \beta_{V,H} \mu_V m_v}_{\text{vertical transmission}} \Big).\end{aligned} \tag{4.14}$$

m is the likelihood of vertical infection, and ϱ_V is the vector-to-vector encounter rate, which primarily depends on the vector group size, whereas the absolute host-vector encounter rate is generally related to $\frac{N_V}{N_H}$, that is, the relative number of vectors per host within reach of each other. Together, this gives us the basic reproductive number of a zoonotic disease with horizontal and vertical transmission as

$$\mathfrak{R}_0 = \frac{\beta_{H,H}}{\gamma + \mu_H} + \frac{\varrho^2 \beta_{H,V} \beta_{V,H}}{\gamma + \mu_V} \left(1 + \frac{\varrho_V \beta_{V,V}}{\mu_V} + \frac{\mu_V m_V}{2}\right). \tag{4.15}$$

Computational Note 4.4 Managing complex models with structures

Compartmental models of zoonotic disease are somewhat notorious for their complexity. Each plane of transmission is a model in its own right, with all its panoply of parameters. It can get quite difficult to keep track of these. Consider the model that yielded Fig. 4.9. Its starting conditions were

```
beta_HH = 0.01
beta_HV = 0.01
beta_VV = 1.225
beta_VH = 0.01

S_H_0 = 1
```

```
S_V_0 = 0.9
I_H_0 = 0
I_V_0 = 0.1

mu_I_V = 0.45
mu_I_H = 0.01

gamma_H, gamma_V = 1/30, 0

mu_V, mu_H = 5/365, 0.05/365
```

Note that this is in fact a simplified version of the model, since we do not track the recovered compartment. The value of `gamma_V = 0` reflects the fact that vectors do not recover. If the lifespan of vectors is sufficiently short, almost always the case for arboviruses and, often enough, the case for most other vectors by a fair degree of approximation, one may consider vectors to be infectious for life.

To manage this array of parameters, it helps to write the derivative function as accepting three tuples:

- `beta`, which comprises each of the values for β,
- `gamma`, which comprises each of the values for γ, and
- `mu`, for the birth and mortality rates in the model.

We thus specify the model as follows:

```
def deriv(t, y, beta, gamma, mu):
    S_H, S_V, I_H, I_V = y
    gamma_H, gamma_V = gamma
    beta_HH, beta_HV, beta_VV, beta_VH = beta
    mu_H, mu_V = mu

    dShdt = mu_H - (beta_HH * S_H * I_H + beta_HV * S_H * I_V) \
            - mu_H * S_H
    dSvdt = mu_V - (beta_VV * S_V * I_V + beta_VH * S_V * I_H) \
            - mu_V * S_V
    dIhdt = (beta_HH * S_H * I_H + beta_HV * S_H * I_V) - \
                gamma_H * I_H - mu_H * I_H - mu_I_H * I_H
    dIvdt = (beta_VV * S_V * I_V + beta_VH * S_V * I_H) - \
                gamma_V * I_V - mu_V * I_V - mu_I_V * I_V

    return dShdt, dSvdt, dIhdt, dIvdt
```

The necessary corollary of this is that when we invoke our solver (in this case, `solve_ivp`, but the point applies equally well for `odeint`), we must provide the tuples we deconstruct in the beginning of our derivative function as such:

```
res = solve_ivp(fun=deriv,
                t_span = (0, 1000),
                y0=y_0,
                max_step=1,
                method="BDF",
                args=(
                    (beta_HH, beta_HV, beta_VV, beta_VH),
                    (gamma_H, gamma_V),
                    (mu_H, mu_V)
                    )
                )
```

A function that takes too many arguments can get clunky and hard to maintain. It is also considered by some to be generally unsightly code. Specifying tuple inputs can clarify the process and ensure more maintainable and legible code.

A notebook implementing the contents of this Computational Note is available on the book's companion Github repository in the folder `/ch04/ross_macdonald`.

In most cases, I_v outnumbers I_h quite significantly. Therefore even where human transmission is quite modest, the disease can persist if there is a significant transmission among reservoir hosts. By way of example, ignoring incidents of laboratory accidents, there have been, between 1975 (the first natural case) and 2021, 11 instances of MARV outbreaks in human populations. These correspond to 11 spillover events over the span of forty-six years, over a population at risk of approximately 800 m (with one instance where both MARV and Ravn virus (RAVV), a related marburgvirus, experienced quasi-simultaneous spillover events over the same population at roughly the same time [132]). This translates to a spillover rate of 3.26×10^{-10} per person per year. Few of these spillover events have caused more than a handful of human cases (with the 2004–2005 Angolan outbreak and the 1998–2000 DRC outbreak, each of which had cases in the hundreds, being more exceptions than the rule). Human-to-human chains of transmission are very short, largely due to the severe course of the disease and rather unambiguous signs of illness. Yet MARV is alive and well, emerging from its reservoir host every few years to infect humans (and, presumably, more often infecting nonhuman animals as well, which however generally goes unrecorded). A serosurvey by Paweska et al. [133] of Egyptian rousettes (*Rousettus aegyptiacus*, the putative reservoir host of MARV [134]) in South Africa showed 71.6% of adult rousettes to be seropositive for MARV. MARV serves as a poignant example of the way a pathogen that is highly prevalent in its reservoir host population can hide safely without human notice, until in some unfortunate accident, hosts and vectors cross paths.

Practice Note 4.2 Looking beyond the host

The wider availability of data—and the tools to process it—has recently reawakened interests in early warning systems of zoonotic diseases. The perspective of One Health (see Practice Note 4.1) is that especially where humans and animals coexist, their disease dynamics are often coupled. Wolking et al. [135] showed, for instance, that from a sample of pastoral communities in Tanzania, almost two-thirds of households with at least one human patient with diarrhoea also had at least one calf with diarrhoea in their herds. *Cryptosporidium spp.* and *Giardia duodenalis* affect bovine populations similarly to humans, and are easily transferred through every-day contact. Focusing not only on the host species but also on vectors or alternate hosts can serve as an early warning system for an emergence of the disease in human populations. Bisson, Ssebide, and Marra [136], for instance, propose the reporting of animal morbidity as sentinels to pathogenic emergence in human populations.

The historical experience in monitoring vectors to anticipate the human impact has not been extensive. Amato et al. [137] describe such an early warning project, noting that it is far from typical. Many disease vectors are particularly difficult to capture and test (bats take pride of place in this respect). Sampling networks are time- and resource-intensive to implement and analyze, and as Herten et al. [138] point out, using animals as bellwethers of human health might raise complex ethical (and economic) questions. As Hussain-Alkhateeb et al. [139]'s scoping review shows, such early warning systems have so far fallen short of the expectations of their promises. With the philosophical underpinnings provided by One Health and the technological solutions that have emerged over the last decades, the hope is that the promise of early warning systems of disease in host populations by examining the vector population or reservoir hosts might come nearer fulfillment.

4.2.4 Seasonal variance of zoonotic transmission

Humans have different needs at different times: most of us want some balance of solitude and privacy on one hand and communality and sharing on the other. We are, like much of the animal world, facultative cooperators.

- Cooperative behavior often has a protective effect for prey and the ability to hunt more effectively for predators. The Allee effect suggests that increasing group size increases individual fitness. Especially in resource-poor times or where there is no benefit in self-interested territoriality (e.g., during winter, where common survival is more important than mating), individuals might opt to cooperate [140].
- On the other hand, the cost of whatever benefits the Allee effect provides is that it has to be shared with a potentially quite large population of others. Thus when the pressures that justify cooperation—the need to hunt cooperatively due to food scarcity or the need to bunch together for defence—are less intense, individuals

might fare better focusing, essentially, on their own good. Wilson, Goodson, and Kingsbury [141], for instance, found evidence of neurobiochemical correlates of this process in sparrows, showing a seasonal shift between territoriality during mating season (summer) and flocking during resource scarcity (winter).

Just as most humans prefer a situation-appropriate mixture of solitude and engagement, the rate at which hosts and vectors interact with each other might vary with time:

- $\beta_{h \to h}$ may generally higher whenever hosts congregate on a seasonal basis. Weather modulates social behavior in humans and nonhuman animals alike. In the animal world, resource scarcity, cold weather, and migration can drive individuals together. Humans tend to congregate during the winter (a phenomenon discussed in more detail in Chapter 7) and, in agricultural societies, during seasons of increased cooperation (sowing and harvest season).
- $\beta_{h \to v}$ and $\beta_{v \to h}$ often change when interaction between the host and vector species changes. In some parts of the world, agriculture brings vectors and human host populations closer (a situation described also in Subsection 4.2.1), and the annual periodicity of agricultural activity often influences transmission rates.
- $\beta_{v \to v}$ is similarly governed by increased density of vectors, as it happens often during breeding periods. Vectors, too, may engage in seasonal flocking periods.

The effect of seasonal variance on a dynamic process in time is referred to as temporal forcing, and is explored in detail in Subsection 7.3. In a temporal forcing model of transmission, we represent one or more values of β not as constants but as time-varying quantities, quite typically as a sinusoidal function that recapitulates rather conveniently how such processes evolve in nature.

Thus we could model a process like the annual seasonality of increased vector-to-host transmission due to

$$\beta(t) = \beta_0(\overbrace{1}^{\text{base } \beta} + \overbrace{\underbrace{\beta_1}_{\text{forcing parameter}} \sin(\underbrace{\omega}_{\text{period}} t + \underbrace{\psi}_{\text{phase}})}^{\text{time-varying } \beta}). \tag{4.16}$$

This should, of course, be quite familiar from high school physics, as it is essentially the sine wave equation. The amplitude of forcing β_1 is bounded by [0, 1], as negative values of β are biologically nonsensical. ω describes the angular frequency, which relates to the ordinary frequency f as $\omega = 2\pi f$. Thus for a process that repeats once a year, ω is 2π. Where ω is supplied as radians per year, care must be taken to supply t and ψ as fractions of the year by dividing them by 365. Compartmental models can easily be expanded to accommodate time-varying rates. For instance, the SI+SIR model from Eq. (4.12) can be parameterized with a time-varying value for $\beta_{v \to h}$ by replacing the term with its time-varying form, then defining it as a function of t:

$$\frac{dS_h}{dt} = \mu_h - (\overbrace{\beta_{h\to h} S_h I_h}^{h\to h} + \overbrace{\beta_{v\to h} S_h I_v}^{v\to h}) - \overbrace{\mu_h S_h}^{\text{mortality}},$$

$$\frac{dI_h}{dt} = \overbrace{\beta_{h\to h} S_h I_h}^{h\to h} + \overbrace{\beta_{v\to h}(t) S_h I_v}^{v\to h} - \overbrace{\gamma I_h}^{\text{recovery}} \overbrace{-\mu_h I_h - \mu_h^I I_h}^{\text{mortality}}, \tag{4.17}$$

$$\beta_{v\to h}(t) = \underbrace{\beta_{v\to h_0}}_{\text{base } \beta_{v\to h}} (1 + \underbrace{\beta_{v\to h_1}}_{\text{forcing parameter}} \sin\Big(\underbrace{\omega}_{\text{frequency}} \frac{2\pi}{365}(\underbrace{t}_{\text{days}} + \underbrace{\psi}_{\text{phase}})\Big).$$

For an annually repeating process, $\omega = 1$, and thus can altogether be omitted. The conversion factor $\frac{2\pi}{365}$ converts days into radians, thus allowing us to specify t and ψ as days.

Computational Note 4.5 Time dependence in ODE solvers

Until now, we have primarily worked with autonomous ODEs, that is, ODEs that describe the right-hand side without reference to t. Temporal forcing requires us to accommodate nonautonomous models, but as we shall find, this is not greatly different from previous cases. All major ODE solvers expose the time of evaluation as a variable to the derivative function. The `odeint` and `solve_ivp` APIs both make this available using the time parameter of the derivative function. Recall that we have defined the derivative as

```
def deriv(t, y,...):
    ...
    return ...
```

for `solve_ivp` (and much the same way for `odeint`, except that in that case, the order of `t` and `y` is reversed). The consequence is that `deriv()` has access to `t` at any time. Thus a time-dependent term is simply an expression over `t`. It sometimes pays to calculate the value of $\beta(t)$ at t and store it as a separate variable within the `deriv()` function's scope (prefixing the current value with `eff_` is a good enough starting point). This reduces ambiguity and prevents accidentally overwriting other variables.

Eq. (4.17) would then be formulated—omitting unchanged lines from Computational Note 4.4—as

```
def deriv(t, y, beta, gamma, beta_1, psi):
    ...
    eff_beta_VH = beta_VH * \
        (1 + (beta_1 * np.sin(2 * np.pi/365 * (t - psi))))
```

```
    dSvdt = mu_v - (beta_VV * S_V * I_V +\
                    eff_beta_VH * S_V * I_H) -  mu_V * S_V

    dIvdt = (beta_VV * S_V * I_V + eff_beta_VH * S_V * I_H) - \
            gamma_V * I_V - mu_V * I_V - mu_I_V * I_V

    return dShdt, dSvdt, dIhdt, dIvdt
```

Note that time-dependent functions in the derivative function can be quite arbitrary, but adding conditional statements and branching processes tends to slow down the integrator considerably.

A notebook implementing the contents of this Computational Note is available on the book's companion Github repository in the folder `/ch04/zoonosis`.

4.2.5 Zoonoses and birth dynamics

We typically conceive of hosts and vectors as being fundamentally different; one long-lived and relatively rare, the other short-lived yet abundant and reproducing rapidly. This is one of the very rare examples of inherent prejudice born of the human experience that is not entirely incorrect. Species with low birth rates and long lifespans generally make very bad vectors. Not only might they live long enough to clear (or, in the case of humans, treat) the pathogen, they might also succumb to the effects of the pathogen all too early. For what they can do for the pathogen in terms of disease transmission, they are much too rare. Ideal vectors have high birth rates, even if it is the result of short lives. For this reason, vectors that have seasonal dynamics of birth are worthy of our special attention.

In 2015, Hayman [142] remarked on the way the bi-annual periodicity of birth in *R. aegyptiacus* allowed a pathogen—in that case, MARV—to persist. In mammals, pregnancy modulates the maternal immune system [143], to prevent rejection of the allogeneous offspring during pregnancy. Thus a vector population, and sometimes, a host population, may be increasingly susceptible to sustaining infection during pregnancy. Where births are seasonal in either the host or the vector population, related effects might emerge on transmission to the relevant compartment.

The consequence of multiple forcings with different periodicities is the evolution of complex time-dependent phenomena, much of which depend not only on the difference of the value between seasons but also on their temporal relationships (phase). Fig. 4.10 shows the effect of phase on a model with seasonally varying births and a seasonally varying transmission term (in that case, $\beta_{v \to v}$). For a relatively stable host population, a birth pulse in the vector population means a quick rise in susceptibles, who will in due course be recruited into the infectious vector compartment, and thus a rise in the relative number of infectious vectors per susceptible host. On the other hand, temporal dynamics of host-vector interactions can change the likelihood of a

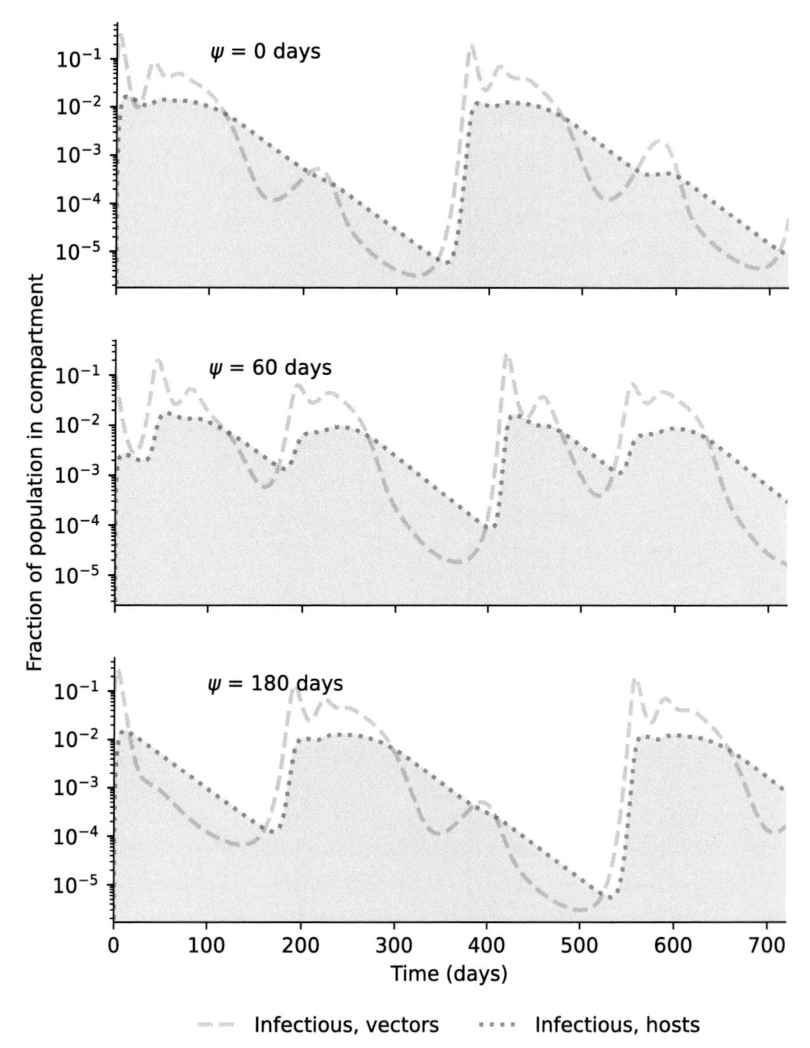

Figure 4.10 A dual temporally forced host-vector system at different phases. In this system, both the birth rate of vectors and intraplane transmission among vectors is temporally forced; the former with a period of 6 months, resembling the biannual births discussed by Hayman [142], the latter at an annual period. The effect of phase considerably changes the number of infections in the host compartment.

host acquiring the infection at times when the transmission coefficient rises. If these two processes are perfectly in phase, resonant phenomena can emerge, resulting in sharp spikes of infection in the host population.

Multipathogen dynamics
Competition, cross-immunity, and cosusceptibility

When a new influenza virus emerges, it is highly competitive, even cannibalistic. It usually drives older types into extinction. This happens because infection stimulates the body's immune system to generate all its defenses against all influenza viruses to which the body has ever been exposed. When older viruses attempt to infect someone, they cannot gain a foothold. They cease replicating. They die out. So, unlike practically every other known virus, only one type—one swarm or quasi species—of influenza virus dominates at any given time. This itself helps prepare the way for a new pandemic, since the more time passes, the fewer people's immune systems will recognize other antigens.

Barry, The Great Influenza: the epic story of the deadliest plague in history, 2005 [144]

In our previous pursuits, we have largely focused on a single pathogen, much to the exclusion of all others. This is a sensible choice when the likelihood of interaction between our pathogen of concern and other pathogens is not particularly significant. Keeling and Rohani [39] note that there is almost always some effect between multiple pathogens that can infect the same organism, a phenomenon they refer to as "interference." This may take place either on a physiological/biochemical level, such as by way of cytokine activation by one pathogen that makes the individual less susceptible to subsequent infection by another in the short term [145]. Alternatively, behavioral factors may be at play: symptomatic illness may reduce an individual's encounter rate, thus reducing the likelihood of infection with another pathogen.

In this chapter, we are concerned with multiple pathogens. Multipathogen modeling is crucial where the interaction is nontrivial. This may be because there is a meaningful biological connection between the pathogens, for instance, the similarity in at-risk groups and the inherent increase in the likelihood of seroconversion due to immunosuppression means Hepatitis C often presents as a coinfection with HIV [146]. Alternatively, there might be an evolutionary concern, such as the development of antibiotic-resistant strains (which is explored in more detail in Subsection 5.1.5). Finally, there are often interactions that are behaviorally defined. Pertussis and measles, which is discussed in more detail in Subsection 5.2.1, is perhaps one of the most prominent examples, where there is a behavioral element to the interaction between two pathogens.

Multipathogen models are challenging, because they comprise, essentially, multiple compartmental models stacked upon each other. The complexity increases rather rapidly with the number of pathogens.

Computational Modeling of Infectious Disease. https://doi.org/10.1016/B978-0-32-395389-4.00014-1

5.1 Multipathogen systems with cross-immunity

This section deals with scenarios where there is partial or total cross-immunity. Cross-immunity has been at the heart of Jenner's vaccination for smallpox: Jenner's vaccination worked, because infection with cowpox generated cross-immunity with smallpox (see Section 6.1). Models of complete cross-immunity are the simplest multipathogen models, and thus we shall begin with this simple scenario.

Because immunity is at the heart of this matter, we will consider two pathogens to be distinct if they are immunologically distinct.

Definition 5.1 (Immunological distinctiveness). Two organisms are immunologically distinct if there is a meaningful difference between the immune response. Thus if the pathogens a and b are immunologically distinct, the immune response evoked by a is less effective in respect of b than it is in respect of a.

Thus we take a principally immunological view of genetic variance: pathogens are "distinct" (i.e., in terms of serovariants) as long as antibodies against one type are less efficient against another. With that in mind, let us explore the effects of serovariance and serological diversity on infectious dynamics.

5.1.1 Case study: too much of a good thing?

Dengue fever is an unpleasant, but rarely life-threatening, illness caused by the dengue virus (DENV), an RNA flavivirus, and hence a distant relative of yellow fever. The majority of patients experience an approximately week-long illness, comprising fever, arthralgia, and a rather peculiar rash. A small minority, however, go on to develop dengue haemorrhagic fever, which may be life-threatening. Similarly to other haemorrhagic fevers, it manifests as thrombocytopaenia, haemorrhage, and hypovolaemic shock.

In 2016 a vaccine became available for dengue fever. CYD-TDV is a chimeric vaccine that uses an attenuated yellow fever strain altered to express the envelope (E) and premembrane (PrM) proteins from selected dengue serotypes [147]. Following a vaccination campaign across the Philippines, it emerged in a 2018 analysis that seronegative individuals who were vaccinated had a paradoxically elevated risk of severe illness [148]. Though those who have had dengue before have benefited greatly from the vaccine, for dengue seronegative vaccine recipients aged 9 to 16, the hazard ratio of virologically confirmed dengue infection requiring hospitalization was 1.57%, as opposed to only 1.09% among unvaccinated controls, a hazard ratio of 1.41.

An explanatory hypothesis was presented by Halstead in 2017, arguing that the vaccine created antibodies that contributed to the development of antibody-dependent enhancement (ADE) [149]. ADE occurs when a vaccine elicits antibodies that either do not bind at a neutralizing epitope (i.e., their binding does not preclude the virion's ingress into a host cell) or have low binding affinity, and thus, paradoxically, antibodies end up facilitating viral entry. The detailed mechanics of ADE are discussed in Subsection 5.1.4, and its handling in the vaccine context in Practice Note 5.2.

In the case of the dengue vaccine, the indiscriminate administration regardless of serostatus has resulted in exposing individuals to a higher risk of severe disease from dengue infection than they would have experienced in the absence of immunization. Too much of a good thing, alas, can thus have detrimental outcomes. Revelations about the potential for ADE arising from the dengue vaccine have led to a widespread crisis of confidence in vaccinations in the Philippines, which compounds the damage [150]. When considering immunizations for macrophagotrophic or respiratory pathogens, the low but nonzero risk of ADE must not be discounted.

5.1.2 Simple multipathogen models with perfect cross-immunity

The simplest multipathogen models ordinarily pertain to two strains or pathogenic species, which we shall denote using numbers. Simple models follow the principle known as Gause's law of competitive exclusion: competition between two organisms occupying the same ecological niche is always "absolute," in that the stable equilibrium of such competition is the absolute dominance of one organism [151].

Definition 5.2 (Gause's law of competitive exclusion). If two or more organisms inhabit the same ecological niche, the stable equilibrium will be the complete dominance of the more advantaged species and the complete extinction of the disadvantaged species.

Our simple models reflect this by allowing for one and only one course of infection, after which individuals return to a recovered compartment. This is the case when there is perfect cross-immunity, i.e., recovery from pathogen 1 (R_1) immunizes against pathogen 2 as well, and vice versa. The model can thus be described as

$$\begin{aligned}
\frac{dS}{dt} &= -\overbrace{\overbrace{\beta_1 S I_1}^{\text{from strain 1}} - \overbrace{\beta_2 S I_2}^{\text{from strain 2}}}^{\text{mass action}}, \\
\frac{dI_1}{dt} &= \beta_1 S I_1 - \gamma_1 I_1, \\
\frac{dI_2}{dt} &= \beta_2 s I_2 - \gamma_2 I_2, \\
\frac{dR}{dt} &= \underbrace{\underbrace{\gamma_1 I_1}_{\text{from strain 1}} + \underbrace{\gamma_2 I_2}_{\text{from strain 2}}}_{\text{recoveries}}.
\end{aligned} \tag{5.1}$$

This model generalizes for n pathogens as

$$\frac{dS}{dt} = -\overbrace{\sum_{i=1}^{n} \overbrace{\beta_i S I_i}^{\text{mass action for } i}}^{\text{total mass action}}$$

$$\frac{dI_i}{dt} = \overbrace{\overbrace{\beta_i S I_i}^{\text{mass action}} - \overbrace{\gamma_i I_i}^{\text{recoveries}}}^{\text{for } i}, \tag{5.2}$$
$$\frac{dR}{dt} = \underbrace{\sum_{i=1}^{n} \underbrace{\gamma_i I_i}_{\text{recoveries for } i}}_{\text{recoveries across all } i},$$

and in line with the equation in Subsection 2.1.5, the pathogen-wise basic reproduction number $\mathfrak{R}_0^i$ would be $\frac{\beta_i}{\gamma_i}$ for all i in n.

The dynamics of multiple pathogens with different values of $\mathfrak{R}_0$ and providing complete cross-immunity are somewhat nontrivial. In the long run, competitive exclusion means that the pathogen with the larger value of $\mathfrak{R}_0$ will become absolutely dominant (i.e., the pathogen with the smaller value of $\mathfrak{R}_0$ will become extinct). A pathogen i is at an equilibrium if $S = \mathfrak{R}_0^i$ (i.e., if $S\mathfrak{R}_0^i = 1$). This results in n equilibrium points for n pathogens. For the simple case of two pathogens, the equilibrium points will be $S = \frac{1}{\mathfrak{R}_0^1}$ and $S = \frac{1}{\mathfrak{R}_0^2}$. If $\mathfrak{R}_0^2 > \mathfrak{R}_0^1$, then the equilibrium point at $S = \frac{1}{\mathfrak{R}_0^1}$ will not be stable for pathogen 2, because the growth rate of pathogen 2 at that point is still positive. On the other hand, at $S = \frac{1}{\mathfrak{R}_0^2}$, the growth rate of pathogen 1 is negative. Thus pathogen 2 can reach an equilibrium that eventually stabilizes into the extinction of pathogen 1, which is the long-term evolutionary destiny of multipathogen systems with complete cross-immunity and different values of $\mathfrak{R}_0$.

Practice Note 5.1 The long and short (term) of it

Gause's law tells us what happens in the long run. However, short-term processes might be quite different.

Fig. 5.1 describes just such a scenario. With pathogen b having a higher $\mathfrak{R}_0$ than a, its eventual dominance is all but assured. On the other hand, in the short term, the more rapid lifecycle of a makes it temporarily dominant in the short term. In short: the long-term evolutionary destiny is determined by $\mathfrak{R}_0$, whereas short-term dynamics are determined by the values that make up $\mathfrak{R}_0$ (i.e., β and γ).

The consequence of that in practice is that when two pathogens emerge that induce cross-immunity, $\mathfrak{R}_0$ alone might not account for short term processes. Rather, its constituent elements must be taken into account. Epidemiological modeling, such as has been discussed in Computational Note 4.4, can assist in preparing a response that allocates resources accordingly. This is particularly important where the two pathogens produce sufficiently different presentations that might require different treatments or are associated with different severity.

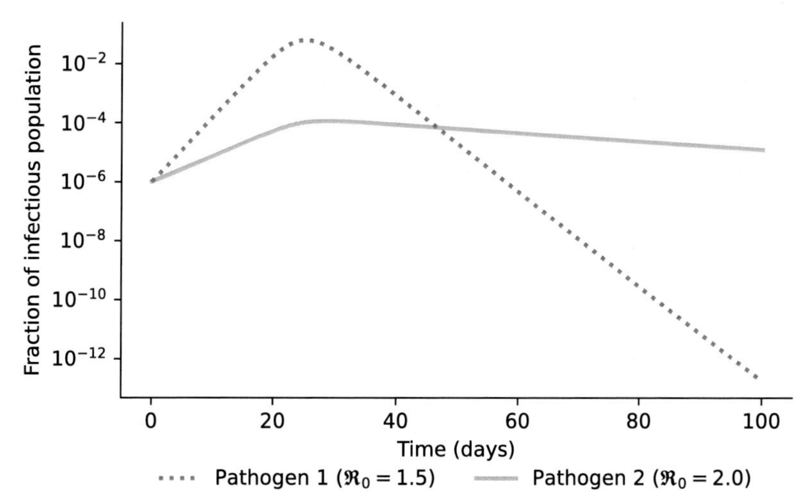

Figure 5.1 Short and long-term dynamics of a two-pathogen system with complete cross-immunity, with $\beta_1 = 1.5, \gamma_1 = 1.0, \beta_2 = 0.4$, and $\gamma_2 = 0.2$ and $I_1(0) = I_2(0) = 10^{-6}$. It follows from the above that whereas pathogen 2 is dominant in terms of $\mathfrak{R}_0$, pathogen 1 follows a much faster dynamics with a shorter lifecycle. The result of this is that whereas pathogen 1 eventually becomes extinct, pathogen 2 is briefly dominant in the short term.

5.1.3 Incomplete cross-immunity

In incomplete cross-immunity, exposure to a pathogen produces a nonzero level of cross-immunity against another pathogen. Fig. 5.2 demonstrates such a scenario with two pathogens. Recovery from either pathogen provides perfect immunity, i.e., there is no "loopback" from R_1 to I_1 and R_2 to I_2, respectively. We denote the compartment of previously recovered individuals who sustain an infection from the other pathogen as the heterotypic infectious compartment (H). The flows are strictly "crosswise," i.e., they go from R_1 to H_2, and vice versa. Recovery from heterotypic infection results in dual immunity, denoted as $R_{1,2}$. Because heterotypic infections are part of the infectious subsystem, they must be considered part of the mass action term of the infectious compartment (since a susceptible individual may be infected by an initial or a heterotypic infection). Therefore the susceptible subsystem is

$$\frac{dS}{dt} = -S(\beta_1(I_1 + H_1) + \beta_2(I_2 + H_2)). \tag{5.3}$$

We denote the heterotypic penalty of each strain as ψ_1 and ψ_2 for strains 1 and 2, respectively. ψ_1 would represent the reduction to β_2 vis-a-vis an individual who has recovered from strain 1, and vice versa. The infectious subsystem (I and H) is then

$$\frac{dI_1}{dt} = \beta_1 S \underbrace{(I_1 + H_1)}_{\text{all infections with strain 1}} - \gamma_1 I_1,$$

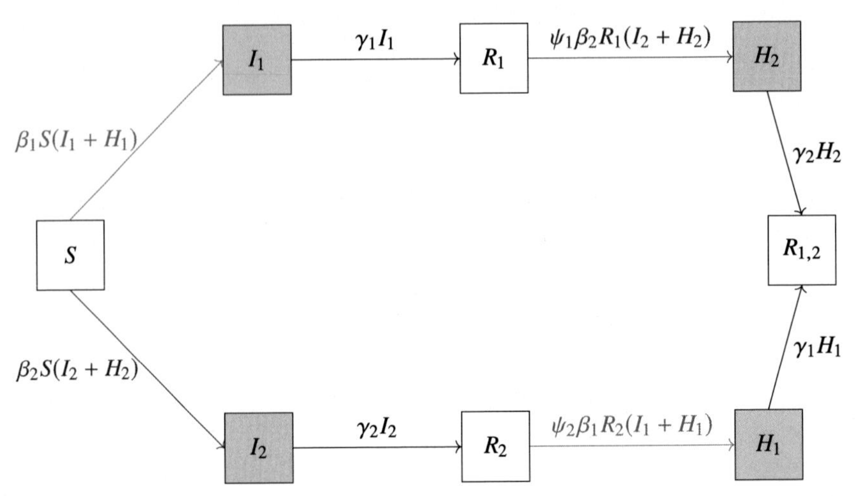

Figure 5.2 Interaction between two pathogens with incomplete cross-immunity. Individuals who recover from one pathogen are still susceptible to heterotypic infection with the other, albeit subject to a penalty ψ.

$$\begin{aligned}
\frac{dI_2}{dt} &= \beta_2 S \underbrace{(I_2 + H_2)}_{\text{all infections with strain 2}} - \gamma_2 I_2, \\
\frac{dH_1}{dt} &= \underbrace{\psi_2 \beta_1 R_2 (I_1 + H_1)}_{\text{heterotypic infections with strain 1}} - \gamma_1 H_1, \\
\frac{dH_2}{dt} &= \underbrace{\psi_1 \beta_2 R_1 (I_2 + H_2)}_{\text{heterotypic infections with strain 2}} - \gamma_2 H_2.
\end{aligned} \tag{5.4}$$

This model for two competing pathogens with some cross-immunity can be expanded to arbitrary numbers of pathogens, although model complexity does increase significantly.

5.1.4 Antibody-dependent enhancement

Antibody-dependent enhancement (ADE) is a phenomenon observed in a small number of pathogens (most significantly perhaps in dengue virus, see, e.g., Subsection 5.1.1) that have multiple circulating subtypes or serotypes. Normally, antibodies bind to virions, then to the fragment crystallizable region gamma IIa receptors (FcγRIIa) on cells that express it (commonly known as CD32+ cells), such as macrophages, dendritic cells, and B-cells. If the antibodies do not neutralize the pathogen, but facilitate its phagocytosis by binding to the FcγRIIa receptor, infection-competent virions can enter these cells with relative ease. Thus, paradoxically, the presence of antibodies work in favor of the pathogen, rather than against it.

There has been relatively little quantitative modeling of ADE in general. Ferguson, Anderson, and Gupta [152] examined a model of ADE in the context of dengue, concluding that ADE can result in temporary oscillatory dynamics, in which the heterotypic serotype that benefits from ADE outpaces its competitors until it burns through the available susceptible population and once again recedes. In the context of COVID-19, Adil Mahmoud Yousif et al. [153] proposed a compartmental approach, in which a certain percentage of the vaccinated cohort (between 1 to 2%) develop a predisposition for ADE. A more effective compartmental approach for the modeling of ADE is to separate serotypes and exposures. For n serotypes, the cross-ADE matrix $\mathbf{A}$ is the $n \times n$ hollow matrix

$$\mathbf{A} = \begin{pmatrix} 0 & \alpha_{1,2} & \cdots & \alpha_{1,n} \\ \alpha_{2,1} & 0 & \cdots & \alpha_{2,n} \\ \vdots & \vdots & \ddots & \vdots \\ \alpha_{n,1} & \alpha_{n,2} & \cdots & 0 \end{pmatrix}, \tag{5.5}$$

where $\alpha_{p,q}$ is the likelihood of developing ADE when exposed to serotype q, given a past exposure to serotype p.

The values of this matrix may be ascertained empirically from *in vivo* studies or from *in vitro* research, since $\alpha_{p,q}$ is likely to relate inversely to the ability of anti-p antibodies to neutralize q (in other words, $\alpha_{p,q}$ is a metric of the deficiency of anti-p antibodies in neutralizing q, i.e., the specificity of anti-p vis-a-vis q). The matrix $\mathbf{A}$ is hollow, because only heterotypic exposures can result in ADE. This is an assumption that is, of course, subject to perfect serological categorization (i.e., it assumes that all of the serotypes are perfect partitions, i.e., every infected individual is infected with one serotype, and the sum of each serotype's infectious subcompartment equates to the entire population, and binding is a binary property of serotypal identity). As such, it is a somewhat idealized picture of reality, where antibodies to a particular serotype may nonetheless bind less than perfectly to their target epitopes. Nonetheless, it is analytically helpful for the purpose of modeling ADE. For the compartment C_p, comprising all individuals who were exposed to, and recovered from, serotype p, ADE on exposure to q accounts for $C_p I_q \mathbf{A}_{p,q}$, so that

$$\frac{dI_q}{dt} = \beta S(1 + \mathbf{A}_{p,q} C_p \epsilon) I_q - \gamma I_q, \tag{5.6}$$

where ϵ is a measure of the impact of ADE, expressed as the increase in β, where ADE occurs (i.e., $\mathbf{A}$ gives the likelihood, ϵ gives the impact). This generalizes for all n serotypes as

$$\frac{dI_q}{dt} = \beta S\Big(1 + \sum_{i=1}^{n} A_{i,q} C_i \epsilon\Big) I_q - \gamma I_q. \tag{5.7}$$

Practice Note 5.2 ADE and vaccination

It bears repeating that ADE is a marginal phenomenon, and must be represented in context. The study by Adil Mahmoud Yousif et al. [153], cited above, found that at most 2% of COVID-19 vaccinations would result in ADE (and in all likelihood, real figures are much lower). ADE figures must be presented in the context of the risk averted by vaccination, which in the overwhelming majority of cases is going to be much higher. During the COVID-19 pandemic, ADE was identified as a significant driver behind vaccine hesitancy [154–156], when in reality the least effective vaccines were quite significantly beneficial, even when accounting for ADE.

It is important not to obscure the reality of ADE, and where appropriate, encourage understanding of the phenomenon to assist physicians in appreciating the sometimes surprisingly severe presentation of infectious diseases in individuals who ought to have acquired or induced immunity [157–160]. At the same time, in the vast majority of circumstances, the benefits from vaccination and the protective effect of recent prior infection are going to be vastly more beneficial than the detriment incurred by ADE. In addition, the effect of ADE depends on changes or diversity in the serotype that renders the previous immunity, acquired or induced, to be no longer effective in neutralizing. Therefore communicating the effects of ADE must take into account the counterfactual risk averted by reducing the risk of contracting the illness in the first place.

5.1.5 Antimicrobial resistance as a multipathogen problem

We may conceptualize antimicrobial resistance as the competition of pathogens s (susceptible or "wild type") and r (resistant). If r and s exhibit complete cross-immunity, then, as we have discerned in Subsection 5.1.2, only one can survive, and the evolutionary dominance will be determined by the relative values of $\mathfrak{R}_0^r$ and $\mathfrak{R}_0^s$.

Because antimicrobial treatment shortens the course of disease, γ_s will be greater than γ_b by a variable we shall call treatment effect or ϵ_T for short, and by the fraction of the population of people who have a susceptible infection and receive antibiotic treatment. The latter is the treatment fraction θ. Thus the infectious subsystem of this model is

$$\begin{aligned} \frac{dI_s}{dt} &= \beta_s S I_s - (\gamma - \overbrace{\theta\epsilon_T}^{\text{penalized recovery for susceptible strain}}) I_s, \\ \frac{dI_r}{dt} &= \beta_r S I_r - \underbrace{\gamma I_r}_{\text{unpenalized recovery for resistant strain}}. \end{aligned} \tag{5.8}$$

It follows from this that s is dominant if $\mathfrak{R}_{0_s}$ is greater than $\mathfrak{R}_{0_r}$, otherwise r becomes the dominant strain, which means that most infections will be caused by the type less susceptible to antimicrobials. The inflection point depends, of course, on the parameters, but most crucially, on θ. Setting the $\mathfrak{R}_0$s equal, we can define the critical value of θ, so that treating a larger proportion of the population would lead towards dominance by the resistant pathogen. This is the critical value ${}^{\Delta}\theta$, which is defined as

$$ {}^{\Delta}\theta = \frac{\gamma}{\epsilon_T}\left(1 - \frac{\beta_s}{\beta_r}\right). \tag{5.9} $$

Subject to the caveat in Practice Note 5.3, we can conceive of ${}^{\Delta}\theta$ as the maximum safe proportion of the infected population that can be treated with antimicrobials of efficacy ϵ_T without enabling the dominance of the resistant strain. This model can, of course, be expanded to multiple antimicrobials, where the pathogenic model would comprise the 2^n permutations of n potential susceptibilities. This is particularly useful where there are multiple dimensions of drug resistance, i.e., multiple drug groups against which a pathogen can develop immunity.

Practice Note 5.3 Compensatory mutations and a caveat about critical values

A corollary of Eq. (5.9) is that what percentage of the population can be treated without handing over dominance to the drug-resistant strain depends on the fitness cost, in terms of relative $\mathfrak{R}_0$s, of drug resistance (along with the effectiveness of treatment, ϵ_T). In general, resistance has a fitness cost, so that drug resistant mutants have a typically lower β (and thus, lower $\mathfrak{R}_0$) [161]. The difference between the respective values of $\mathfrak{R}_0$ is the mathematical expression of this fitness cost. However, evolution is a continuous process. Compensatory mutation refers to the process of subsequent mutations "offsetting" the fitness cost of the drug resistance mutation [162].

The critical value ${}^{\Delta}\theta$ is a theoretical estimate of what percentage of the population can be treated without benefiting the drug-resistant strain; the critical value of ${}^{\Delta}\theta$ in this case may drop over time due to compensatory mutations [163,164]. It is therefore important to treat ${}^{\Delta}\theta$ as an indicative upper bound and approach it with considerable caution. Pathogenic evolution does not stop, and if a pathogen were to overcome its innate deficiencies through compensatory mutations, reasonable treatment rates well below ${}^{\Delta}\theta$ could prove excessive. Vigilant tracking of pathogenic dynamics is crucial for data-driven antibiotic stewardship.

5.1.6 Microscale models of antimicrobial treatment and immunity

The compartmental models we have encountered are mean-field models that represent population flows by mapping the overall rate of processes. We may not be able to account for every individual, but at sufficiently large populations, even relatively simple

Table 5.1 Phenotype-specific parameters of the phenotypes for the model described in Subsection 5.1.6.

Phenotype	Population	Growth rate	Killing rate by antimicrobials
Susceptible	B_r	r_r	κ_r
Partially resistant	B_s	r_s	κ_s

systems of ODEs result in a good enough approximation of reality. The same principle can be applied on the microscale to quantities of bacteria and immune cells to model the effect (and the lack thereof, in some cases) of antibiotic resistance.

Such models, which Hellriegel [165] aptly described as "immuno-epidemiological" models, are microscale in relation to population models, but macroscale models with respect to populations of pathogens and immune cells, which make up the equivalent of the population. Typically, an immuno-epidemiological model would be somewhat similar to the models that deal with different planes of transmission in the context of zoonotic and vector-borne disease (see Chapter 4):

- There is a population of pathogens, potentially of multiple strains, which is being killed at a certain rate by some (but not all) immune cells, and potentially by external influences (such as the administration of an antimicrobial).
- There is a separate population of immune cells, which go through a cycle of maturation, typically from some form of precursor cell to effector cells and eventually, some fraction matures into memory cells. The immune systems of higher organisms are fairly complex, and it is often a matter for the judgment of an infectious disease modeler to determine what detail is required to adequately represent the host population's immune topology. Studying the peculiarities of the host population's immune system tends to pay abundant dividends.

The model of antimicrobial therapy developed by Gjini and Brito [161] is a good example of such a model. It represents two concurrent planes:

1. In the pathogenic plane, a susceptible and a partially resistant strain compete.
2. In the immune plane, immune cells differentiate, mature, and divide from naive immune cells to effector cells to memory cells.

Both of these processes take place in the presence of antimicrobial treatment. Fig. 5.3 describes the structure of such a model, describing a typical CD8+ mediated immune response in the pathogenic and the immune plane. The pathogenic plane contains two phenotypes: B_s, susceptible to the antibiotic, and B_r, which is partially resistant. The properties of the phenotypes are described in Table 5.1.

We may characterize the dynamics of each plane, following Gjini and Brito [161], by a system of two sets of differential equations: one pair of ODEs corresponding to B_r and B_s, and three ODEs corresponding to naive precursor cells (N), effector cells (E), and memory cells (M). Antimicrobial treatment is reflected by the pathogen-specific antibiotic kill rate κ, the treatment indicator function $\eta(t)$ determining whether the person is receiving treatment at time t, and the mean serum antibiotic concentration, $\overline{c_a}$. In addition, both pathogens have a pathogen-nonspecific kill rate κ_ℓ, attributable to the leukocytes.

$$\frac{dB_s}{dt} = \overbrace{r_s B_s}^{\text{growth}} - \overbrace{\kappa_\ell B_s (M+E+N)}^{\text{killing by leukocytes}} - \overbrace{\kappa_s B_s \eta(t) \overline{c_a}}^{\text{killing by antibiotics}} ,$$
$$\frac{dB_r}{dt} = \underbrace{r_r B_r}_{\text{growth}} - \underbrace{\kappa_\ell B_r (M+E+N)}_{\text{killing by leukocytes}} - \underbrace{\kappa_r B_r \eta(t) \overline{c_a}}_{\text{killing by antibiotics}} . \tag{5.10}$$

We note that the rate of pathogens killed by the antibiotic is the product of the treatment function $\eta(t)$, which determines whether the patient is receiving treatment at t, the susceptibility of the strain to the antibiotic (expressed as the kill rate κ), and the amount of the antibiotic administered (which we describe by reference to the mean serum antibiotic concentration, $\overline{c_a}$).

The second half of this system represents the immune cell populations. Key to this are three factors: the rate of recruitment of naive immune cells $\mu_{\max}$, the rate at which effector cells die or become memory cells ($\frac{1}{\tau_E}$), and a third factor, which links the immune system's dynamics to that of the pathogenic population's, that is, the pathogen load dependent immune growth rate σ_B. We model the growth of the immune system as a Monod equation (a special case of the Langmuir–Hill equation for $n = 1$) [166], so that the growth rate at a given level of total pathogen load $B = B_s + B_r$ is

$$\mu(B) = \mu_{\max} \frac{B}{K_B + B}, \tag{5.11}$$

where K_B is the half-velocity coefficient, that is, the value of B so that $\frac{\mu(B)}{\mu_{\max}} = \frac{1}{2}$.

We also account for the fact that effector cells may either die off or become memory cells. We assume that after τ_E time, an effector cell either dies, becomes a memory cell with p_M probability. Thus

$$\frac{dN}{dt} = -N \underbrace{\mu_{\max} \frac{B}{K_B + B}}_{\substack{\text{recruitment of} \\ \text{naive T cell precursors}}} ,$$
$$\frac{dE}{dt} = \underbrace{2N\mu_{\max} \frac{B}{K_B + B}}_{\substack{\text{division and differentiation of} \\ \text{naive T cell precursors}}} + \underbrace{E\mu_{\max} \frac{B}{K_B + B}}_{\substack{\text{proliferation of} \\ \text{effector cells}}} - \underbrace{\frac{1}{\tau_E} E (1 - \frac{B}{K_B + B})}_{\substack{\text{apoptosis and differentiation} \\ \text{into memory cells}}} , \tag{5.12}$$
$$\frac{dM}{dt} = \underbrace{p_M \frac{1}{\tau_E} E (1 - \frac{B}{K_B + B})}_{\substack{\text{differentiation into} \\ \text{memory cells}}} .$$

The benefit of microscale immunoepidemiological models is that they allow us to reason quantitatively about antimicrobial treatment. Gjini and Brito [161], for instance, utilized this model to evaluate different treatment strategies by different for-

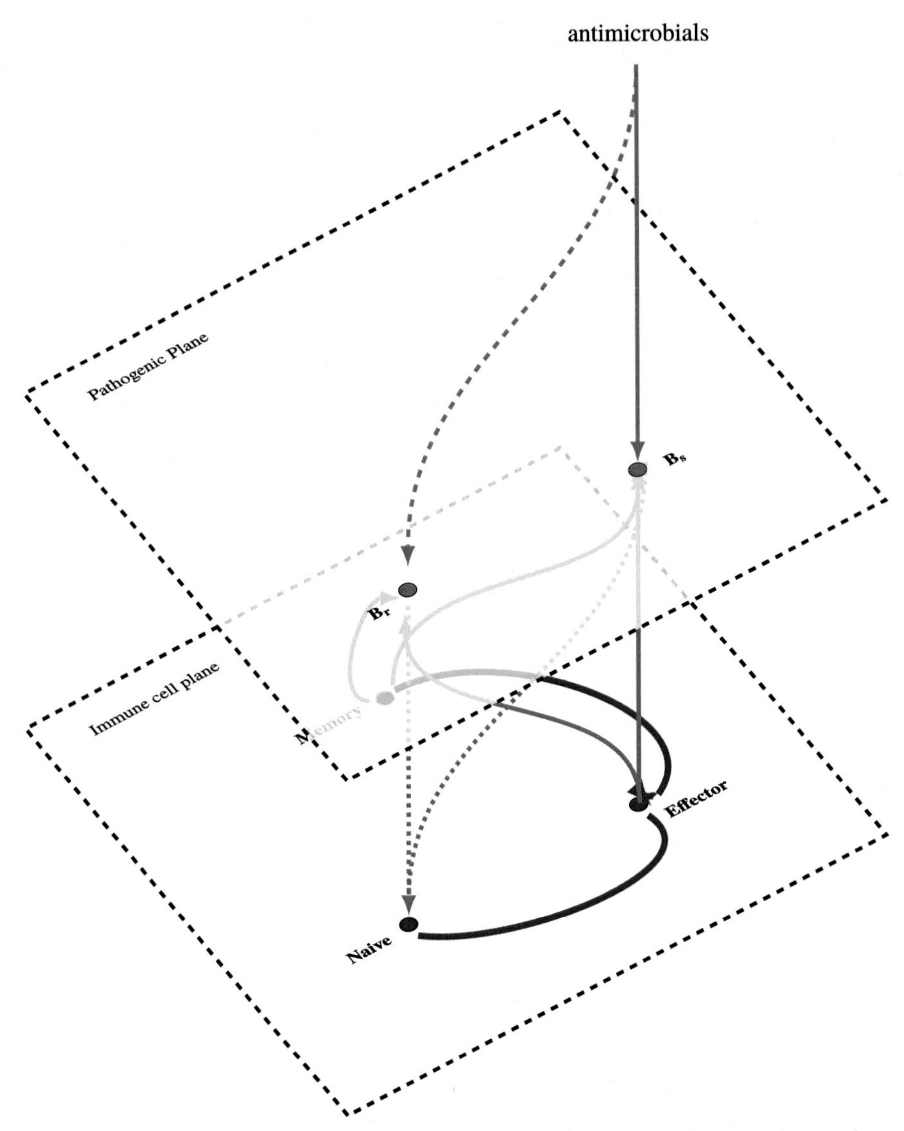

Figure 5.3 Outline of an immunoepidemiological model with two pathogens, B_s (susceptible to the antimicrobial) and B_r (partially resistant). Dotted red lines describe the influence of pathogens on differentiation and proliferation of immune cells. Solid lines describe the effect of immune cells on the pathogen. The dashed lines represent the partial efficiency of antimicrobials against B_r.

mulations of the treatment indicator function $\eta(t)$, contrasting a fixed-time treatment regimen with a critical pathogen density-dependent treatment strategy.

It is worth reflecting on a particular feature of immunity that this model illuminates. The rate at which long-term immunity is incurred (through memory cells) depends on the rate at which the immune system is activated. We have seen in Eq. (5.11) that this is a function of pathogenic load. For an n-day illness, the total pathogenic load for the entire illness is

$$\begin{aligned} B_{\text{total}} = \int_0^n \Big(& r_s B_s(t) + r_r B_r(t) \\ & - \kappa_\ell (B_s(t) + B_r(t))(M(t) + E(t) + N(t)) \\ & - (\kappa_s B_s(t) + \kappa_r B_r(t)) \eta(t) \overline{c_a} \Big) \, dt. \end{aligned} \tag{5.13}$$

This is, of course, a strictly monotonically decreasing function of the total antimicrobial burden

$$\text{TAB} = \int_0^n \eta(t) \overline{c_a} \, dt.$$

On the other hand, it follows from Eq. (5.12) that the overall memory T cell gain M^+ by the end of infection,

$$M^+ = \int_0^n p_m \frac{1}{\tau_E} E(t) \Big(1 - \frac{B_r(t) + B_s(t)}{K_B + B_r(t) + B_s(t)} \Big) \, dt \tag{5.14}$$

is a strictly monotonically increasing function of B_{total}. The consequence of this is that higher antimicrobial loads will result in a lower total pathogenic load, but also a lower involvement of the immune system, and therefore less immunity in the long run (as indeed has been empirically demonstrated in a number of experiments summarized in a sweeping review by Benoun, Labuda, and McSorley [167]). Thus though rapid and aggressive antimicrobial treatment is sometimes appropriate, the long-term absence of ensuing CD4+ immunity is its cost. The overall effectiveness of the same TAB decreases as the fitness benefit of resistance $\frac{\kappa_r}{\kappa_s}$ increases. The consequence is that antibiotic treatment exerts a selection pressure in favor of resistant strains, while resulting in a reduction of developing long term T cell immunity that would not be affected by the pathogen's antimicrobial resistance, that is, trading long term protection for immediate elimination and the reduction of early infectious potential. Microscale models resting on empirical data are useful in quantifying the cost-benefit ratio of early and aggressive antimicrobial interventions in view of the risks posed by the pathogen and the prevalence of resistant strains.

5.2 Multipathogen systems without cross-immunity

In this section, we turn to the study of multipathogen systems without cross-immunity. Pathogens are no exception to the fundamental imperatives of biology: all life exists in competition for finite resources. For pathogens, the crucial resource is, of course, the availability of suitable hosts.

An inherent bias of epidemiology is that we observe all things from the perspective of the host. In reality, looked at it from the perspective of the "pathogen," much (but not all) of the pathogenic effect is quite epiphenomenal to the pathogen's "interests." A hypothesis commonly known as the "trade-off theory" or "avirulence theory" argues that pathogenic evolution moves inevitably towards higher infectiousness but lower lethality [168–170]. Even if a strict trade-off theory, whereby evolution implies monotonously decreasing virulence over time, does not necessarily hold in that simple formulation [171–173], our mathematical models clearly show that altogether too high a mortality leads to rapid epidemic burnout, whereas successful pathogens converge to be less virulent but more infectious.

The risk of this inherent bias is that we might not realize the pressure under which any pathogen finds itself, that is, not from the host's defenses but from other pathogens. When multiple pathogens affect the same host, they may act synergistically or they may act adversarially. This section examines cases where no cross-immunity is elicited, i.e., where immunity to one pathogen does not confer immunity to the other.

5.2.1 Case study: pertussis and measles

Pertussis, more commonly known as whooping cough, is a bacterial disease caused by the bacterium *Bordetella pertussis*. Although quite rare in the developed world due to near-universal vaccination (typically as part of the TDaP combination vaccine), pertussis was once a major cause of death among children in particular. Even today, the death toll from pertussis is not insignificant: Koenig and Guiso [174] note that in 2014, pertussis claimed the lives of 160,000 children under 5.

Measles is a viral disease caused by the measles morbillivirus (MeV), the single most infectious naturally occurring agent known to man [175]. Fortunately, measles, too, is rarely seen in the developed world, thanks to vaccination programmes, although there has been a worrying increase in cases following the surge in vaccine hesitancy in the aftermath of widely spread misinformation of a link between the MMR vaccine and autism [176,177].

On the surface, the two seem to have little to do with each other. However, in a 2015 article, Coleman [178] noted an interaction between the temporal dynamics of measles and pertussis. Examining state-level data on incidence for both conditions in the United States between 1938 and 1954, Coleman found that states with a higher incidence of measles would also go on to have a higher incidence of pertussis. More poignantly, he also noted that the dynamics of pertussis lagged behind measles by about 3–4 weeks. Coleman hypothesizes that measles infection might result in recovery into an immunosuppressed state, which increases the likelihood of contracting pertussis. Recent studies on the immunological effect of measles have identified a postinfectious state of "immune amnesia," which included not merely immunosuppression through the destruction of T and B lymphocytes, but also the destruction of memory B cells that serve as the mainstay of immunological memory in humans [179,180]. A 2020 study by MacIntyre, Costantino, and Heslop [181] following the 2019 Samoa measles outbreak seems to support this hypothesis.

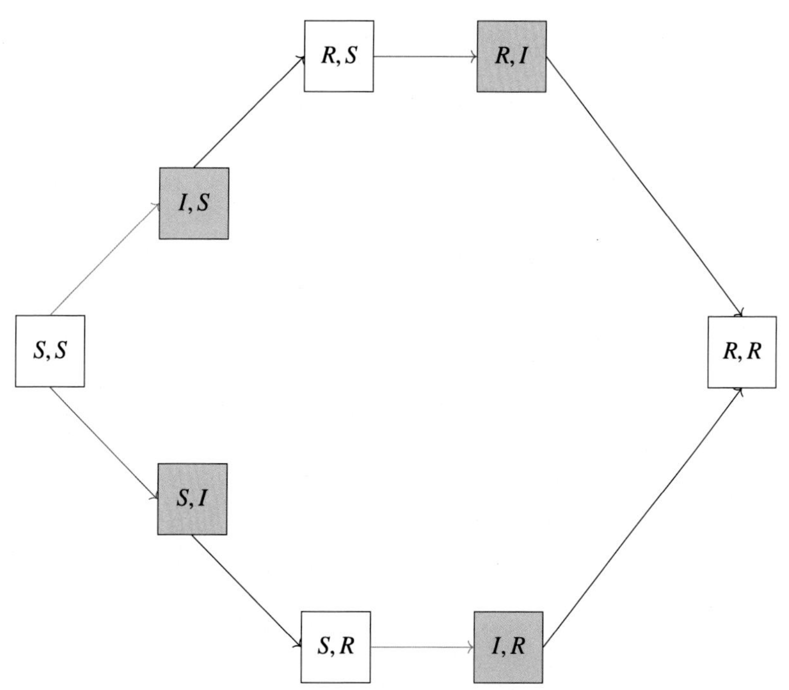

Figure 5.4 Interaction between two pathogens with no cross-immunity. Red boxes denote the infectious subsystem of pathogen 1 and blue boxes that of the pathogen 2, respectively. Arrows that transition into infectious states are colored based on which infectious subsystem governs the mass action term.

Fortunately, for much of the world's population, pertussis is a thing of the past, and though measles has regained an odd stature as a "reemerging" pathogen, it remains quite rare. Nevertheless, the dynamics observed by Coleman [178] show the complex way in which pathogens exist in a complex microbiome, competing but also, from time to time, facilitating each other.

5.2.2 Simple multipathogen systems without cross-immunity and without coinfection

The absence of cross-immunity makes models rather more complex, since we have to account for each person's status in respect of two pathogens. Simple models assume the absence of coinfection, i.e., one individual can be in no more than one infectious system at any given time. As Keeling and Rohani [39] note, this is epidemiologically plausible, since infection tends to reduce encounters sufficiently to make infectious interactions less frequent and may, through mortality, permanently render individuals unavailable to the competitor infection.

Fig. 5.4 describes the structure of a model without cross-immunity and without coinfection. We may, given the infectious subsystems $I_1 = N_{I,S} + N_{I,R}$ and

$I_2 = N_{S,I} + N_{R,I}$, express this model as the following system:

$$\begin{aligned}
\frac{dN_{S,S}}{dt} &= -(\beta_1 N_{S,S} I_1 + \beta_2 N_{S,S} I_2), \\
\frac{dN_{I,S}}{dt} &= \beta_1 N_{S,S} I_1 - \gamma_1 N_{I,S}, \\
\frac{dN_{S,I}}{dt} &= \beta_2 N_{S,S} I_2 - \gamma_2 N_{S,I}, \\
\frac{dN_{R,S}}{dt} &= \gamma_1 N_{I,S} - \beta_2 N_{R,I} I_2, \\
\frac{dN_{S,R}}{dt} &= \gamma_2 N_{S,I} - \beta_1 N_{I,R} I_1, \\
\frac{dN_{R,I}}{dt} &= \beta_2 N_{R,I} I_2 - \gamma_2 N_{R,I}, \\
\frac{dN_{I,R}}{dt} &= \beta_1 N_{I,R} I_1 - \gamma_1 N_{I,R}, \\
\frac{dN_{R,R}}{dt} &= \gamma_2 N_{R,I} + \gamma_1 N_{I,R}.
\end{aligned} \tag{5.15}$$

From a computational perspective, it may often be useful in such situations to make use of the techniques discussed in Computational Note 5.1 to handle vector-valued calculations.

Computational Note 5.1 Modeling the no-coinfection no-cross immunity interaction

One might be tempted to transpose the system described in Eq. (5.15) directly into a derivative function:

```
def deriv(t, y, beta_1, beta_2, gamma_1, gamma_2, mu, nu):

    SS, IS, RS, SI, RI, SR, IR, RR = y

    I_1 = IS + IR
    I_2 = SI + RI

    dNSSdt = nu - beta_1 * SS * I_a + beta_2 * SS * I_2 - mu * SS
    dNISdt = beta_1 * SS * I_a - gamma_1 * IS - mu * IS
    dNRSdt = gamma_1 * IS - beta_2 * RS * I_2 - mu * RS
    dNSIdt = beta_2 * SS * I_2 - gamma_2 * SI - mu * SI
    dNRIdt = beta_2 * RS * I_2 - gamma_2 * RI - mu * RI
    dNSRdt = gamma_1 * IS - beta_1 * SR * I_1 - mu * SR
    dNIRdt = beta_1 * SR * I_1 - gamma_1 * IR - mu * IR
```

```
    dNRRdt = gamma_1 * IR + gamma_2 * RI - mu * RR

    return dNSSdt, dNISdt, dNRSdt, dNSIdt, dNRIdt, dNSRdt, dNIRdt,
           dNRRdt
```

There is, strictly speaking, nothing wrong with this, but we can turn this into better, more elegant and more performant code. First, note that there are four terms that describe recovery: those that describe recovery from the pathogen 1 subsystem (`gamma_1 * I[S/R]`) and those that describe recovery from the pathogen 2 subsystem (`gamma_2 * [S/R]I`). We can turn this into a matrix quite easily, by constructing a matrix in which each column corresponds to a subsystem, then multiplying it by a vector of both values of γ:

$$\begin{pmatrix} IS & SI \\ IR & RI \end{pmatrix} \vec{\gamma} = \begin{pmatrix} \gamma_1 IS & \gamma_2 SI \\ \gamma_1 IR & \gamma_2 RI \end{pmatrix}. \tag{5.16}$$

We do so by constructing a matrix on the fly with `np.array` and multiplying it:

```
gmat = gamma * np.array([[IS, SI], [IR, RI]])
```

Next, we note that the first equation involves subtracting $\beta_1 N_{S,S} I_1$ and $\beta_2 N_{S,S} I_2$. If we create a vector $\vec{I}$ for the infectious subsystems, where $\vec{I} = \begin{pmatrix} I_1 \\ I_1 \end{pmatrix}$, we can simplify this as $\vec{\beta} N_{S,S} \vec{I}$. We solve this by first defining the infectious subsystem vector:

```
I = np.array([[IS + IR], [SI + RI]])
```

Next, we substitute the dot product form into the derivative functon's calculation of $\frac{dN_{S,S}}{dt}$:

```
dNSSdt = nu - (beta * SS).dot(I) - mu * SS
```

Finally, we note that the force of infection in the mass action terms can be vectorized out as $\vec{\beta}\vec{I}$. The element of the vector that corresponds to subsystem a is multiplied by $N_{S,S}$ and $N_{S,R}$, whereas the element of the vector that corresponds to b is multiplied by $N_{S,R}$ and $N_{R,S}$. This gives us

```
bmat = beta * I * np.array([[SS, SS], [SR, RS]])
```

which is of course a matrix representation of

$$\begin{pmatrix} \beta_a I_a SS & \beta_b I_b SS \\ \beta_a I_a SR & \beta_b I_b RS \end{pmatrix}. \tag{5.17}$$

Thus our derivative function in its final glory, will look something rather like this:

```
def deriv(t, y, beta, gamma, mu, nu):

    SS, IS, RS, SI, RI, SR, IR, RR = y

    I = np.array([[IS + IR], [SI + RI]])
    gmat = gamma * np.array([[IS, SI], [IR, RI]])
    bmat = beta * I * np.array([[SS, SS], [SR, RS]])

    dNSSdt = nu - (beta * SS).dot(I) - mu * SS
    dNISdt = bmat[0,0] - gmat[0,0] - mu * IS
    dNRSdt = gmat[0,0] - bmat[1,1] - mu * RS
    dNSIdt = bmat[0,1] - gmat[0,1] - mu * SI
    dNRIdt = bmat[1,1] - gmat[1,1] - mu * RI
    dNSRdt = gmat[0,0] - bmat[1,0] - mu * SR
    dNIRdt = bmat[1,0] - gmat[1,0] - mu * IR
    dNRRdt = gmat[1:,].sum() - mu * RR

    return dNSSdt, dNISdt, dNRSdt, dNSIdt, dNRIdt, dNSRdt, dNIRdt,
           dNRRdt
```

The output of integrating this ODE is presented as Fig. 5.5.

Concise code is almost always better code, and the use of matrices and vectorization can greatly assist with improving code quality and speed of execution. A caveat is that since there are design choices involved in the use of matrices (e.g., what to represent in rows and what in columns), such code should be assiduously documented (as is indeed good practice in all cases).

A notebook implementing the contents of this Computational Note is available on the book's companion Github repository in the folder `/ch05/transition_matrix`.

5.2.3 Multipathogen systems without cross-immunity and with coinfection

In the previous subsection, we primarily examined systems that excluded coinfection; every individual was at most in a single infectious compartment. In this section, we

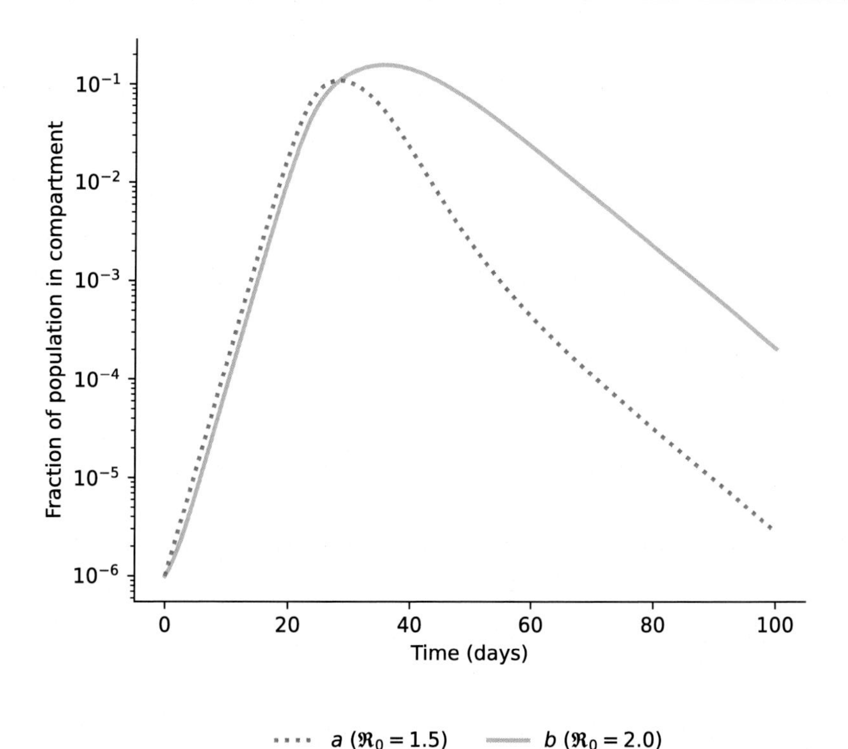

Figure 5.5 Short and long-term dynamics of a two-pathogen system with no cross-immunity and no coinfection, with $\beta_1 = 1.5$, $\gamma_1 = 1.0$, $\beta_2 = 0.4$, and $\gamma_2 = 0.2$ and $I_1(0) = I_2(0) = 10^{-6}$. The calculations leading up to this figure are discussed in Computational Note 5.1.

turn to situations in which the infectious compartment of one pathogen is also susceptible to infection by the other. It differs little from the model without coinfection discussed in Subsection 5.2.2, except that it also permits for a coinfected compartment I, I:

$$\frac{dN_{S,S}}{dt} = -(\overbrace{\beta_1 N_{S,S} I_1}^{\text{infections with 1}} + \overbrace{\beta_2 N_{S,S} I_2}^{\text{infections with 2}}),$$

$$\frac{dN_{I,S}}{dt} = \underbrace{\beta_1 N_{S,S} I_1}_{\text{new infections}} - \underbrace{\gamma_1 N_{I,S}}_{\text{recoveries}} - \underbrace{\beta_2 N_{I,S} I_2}_{\text{superinfection with 2}},$$

$$\frac{dN_{S,I}}{dt} = \underbrace{\beta_2 N_{S,S} I_2}_{\text{new infections}} - \underbrace{\gamma_2 N_{S,I}}_{\text{recoveries}} - \underbrace{\beta_1 N_{S,I} I_1}_{\text{superinfection with 1}},$$

$$\frac{dN_{I,I}}{dt} = \underbrace{\beta_1 N_{S,I} I_1}_{\text{superinfections with 1}} + \underbrace{\beta_2 N_{I,S} I_2}_{\text{superinfections with 2}} - \underbrace{\gamma_c N_{I,I}}_{\text{superinfection recoveries}},$$

$$\frac{dN_{R,S}}{dt} = \underbrace{\gamma_1 N_{I,S}}_{\text{recoveries}} - \underbrace{\beta_2 N_{R,I} I_2}_{\text{infections with 2}}, \tag{5.18}$$

$$\frac{dN_{S,R}}{dt} = \underbrace{\gamma_2 N_{S,I}}_{\text{recoveries}} - \underbrace{\beta_1 N_{I,R} I_1}_{\text{infections with 1}},$$

$$\frac{dN_{R,I}}{dt} = \underbrace{\beta_2 N_{R,I} I_2}_{\text{infections with 2}} - \underbrace{\gamma_2 N_{R,I}}_{\text{complete recoveries}},$$

$$\frac{dN_{I,R}}{dt} = \underbrace{\beta_1 N_{I,R} I_1}_{\text{infections with 1 from } S,R} - \underbrace{\gamma_1 N_{I,R}}_{\text{recoveries}},$$

$$\frac{dN_{R,R}}{dt} = \underbrace{\gamma_2 N_{R,I}}_{\text{recoveries from } R,I} + \gamma_1 \underbrace{N_{I,R}}_{\text{recoveries from } I,R} + \underbrace{\gamma_c N_{I,I}}_{\text{recoveries from coinfections}}.$$

The underlying assumption here is that β_1 and β_2 are the probabilities of infection with pathogens 1 and 2, respectively, regardless of status with respect to the other pathogen. This is appropriate where there is neither significant superinfection exclusion or a significant multipathogen facilitation, i.e., where the coinfection coefficient is around zero.

5.2.4 The coinfection matrix

When dealing with multipathogen systems, we are sometimes interested in a simplified picture of relative transmission risk as a function of infection status. The state transition matrix is a useful tool that tells us about the quantities in the compartments, but not all that much about relative risk. The coinfection matrix **C** is analogous to a WAIFW matrix (see Subsection 3.1.2), except that it describes the transmission between groups with respect to their pathogenic status over the number of pathogens. The coinfection matrix for two pathogens, $\mathbf{C}^2$, for instance, is

$$\mathbf{C}^2 = \begin{pmatrix} 0 & \beta_{S,S\to S,I} & \beta_{S,S\to I,S} & 0 \\ 0 & 0 & 0 & \beta_{S,I\to I,I} \\ 0 & 0 & 0 & \beta_{I,S\to I,I} \\ 0 & 0 & 0 & 0 \end{pmatrix}. \tag{5.19}$$

Coinfection matrices are strictly upper triangular, because the number of pathogens for which the individual is infectious increases monotonically along the axis, and only infectious individuals can transmit the pathogen. The relevance of the coinfection matrix lies in its usefulness to understand how infectious status with respect to one pathogen modulates sustaining infection by another. From this, we can calculate the coinfection coefficient $\zeta_{p,q}$, which represents the relative effect that infection with the p-th pathogen has on infection with the q-th pathogen. For two pathogens, infection by the first pathogen changes the relative transmission coefficient with respect to the second by

$$\frac{\beta_{S,S\to S,I}}{\beta_{I,S\to I,I}}. \tag{5.20}$$

We may express this more generally.

Definition 5.3 (Coinfection coefficient). Let there be a system of n pathogens, in which we distinguish a set $\mathbb{P}$ of 2^n distinct combinations of states with respect to each pathogen. Any two elements in $\mathbb{P}$ are associated with a transmission coefficient $\beta_{i,j}$ for $i, j \in \mathbb{P}$. The coinfection coefficient with respect to the p-th and the q-th pathogen is defined as

$$\zeta_{p,q} = \log\left(\frac{1}{|\mathbb{P}|}\frac{\sum_{i\in\mathbb{P}[p=S,q=S]}\sum_{j\in\mathbb{P}[p=S,q=I]}\beta_{i\to j}}{\sum_{k\in\mathbb{P}[p=I,q=S]}\sum_{m\in\mathbb{P}[p=I,q=I]}\beta_{k\to m}}\right). \tag{5.21}$$

The coinfection coefficient of two pathogens determines the impact of p on subsequent infections with q:

- If $\zeta_{p,q}$ is negative, infection with p makes infection with q less likely. This is seen in certain cases of superinfection exclusion (SIE) (see Subsection 5.1.2).
- If $\zeta_{p,q}$ is zero, infection with p does not make infection with q more or less likely.
- If $\zeta_{p,q}$ is positive, infection with p increases the likelihood of sustaining infection with q upon an encounter. We say that p facilitates q.

5.2.5 Multipathogen facilitation

In facilitation, infection with one pathogen results in a greater susceptibility for infection by another. This facilitation may be one-sided or two-sided. For instance, HIV facilitates the virulence of *Candida albicans*, a normally rather commensal opportunistic fungus, but the extent of *C. albicans* infection does not affect HIV itself, thus resulting in a one-sided facilitation [182]. On the other hand, not only is infection with *Mycobacterium tuberculosis* facilitated with the immune suppression that accompanies HIV infection, but through a mechanism that at this time remains poorly understood, *M. tuberculosis* infection facilitates the replication of HIV in turn [183]. This makes HIV and *M. tuberculosis* an example of two-sided facilitation.

The coinfection matrix and the coinfection coefficient (see Subsection 5.2.4) are our primary tools to explore multipathogen facilitation quantitatively. We speak of pathogen p facilitating q if $\zeta_{p,q}$ is positive. In such cases, an encounter between an individual susceptible to both p and q with a q-infectious individual is less likely to produce infection than it would if the individual were susceptible to q but infectious with respect to p. In general, the coinfection coefficient is not commutative, i.e., $\zeta_{p,q}$ does not necessarily equal $\zeta_{q,p}$. This reflects situations such as the above-mentioned example of HIV and *C. albicans*.

One-sided facilitation is adequately modeled using a differential-β approach, similarly to the way we have tackled risk in host heterogeneities (see Subsection 3.1.2), with the difference that in this case, the transmission coefficient will change as a function of infectious compartment status.

It is worth noting that in general, only symptomatic disease is capable of facilitating infections. This is *a fortiori* the case for HIV, where pathogenic facilitation does not emerge until CD4+ counts are sufficiently depleted, signaling full-blown AIDS. Similarly, symptomatic infection may often be forestalled by early initiation of aggressive treatment. For this reason, where multipathogen facilitation is modeled, the phase of infection needs to be accounted for. This is often best accomplished by creating two subcompartments of the infectious compartment: nonfacilitating disease (e.g., HIV infection without significant depression of CD4+ counts) and facilitating disease (HIV infection with immune compromise).

Modeling the control of infectious disease

Pharmacological and nonpharmacological interventions

In 1736 I lost one of my sons, a fine boy of four years old, by the small-pox, taken in the common way. I long regretted bitterly, and still regret that I had not given it to him by inoculation. This I mention for the sake of parents who omit that operation, on the supposition that they should never forgive themselves if a child died under it; my example showing that the regret may be the same either way, and that, therefore, the safer should be chosen.

From Benjamin Franklin's Autobiography, quoted by Best, Katamba, and Neuhauser [184]

6.1 Modeling vaccination

The word "vaccination" comes from *vacca*, the Latin word for "cow." This is a poignant recapitulation of the history of vaccines. The first vaccine properly so called had, as its active ingredient, the cowpox virus, a close relative of smallpox that however was much less likely to cause severe, disfiguring, or lethal disease. Edward Jenner (1749–1823) observed that milkmaids, who were often exposed to cowpox, suffered a relatively mild disease, but would be immune to the much more serious smallpox. In an experiment that would unlikely pass muster in the modern world, he infected James Phipps, then an 8-year-old, with cowpox. He suffered a mild and transient illness, but when he was later exposed to scabs from a smallpox patient, he proved immune [185]. Unlike the earlier practice of variolation, which was practised in late Song dynasty China [186] as a way to induce the cutaneous form of smallpox, *variola minor*, to protect against the more severe forms of smallpox (*variola major*), Jenner's vaccination used a different and less pathogenic virus. He relied on what would later be called "antigenic similarity," but which was at the time hardly understood.

Much has changed since Jenner's inoculations with cowpox. Vaccination has made smallpox extinct in the wild, as well as rinderpest, a relative of measles that affects cattle and buffalo, among others. Poliomyelitis, which has in its heyday killed and maimed millions of children and adults alike, is close to eradication, with fewer than 200 wild-type cases documented in 2020. Vaccines are some of the most effective public health interventions against infectious disease. For this reason, we shall devote this section to modeling vaccination against infectious disease.

Computational Modeling of Infectious Disease. https://doi.org/10.1016/B978-0-32-395389-4.00015-3

6.1.1 Case study: measles vaccination over the years

Measles is the Comeback Kid of infectious diseases. The measles morbillivirus (MeV), which causes measles, is the most infectious known pathogen, with conservative estimates of its $\mathfrak{R}_0$ around 11–18 [175]. It emerged sometime between the 6th century BCE and the mid-11th century CE from rinderpest, a morbillivirus of even-toed ungulates [187]. Unlike its ancestor, which holds the distinction of being the first animal disease eradicated using vaccination [188], measles is making a comeback. The recent years have seen increasing vaccine hesitancy specifically concerning the MMR vaccine [189], and consequently a number of localized outbreaks in the United States [190].

The overall vaccination rate in the United States is approximately 92%, although this figure is much lower in some communities [191]. Studies have estimated the effectiveness of the vaccine to prevent clinical measles at approximately 95%, and intrahousehold transmission in about 92% of cases [192].

The concern raised by localized outbreaks is that the protection that flows from collective immunity may be lost. Recall that a pathogen requires at least $\mathfrak{R}_0^{-1}$ susceptible fraction of population to be able to persist. For the higher estimates of $\mathfrak{R}_0$ for measles, this would correspond to around 5.5% of the population, in other words, as long as more than 5.5% of the population are susceptible, there is a constant risk of outbreaks. Susceptibility includes not only those who are not, or cannot be, vaccinated, but also those in whom the vaccination has, for whatever reason, failed to induce sufficient immunity. As such, $\mathfrak{R}_0^{-1}$ represents an optimistic upper bound, premised on the absence of vaccine failure, and thus, collective immunity might be lost at a much lower proportion of nonvaccinated individuals.

This section deals with modeling the effect of vaccines on populations. We will, in sequence, look at vaccines that are equally effective against illness and transmission, then move on to vaccines, where the effect on transmission is, as is the case for measles vaccines, different. Finally, we shall consider alternatives to mass vaccination that have proven their worth historically.

6.1.2 Modeling vaccines effective against both illness and transmission

6.1.2.1 Initial vaccination

Where a vaccine is equally effective against both illness and transmission, it is typically best modeled as a separate compartment that has, under the assumption of unlimited duration and full effectiveness of immunity, no outflows towards the infectious compartments. A simple model might assume a preexisting and static cohort of vaccinated.

Definition 6.1 (Initial vaccination). In **initial vaccination**, also called **vaccination at recruitment**, vaccinated individuals are present at the beginning of the model. The vaccination fraction ν is the fraction of susceptibles who are vaccinated at the initial point in time, and there are no accruals to the vaccinated compartment.

Given the vaccination fraction ν, we initialize the model for n_I initially infected as follows:

$$\begin{aligned} S(0) &= 1 - R(0) - \nu, \\ V(0) &= \nu, \\ R(0) &= 0, \\ I(0) &= 1 - S(0) - \nu. \end{aligned} \tag{6.1}$$

In other respects, the model is quite identical to a regular SIR model: the impact of the vaccination takes place before the infectious process begins, by way of "shielding" the vaccinated individual from being affected by the infectious process.

6.1.2.2 Fixed-rate vaccination

Process vaccination refers to time-dependent changes in the number of vaccinated individuals, by a fixed rate. Typically, a fixed vaccination rate is most appropriate when examining a relatively short period of time during which the vaccination rate is relatively steady, e.g., examining a single influenza epidemic. Another assumption is that the vaccine is given only to susceptible individuals.

$$\begin{aligned} \frac{dS}{dt} &= -\underbrace{\beta SI}_{\text{mass action}} - \underbrace{\nu S}_{\text{new vaccinations}}, \\ \frac{dV}{dt} &= \underbrace{\nu S}_{\text{new vaccinations}}, \\ \frac{dI}{dt} &= \underbrace{\beta SI}_{\text{mass action}} - \underbrace{\gamma I}_{\text{recovery}}, \\ \frac{dR}{dt} &= \underbrace{\gamma I}_{\text{recovery}}. \end{aligned} \tag{6.2}$$

In practice, fixed-rate vaccination is often applied in an age-regularized manner. Typically, most school-age children receive their vaccinations at the start of the school term, and in the Northern hemisphere, influenzavirus vaccination campaigns typically start around late September, when the annual vaccine is released. Where the rate of vaccination is deterministic but time-varying, it is possible to apply a time-varying forcing model, similar to the way we modeled annual peaks of epidemic processes in the vector-borne context Subsection 4.1.6.

6.1.2.3 Pulse vaccination

Vaccinating large numbers of individuals is expensive, time-consuming, and may at times not be practicable. Pulse vaccination is a "shortcut" of sorts to attain immunity at relatively lower rates of vaccination, carried out periodically in the form of "pulse" vaccination campaigns [193]. Barik, Chauhan, and Bhatia [194] demonstrated that in certain circumstances, pulse vaccination may in fact be more effective than continuous vaccination.

At the heart of pulse vaccination is the fact that once $1 - \mathfrak{R}_0^{-1}$ of all susceptibles have been vaccinated, the growth rate of the disease can only be negative. Thus pulse vaccination aims at limiting the fraction of S (as a fraction of the total population) below $\mathfrak{R}_0^{-1}$ through periodic vaccination campaigns. This may be difficult to achieve in a single campaign, but feasible by multiple pulses.

Let us assume we vaccinate the proportion ν of the unvaccinated population every τ_ν time (the pulse interval). Then, the vaccinated part of the population after n pulses τ_ν apart is

$$V(n) = \nu \sum_{i=0}^{n} S(i\tau_\nu - \epsilon)\delta(t - n\tau_\nu), \tag{6.3}$$

where $S(n\tau_\nu - \epsilon)$ is the number of susceptible individuals an instant before the n-th pulse (ϵ here stands for the infinitesimal time difference), and δ is the Dirac delta function. Then, given an equal rate of births and deaths μ, we may represent this system's dynamics in the long run as

$$\begin{aligned}
\frac{dS}{dt} &= \mu - \beta SI - \nu \sum_{i=0}^{\infty} S(i\tau_\nu - \epsilon)\delta(t - n\tau_\nu) - \mu S, \\
\frac{dI}{dt} &= \beta SI - \gamma I - \mu I, \\
\frac{dR}{dt} &= \gamma I - \mu R, \\
\frac{dV}{dt} &= \nu \sum_{i=0}^{\infty} S(i\tau_\nu - \epsilon)\delta(t - n\tau_\nu) - \mu V.
\end{aligned} \tag{6.4}$$

The disease-free mean numbers of I and S during the n-th pulse cycle, which is defined as $[(n-1)\tau_\nu, n\tau_\nu]$, are

$$\begin{aligned}
\overline{S}(t) &= 1 - \frac{\nu e^{\mu\tau_\nu}}{e^{\mu\tau_\nu} - (1-\nu)} e^{-\mu(t-(n-1)\tau_\nu)} \\
&\quad - \nu\left(1 - \frac{\nu e^{\mu\tau_\nu}}{e^{\mu\tau_\nu} - (1-\nu)} e^{-\mu\tau_\nu}\right) \int_{(n-1)\tau_\nu}^{t} \delta(t - n\tau_\nu) \\
\overline{I}(t) &= 0.
\end{aligned} \tag{6.5}$$

The stability analysis of this model, which is due to Shulgin, Stone, and Agur [195], approaches this problem using Floquet theory, as one of small perturbations on stable level, where the current value of $S(t)$ and $I(t)$ are the sum of the pulse-wise means $\overline{S}(t)$ and $\overline{I}(t)$, respectively, plus a perturbation term s or i. These are obtained by

Taylor expansion of Eq. (6.4), and discounting higher–order terms, it yields

$$\begin{aligned}\frac{ds}{dt} &= -\mu s - \beta\overline{S}(t)i - \nu s(n\tau_\nu - \epsilon)\sum_{n=0}^{\infty}\delta(t - n\tau_\nu),\\ \frac{di}{dt} &= \beta\overline{S}(t)i - \gamma i - \mu i.\end{aligned} \tag{6.6}$$

Given the initial conditions

$$\begin{cases} s_1(0) = 1 \\ i_1(0) = 0 \\ s_2(0) = 0 \\ i_2(0) = 1 \end{cases} \tag{6.7}$$

we obtain the fundamental matrix of the system in Eq. (6.6),

$$\mathbf{F}(t) = \begin{pmatrix} s_1(t) & s_2(t) \\ i_1(t) & i_2(t) \end{pmatrix}. \tag{6.8}$$

The Floquet multipliers are the eigenvalues λ_1 and λ_2 of $\mathbf{F}(\tau_\upsilon)$, namely

$$\lambda_1 = (1-\nu)e^{-\mu\tau_\upsilon}\lambda_2 = e^{\int_0^{\tau_\upsilon}\beta(t)\overline{S}(t)dt - \mu\tau_\upsilon - \gamma\tau_\upsilon}. \tag{6.9}$$

The solution in Eq. (6.5) is stable if both of the Floquet multipliers in Eq. (6.9) have absolute values less than one. This is necessarily the case for λ_1 and true for λ_2 if

$$\frac{1}{\tau_\upsilon}\int_0^{\tau_\upsilon}\overline{S}(t)dt < \frac{\mu+\gamma}{\beta}. \tag{6.10}$$

By integrating the stability condition, we get

$$\frac{(\mu\tau_\upsilon - \nu)(e^{\mu\tau_\upsilon} - 1) + \mu\nu\tau_\upsilon}{\mu\tau_\upsilon(\nu - 1 + e^{\mu\tau_\upsilon})} < \frac{\mu+\gamma}{\beta}. \tag{6.11}$$

For pulses that are much shorter than the mean lifespan and assuming that the duration of the disease, $\frac{1}{\gamma}$ is also shorter than the mean lifespan, we may approximate the critical maximum value of τ_υ, which we will call $\tau_{\max}$, as

$$\tau_{\max} = \frac{\gamma\nu}{\beta\mu(1 - \frac{\nu}{2} - \frac{\gamma}{\beta})}. \tag{6.12}$$

This allows us to calculate the minimum safe pulse frequency. Note that this approach makes several major assumptions, in particular that

- vaccines do not wane in effect, and
- vaccines do not fail.

We know from practice that this is not the case, and it may be necessary to account for these. A common method of accounting for the effect of vaccine failure is to multiply the vaccination rate ν by the vaccine efficacy VE, essentially discounting a fraction of vaccinated individuals to represent those in whom no protective effect is attained.

6.1.2.4 Seasonally rate-varying vaccination

In practice, many diseases for which we vaccinate occur seasonally. Sinusoidal forcing, which we discuss in rather more detail in Subsection 7.3.2, is a technique to model diseases with periodically (seasonally) oscillating transmission rates. We can then rewrite Eq. (6.4) in a time-dependent form:

$$\begin{aligned}
\frac{dS}{dt} &= \mu - \beta(t)SI - \nu\sum_{n=0}^{\infty} S(n\tau_\nu - \epsilon)\delta(t - n\tau_\nu) - \mu S, \\
\frac{dI}{dt} &= \beta(t)SI - \gamma I, \\
\frac{dR}{dt} &= \gamma I, \\
\frac{dV}{dt} &= \nu\sum_{n=0}^{\infty} S(n\tau_\nu - \epsilon)\delta(t - n\tau_\nu),
\end{aligned} \tag{6.13}$$

where $\beta(t)$ denotes the time-varying contact rate, which is composed of a background rate β_0 and a factor β_1, which describes the amplitude of the seasonal variation above the background rate. Specifically,

$$\beta(t) = \beta_0 + \beta_0\beta_1 \cos(2\pi t + \psi_0), \tag{6.14}$$

where ψ_0 is the phase offset between the vaccination and the seasonal variations of β.

Using the same Floquet techniques that yielded us the stability condition in Eq. (6.10), we get the Floquet multipliers

$$\begin{cases} \lambda_1 = (1-\nu)e^{-\mu\tau_v} \\ \lambda_2 = e^{\int_0^{\tau_v} \beta(t)\overline{S}(t)dt - \mu\tau_v - \gamma\tau_v}. \end{cases} \tag{6.15}$$

If $|\lambda_2| < 1$, the periodic infection-free solution is stable. This is the case if

$$\frac{1}{\tau_v}\int_0^{\tau_v} \beta(t)\overline{S}(t)dt < \mu + \gamma. \tag{6.16}$$

Under the Taylor expansion of Eq. (6.16), and the assumptions we have made previously, as well as the assumption that $\mu < 2\pi$, we obtain $\tau_{\max}$ in the presence of forcing as

$$\tau_{\max} = \frac{\gamma\nu}{\beta_0\mu(1 - \frac{\nu}{2} - \frac{\gamma}{\beta_0})} + \frac{\nu\beta_1(\mu\cos\psi_0 - 2\pi\sin\psi_0)}{4\pi^2(1 - \frac{\nu}{2} - \frac{\gamma}{b_0})}. \tag{6.17}$$

The first part, which ought to be rather familiar from Eq. (6.12), represents the maximum safe pulse period in the absence of temporal forcing, whereas the second part represents the effect of seasonal forcing via β_1. The relative value of the temporally forced and the temporally unforced component determines the degree to which a pathogen's temporally forced dynamics affect pulse vaccination. This is the effect of forcing ζ, which we calculate as

$$\zeta = \frac{\beta_0 \beta_1 \mu (\mu \cos \psi_0 - 2\pi \sin \psi_0)}{4\pi^2 \gamma}. \tag{6.18}$$

By evaluating the maximum of ζ, we can calculate the likely effect of phase differences: the greater the maximum of ζ is, the greater the impact of ψ_0 (i.e., the phase difference between peak infections and vaccinations).

Practice Note 6.1 When to schedule vaccinations

Shulgin, Stone, and Agur [195] estimate that for their infectious parameters of measles, ζ would take a value of approximately 0.02, meaning that for all intents and purposes, it does not quite matter when a pulse vaccination is initiated with respect to the peak infectious period. On the other hand, ζ might be quite considerable, especially where γ is small. When advising on pulse vaccination schedules, it is crucial for infectious disease modelers to take into account the expected impact of setting a mathematically more appropriate pulse vaccination schedule vis-a-vis other considerations. Pulse vaccination of school age children, for instance, is difficult outside of term-time, and a suboptimally scheduled vaccination campaign with high vaccination rates (ν) might often be superior to a better timed but less thorough one. In addition, pulse vaccinations may create hard-to-predict long-term nonlinear effects. These are discussed in detail in Subsection 7.3.5.

6.1.3 Risk-targeted vaccination

As we have learned in Chapter 3, in epidemiology, all are not created equal. Vaccine targeting is the conscious determination of particular segments of the population to achieve optimum effects.

It is known, for instance, that vaccines are generally less effective in people with suppressed immune systems, whether as a consequence of immunosuppressive therapy (e.g., for autoimmune disorders), as a consequence of illness (e.g., AIDS or congenital conditions affecting the immune system, such as DiGeorge syndrome) or are too young or too old to mount effective immunity. Given a limited number of vaccines, the strategy of targeting the first-degree contacts (such as household members and caregivers) of persons less likely to respond to the vaccine can create an effect sometimes described as "cocooning," reducing the risk of infection among close contacts to avoid infection [196].

This strategy is widely used to protect newborns from, among others, pertussis. In addition to maternal immunization during pregnancy with the TDaP vaccine, immunizing the child's close contacts can significantly reduce the risk of pertussis [197].

Risk-targeted vaccination, on the other hand, is an *a priori* approach, in which vaccines are targeted at individuals who are more likely to contract the infection in the first place. Looking back at Chapter 3, we recall that populations may be stratified by their risk. One of the consequences of such a stratification is that the overall impact of vaccination on different strata might yield different results. We know that random vaccination will eradicate the infection (under the assumption of perfectly effective vaccines), where $1 - \frac{1}{\mathfrak{R}_0}$ of the population is vaccinated. This is the collective immunity threshold ${}^{\Delta}\nu$.

Definition 6.2 (The critical threshold of collective immunity). A pathogen cannot persist in a population where a fraction of $1 - \frac{1}{\mathfrak{R}_0}$ or more of the population are immune, whether by vaccination or post-infectious immunity. We call this value the **critical threshold of collective immunity** or **herd immunity threshold** ${}^{\Delta}\nu$ for short.

However, where risk is heterogeneous (see Definition 3.2), *who* gets vaccinated may matter more than *how much* of the population ends up vaccinated. The consequence is that the infection can be controlled at less than ${}^{\Delta}\nu$ vaccinated, as long as higher-risk strata are vaccinated to a sufficient degree.

Let us consider the scenario from Subsection 3.1.2, with two strata (see Definition 3.1): high-risk (H) and low-risk (L), with n_H and n_L denoting the fractions of each stratum as part of the population. We account for differential vaccination through the vaccination rate parameters ν_H and ν_L (Fig. 6.1). We can computationally estimate a curve at which the disease is eradicated by determining the area on a coordinate space defined by the total percentage vaccinated and the percentage of high-risk individuals vaccinated, where $\mathfrak{R}_0 \leq 1$. As Keeling and Rohani [39] note, such issues are not quite amenable to analytical solutions in most cases, but computational optimization can often get us the results we need. Computational Note 6.1 lays out a computational approach to solving such problems, creating a nomogram of control.

Computational Note 6.1 Targeted vaccination

Let us consider a population with characteristics identical to the example in Computational Note 3.1, namely a WAIFW matrix of

$$\mathbf{b} = \begin{pmatrix} 10 & 0.5 \\ 0.5 & 2 \end{pmatrix}$$

and 20% of the population belonging to the high-risk category. Furthermore, we shall assume a γ of $\frac{1}{20}$. Recall from Computational Note 3.1 that the infectious subsystem was given by

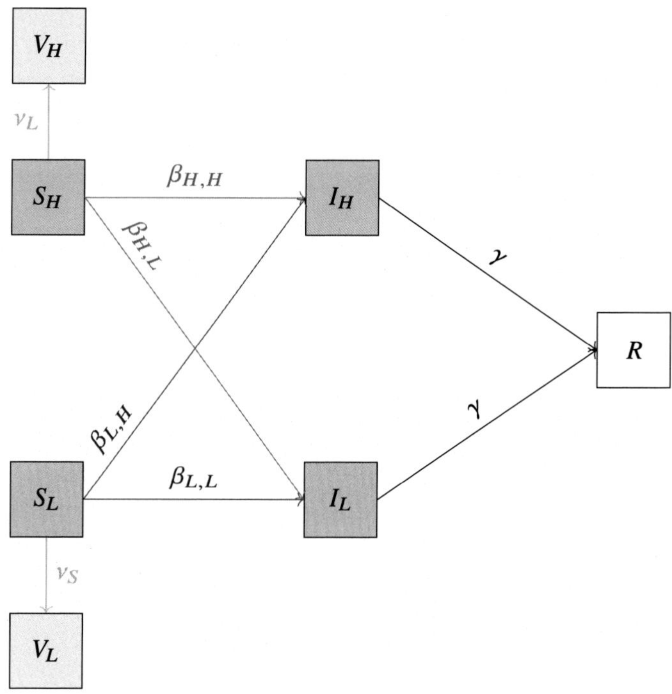

Figure 6.1 A model of differential vaccination in a population with two risk strata. ν_H and ν_L are the vaccination rates of the high and low risk strata, respectively.

```
d_Is = Matrix(np.array([(beta_HH * n_H - gamma_H) * I_H
                         + (beta_HL * n_H) * I_L,
                        beta_LH * n_L * I_H
                         + (beta_LL * n_L - gamma_L) * I_L]).T)
```

We obtained the Jacobian of `d_Is` as

```
coeffs = d_Is.jacobian(infectious_system)
```

namely

$$\begin{bmatrix} \beta_{HH}n_H - \gamma_H & \beta_{HL}n_H \\ \beta_{LH}n_L & \beta_{LL}n_L - \gamma_L \end{bmatrix}$$

and calculated $\mathfrak{R}_0$ by inserting values into this matrix.

We are slightly amending our approach in the present case. What we are specifically interested in is what the expected value of $\mathfrak{R}_0$ would be, given a certain percentage of vaccination of each of the strata. We consider initial vaccination only, and we are simply treating it as removing a fraction of n_H* or n_L* (the initial population fractions), so that $n_H = (1 - \nu_H)n_H*$ and, correspondingly, $n_L = (1 - \nu_L)n_L*$. We then perform a grid search by estimating the value of $\mathfrak{R}_0$ for any permutation of ν_H and ν_L.

We begin by setting up the grid, using NumPy's `meshgrid` function:

```
nx, ny = (100, 100)
x = np.linspace(0.01, 1, nx)
y = np.linspace(0.01, 1, ny)
xx, yy = np.meshgrid(x, y)
```

Next, we use `lambdify` to "bake in" the eigenvalues of the Jacobian subject to certain values that are not going to change. `lambdify` is a function of SymPy that turns a particular expression into a function subject to certain variables.

```
f = lambdify([v_H, v_L], coeffs.subs({"beta_HH": 10,
          "beta_HL": 0.5,
          "beta_LH": 0.5,
          "beta_LL": 2,
          "n_H": 0.2 * (1-v_H),
          "n_L": 0.8 * (1-v_L),
          "gamma_H": 0.05,
          "gamma_L": 0.05}).eigenvals())
```

This essentially turned the eigenvalues of a matrix of coefficients into a function callable with two parameters (ν_H and ν_L, respectively). We turn our code seeking the dominant eigenvalue into a vectorized form that we can then directly apply to array objects from NumPy, like our grid:

```
@np.vectorize
def get_r0(v_H, v_L):
    return max([abs(i) for i in f(v_H, v_L)])
```

Finally, we can call this function in conjunction with Matplotlib's contour plotting functions to create estimates of $\mathfrak{R}_0$ at different levels of high- and low-risk vaccination:

```
fig, ax = plt.subplots(constrained_layout=True, figsize=(10, 6))

origin = "lower"

fill = ax.contourf(xx,
                   yy,
```

```
                    get_r0(xx, yy),
                    15,
                    cmap=plt.cm.rainbow,
                    origin=origin)
critical_optimum = ax.contour(xx,
                              yy,
                              get_r0(xx, yy),
                              levels=[1.0],
                              origin=origin)

ax.set_ylabel("Vaccination rate, low-risk ($\\nu_L$)")
ax.set_xlabel("Vaccination rate, high-risk ($\\nu_H$)")

cbar = fig.colorbar(fill)
cbar.ax.set_ylabel("$\mathfrak{R}_0$")
cbar.add_lines(critical_optimum)
```

A notebook implementing the contents of this Computational Note is available on the book's companion Github repository in the folder `/ch06/rtv`.

Fig. 6.2 shows the most important feature of risk-dependent vaccination. The L-shaped contour separates circumstances where epidemic spread is possible (left-below) from circumstances where $\mathfrak{R}_0$ is pushed below 1, and consequently epidemic spread is not possible. The L-shape of the curve indicates two salient points.

First, once a relatively small percentage of high-risk individuals are vaccinated, the population comes quite close to immunity. Once about half of the high-risk individuals are vaccinated (i.e., 10% of the total population), the epidemic potential is greatly diminished. Targeting a high-risk subpopulation means that only a relatively small part (in this case, about 35%) of the low-risk population need to be vaccinated to eliminate the pathogen's epidemic potential.

Second, as Keeling and Rohani [39] point out, though vaccinating the high-risk group above their threshold yields no further benefits and is in fact an ineffective use of vaccines, overvaccinating may be more advantageous than undervaccinating. This is because, as the steepness of the "stem" of the L-shaped curve (the vertical component) indicates, the critical point, a 5% decrease in vaccination rates of the high-risk subpopulation, can plunge the entire population into the risk of a reemergent epidemic.

6.1.4 Game theoretical perspectives on vaccination

Game theory is the study of individual decision-making in the face of competing boundedly rational actors. We may conceive of vaccination as a sequential multi-player game, with two strategies open to each agent.

- If an agent chooses to vaccinate, they will be immune, at least to some extent, from the consequences of a particular infectious disease, for some time going forward.

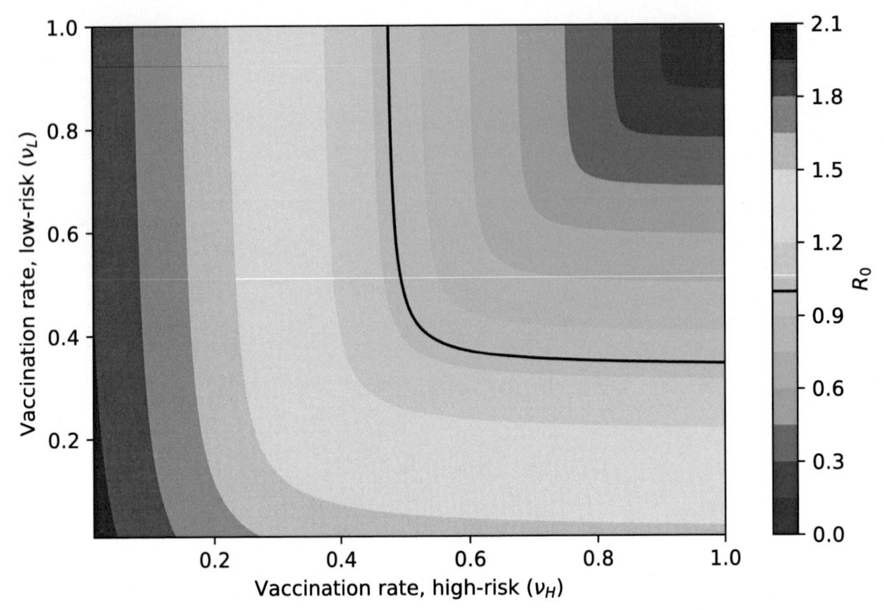

Figure 6.2 Nomogram of control indicating risk-dependent vaccination optima for $\mathbf{b} = \begin{pmatrix} 10 & 0.5 \\ 0.5 & 2 \end{pmatrix}$ and $\gamma = 0.05$. The solid line shows the critical vaccination optimum at $\mathfrak{R}_0 = 1$.

However, they incur some small but nonzero risk of suffering adverse effects from the vaccine. The latter is fixed, whereas the former is a function of what proportion of the population is already vaccinated.

- If an agent chooses not to vaccinate, they do not suffer the fixed cost of the nonzero risk of suffering adverse effects from the vaccine, but risk contracting the disease. Once again, how high that risk is depends on how many people have already been vaccinated.

This leads to a clear paradox: we know, from our study of herd immunity, that it is not necessary to vaccinate an entire population to elicit immunity. Thus once a certain percentage of a population is already immune, nothing is there to gain from vaccination. In fact, not vaccinating becomes the dominant strategy somewhat below that threshold, specifically, at the inflection point, where the risk of disease is lower than the risk of adverse effects. This results in an apparent paradox: at a particular level of vaccination, the dominant individual strategy is to not vaccinate, but if everyone does so, the overall utility is decreased. This notion is widely referred to as the "tragedy of the commons": it is in the interest of each farmer, seen individually, to exploit the common pasture to its maximum extent, but if every farmer adopts that strategy, the overall results are going to be rather unfavorable. Liu et al. [198] note that quite often, the level of vaccination above which it is no longer in an individual's self-interest to vaccinate (Nash vaccination) is well below what is required to elicit collective immunity.

Bauch and Earn [199] propose a general scheme for analyzing the "vaccination game." Individuals adopt a strategy P, which is their general likelihood to vaccinate. For the entire population N, the population-level vaccination fraction ν will be

$$\nu = \frac{1}{|N|} \sum_{i \in N} P_i. \tag{6.19}$$

- If an individual vaccinates, they incur the risk of adverse effects, J_υ.
- If an individual does not vaccinate, they incur the risk of infection, J_i. Since this depends on the mass action term, i.e., the number of infectious (nonvaccinated) individuals, we multiply this by $\pi_{\nu'}$, the likelihood of sustaining infection if the population-level vaccination fraction is ν'.

If we conceptualize the relative risk J as $\frac{J_\upsilon}{J_i}$, the expected payoff for a strategy P given the population-level vaccination fraction ν may be expressed as

$$E(P, \nu) = -JP - \pi_{\nu'}(1 - P), \tag{6.20}$$

since the strategies of vaccinating and not vaccinating are partitions (i.e., mutually exclusive and jointly exhaustive). If a large part of the population adopts P, we call P a Nash equilibrium if anyone choosing any strategy other than P (which we shall call $\neg P$) will receive a lower payoff. If, for any value of ν, $E(P, \nu) > E(\neg P, \nu)$, P is a convergently stable Nash equilibrium (CSNE).

Practice Note 6.2 Game theory, rationality and information asymmetry

Notably, agents make decisions based on their own understanding of the risk. This may not, actually, correspond to the true risk, either of the vaccine or of the illness. "Vaccine scares" are episodic events during which, owing often to exaggerated media interest, the public perception of the risks associated with a vaccine exceed the actual risk. Similarly, agents may overestimate or underestimate the risk from the disease. Incidents such as that following the unexpected adverse effects of a dengue vaccine in the Philippines [200] or the vaccine scare in France following the media interest in a now disproven association between the hepatitis B vaccine and multiple sclerosis [201] affect the perception of risk. As such, it is crucial to understand not only real risks but also the risks as they are understood and evaluated by boundedly rational agents, whose understanding of risks and benefits might be colored by subjective perception and by individual interpretations of the concept of risk.

For simplicity's sake, let us assume a population split into two segments: those vaccinating with probability P and those vaccinating with probability Q, with ρ being the fraction of those adopting P and, since the strategies are mutually exclusive, $1 - \rho$ being the fraction of those who adopt Q. Based on these strategies, the population's vaccination fraction is the product of the proportions and their individual vaccination properties:

$$\nu = \rho P + (1-\rho)Q. \tag{6.21}$$

This gives us the payoffs for P and Q strategists, E_P and E_Q respectively, as

$$\begin{cases} E_P(P,Q,\rho) = E(P, \rho P + (1-\rho)Q) \\ E_Q(P,Q,\rho) = E(Q, \rho P + (1-\rho)Q). \end{cases} \tag{6.22}$$

For any P strategist, the benefit of switching to Q would be

$$\Delta_E(P \to Q) = E_P(P,Q,\rho) - E_Q(P,Q,\rho) = (P-Q)(\pi_{\nu P+(1-\nu)Q} - J). \tag{6.23}$$

Bauch and Earn [199] proved that for any relative risk J, there exists a strategy P^* so that for any strategy $\neg P^*$ and any value ρ, $\Delta_E P \to Q > 0$. The term $\pi_{\rho P+(1-\rho)Q}$ describes the likelihood of infection if ρ part of the population adopt P and $1-\rho$ adopt Q. There are three possible Nash equilibria for the above: the pure equilibria $P^* = 0$ and $P^* = 1$, and a mixed equilibrium of $0 < p^* < 1$. If $J \geq \nu P + (1-\nu)Q$, then the Nash optimal response is to never to vaccinate. If $J < \nu P + (1-\nu)Q$, the Nash optimal strategy is to vaccinate at a nonzero rate of ν^* so that $\pi_{\nu^*} = J$.

It is crucial to examine the ramifications of this last statement. A Nash rational agent keeps vaccinating until the $\pi^{\nu^*} = J$, that is, until the risks of sustaining the illness no longer outweigh the risks of the vaccination itself. If $\nu = 1 - \frac{1}{\mathfrak{R}_0}$, the disease has been effectively eradicated, and π_ν is by definition zero. Recall that we defined J as the fraction of risks from the vaccine J_υ, and risks from infection J_i. It follows from the existence of the mixed equilibrium $0 < \nu^* < 1$ that

$$\lim_{\nu \to \nu^*} \frac{J_\upsilon}{J_i(\nu)} = \infty. \tag{6.24}$$

It may be assumed that $\nu^* = 1 - \frac{1}{\mathfrak{R}_0}$, i.e., the threshold of collective immunity, would be a Nash equilibrium, since there would be no additional benefit to vaccinating (or "unvaccinating," a mathematically possible but biologically nonsensical strategy). However, this is not the case for any case where J_υ is nonzero. For all nonzero values of J_υ, as ν approaches $1 - \frac{1}{\mathfrak{R}_0}$, J_i approaches zero and the risk approaches infinity. Consequently, for all nonzero J_υ, there exists a value ϵ so that $1 - \frac{1}{\mathfrak{R}_0} = \nu^* + \epsilon$.

The consequence of ϵ is to denote a "band," in which Nash strategy and prosocial vaccination, which aims at providing the greatest protection for the population, are different: because ϵ is nonzero if J_υ is nonzero, in every such case, the CSNE at ν^* is attained before collective immunity. The value of ϵ is solely a function of J_υ, the risks of the vaccine, or more specifically, their perception (on which see Practice Note 6.2). It is not, however, a function of the risks from the disease or its severity. Once ν^* has been reached, the impact of the disease is already nullified for the individual, vis-a-vis the risks from the vaccination, and the vaccine does not provide a direct benefit for the individual, only for the collective. As such, effective communication about vaccination needs to shift, as ν^* is approached, away from individual benefit and towards prosocial and altruistic objectives.

6.2 Duration and effectiveness of vaccine-induced immunity

Similarly to post-infectious immunity, vaccine-induced immunity may decrease over time. Some vaccinations provide a lifelong protective effect, whereas others need to be boosted periodically, reinforcing immunity. Equally, the degree of immunity created by vaccination may be variable between individuals, between vaccines, and over time. This section deals with the way we conceptualize the limits on the duration and effectiveness of vaccine-induced immunity.

1. A vaccine may prevent severe illness, but not necessarily mild illness. This is sometimes referred to as a nonsterilizing vaccine. Such vaccines nevertheless affect infectiousness indirectly, by reducing the pathogenic load in the body and/or alleviating the symptoms that contribute to the pathogen's spread. For instance, the rotavirus vaccine does not eliminate the pathogen completely, but prevents severe illness. A secondary effect of that is that milder (and shorter) infection generally results in fewer secondary cases. In the same vein, the COVID-19 vaccines' reduction of severe symptomatic illness has resulted in improved control.
2. A vaccine may be ineffective in an individual, by which we mean to say it fails to obtain the protective effect that it has in the population at large. In certain subpopulations, vaccine failure is more common due to decreased immunogenicity. This includes, in particular, the elderly. For certain vaccines, a more immunogenic preparation is available for such subpopulations, which is typically an adjuvanted version of a nonadjuvanted vaccine.
3. Finally, a vaccine typically induces temporary immunity. Similarly to post-infectious immunity, vaccine-induced immunity may not last indefinitely, and typically wanes in a way that can be approximated as a deterministic process.

Practice Note 6.3 "Natural" immunity

The term "natural immunity" has been often used to express post-infectious immunity and differentiate it from vaccine-induced immunity. In practice, this is not necessarily helpful. There is nothing fundamentally "unnatural" in vaccine-induced immunity, and whereas the minutiae of natural infection and vaccine-induced immunity might differ, this is a quintessentially unhelpful notion.

In addition to encouraging the naturalistic fallacy, whereby "natural" immunity is seen as less risky (when in practice, surviving an infection is almost always *more* dangerous than a vaccine) and more "appropriate," it is also bound to create public misperceptions, e.g., confusion with passive immunization via convalescent plasma or antibodies. A preferable terminology is "post-infectious" or "post-infection" immunity, which highlights that the process leading to immunity was infection rather than immunization.

6.2.1 Case study: Marek's disease

Marek's disease is an alphaherpesvirus-mediated oncogenic disease in chickens caused by the Marek's disease virus (MDV). Infected animals present with T cell lymphomas and lymphocytic tumors [202]. In chicken flocks, Marek's disease has been largely controlled using vaccines. However, the vaccine is not perfect, in fact, because the vaccine does not provide sterilizing immunity [203], vaccinated animals continue to spread the infection. As a result, vaccines have actually increased the virulence of Marek's disease [203].

A vaccine is, in the end, an evolutionary pressure on a pathogen. If the vaccine does not eliminate the pathogen's ability to infect new hosts (i.e., if it is not a sterilizing vaccine), vaccination may increase overall virulence by selecting for the most virulent strain of the pathogen, where virulence and transmissibility are related in any way (as it is where herpesviruses are concerned). The result of the vaccine against Marek's disease has been clearly beneficial in reducing disease from an economically costly disease of poultry, but its evolutionary pressure has made it a much more lethal pathogen.

This is, of course, not an argument against vaccination. Contributing to pathogenic evolution towards higher virulence is, even in extreme cases as Marek's disease, the lesser of two evils. It does, however, call attention to the need to understand the impact that interventions to control disease have on the ecology and evolution of the disease itself. Pathogens do not exist in isolation but exist in a space of competition for hosts and the ability to spread.

6.2.2 Incomplete effect of vaccines

For modeling the incomplete effect of vaccines, preventing severe illness but not mild disease, it is helpful to conceptualize it as the interplay of three variables:

- p_s is the likelihood of severe disease, i.e., the probability that an infected case, from S or V alike, flows into I_s rather than I_m.
- ϵ_I is the vaccination penalty on infection. It reflects the reduction in β as a consequence of vaccination, and is confined to the range [0, 1].
- ϵ_S is the vaccination penalty on virulence. This reflects the reduction that vaccination confers to the likelihood of developing serious illness, and is equally confined to the range [0, 1].

Using the shorthand I for $I_m + I_s$, we may write

$$\begin{aligned}
\frac{dS}{dt} &= -\underbrace{\beta SI}_{\text{infections}} - \underbrace{\nu}_{\text{vaccinations}}, \\
\frac{dV}{dt} &= \underbrace{\nu}_{\text{vaccinations}} \underbrace{-\beta V \underbrace{\epsilon_i}_{\text{infection penalty}} I}_{\text{infections of vaccinated}}, \\
\frac{dI_m}{dt} &= \beta(1-p_s)(\underbrace{SI}_{\text{from } S} + \underbrace{V(1-\epsilon_s)\epsilon_i I}_{\text{from } V}) - \gamma_m I_m, \\
\frac{dI_s}{dt} &= \beta p_s(\underbrace{SI}_{\text{from } S} + \underbrace{V\epsilon_i\epsilon_s I}_{\text{from } V}) - \gamma_s I_s.
\end{aligned} \tag{6.25}$$

The beauty (and utility) of this representation is that ϵ_I is the relative risk (RR) between vaccinees and nonvaccinated with respect to infection, and ϵ_S is the same quantity with respect to severe disease. They are also related to the vaccine efficacy α, the proportional reduction in the attack rate of the disease, by way of $\alpha = 1 - \epsilon$. This connection to the population impact numbers that are typically obtained during clinical and early post-marketing testing of a vaccine enable us to reason about infections in view of the vaccine. Moreover, it allows us to determine the requisite vaccination rate ν to achieve a certain maximum number of severe cases, which is critical for planning vaccinations to protect hospital capacity.

6.2.3 Waning immunity

As we have seen in Subsubsection 2.3.3, post-infectious acquired immunity often wanes after a given period of time. The same is true for vaccine-induced immunity, which is the rationale behind boosting immunizations periodically (see Subsection 6.2.5). In the SIRS model, persons who recovered into an immune state relapsed into susceptibility at the waning rate $\omega = \frac{1}{\tau_R}$, where τ_R is the mean duration of immunity. SVIRS is a variation on the theme of the SIRS model that takes account of vaccination rates ν, as well as waning immunity of the vaccinated. (See Fig. 6.3.)

We may express this as the set of differential equations:

$$\begin{aligned}
\frac{dS}{dt} &= -\underbrace{\beta SI}_{\text{mass action}} + \underbrace{\omega R}_{\text{reversion from } R} + \underbrace{\omega V}_{\text{reversion from } V} - \underbrace{\nu S}_{\text{new vaccinations}}, \\
\frac{dV}{dt} &= \underbrace{\nu S}_{\text{new vaccinations}} - \underbrace{\omega V}_{\text{waning of vaccinated}}, \\
\frac{dI}{dt} &= \underbrace{\beta SI}_{\text{mass action}} - \underbrace{\gamma I}_{\text{recovery}},
\end{aligned} \tag{6.26}$$

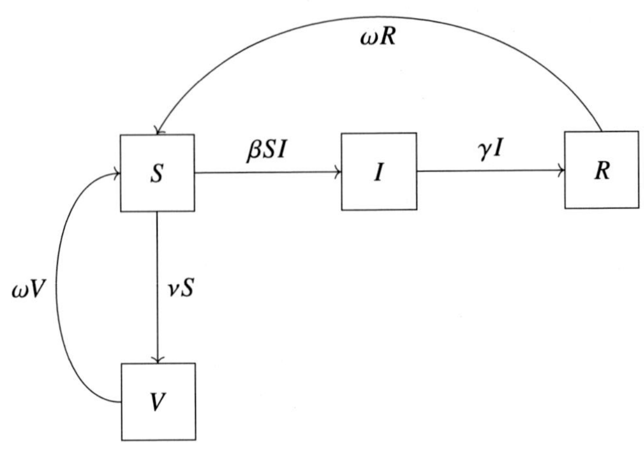

Figure 6.3 SVIRS model accounting for homogeneous waning immunity of the vaccinated. Homogeneity of waning means that post-infectious immunity and post-vaccination immunity are subject to the same waning rate ω.

$$\frac{dR}{dt} = \underbrace{\gamma I}_{\text{recovery}} - \underbrace{\omega R}_{\text{reversion from } R} .$$

This presupposes, of course, that vaccine-induced and post-infectious immunity (on which see Practice Note 6.3) wane at an equivalent rate of ω.

This model's main shortcoming is that waning is not ordinarily a continuous process. Immunity typically wanes after a given amount of time, and not at all before then. Delay differential equations can help us model this situation. Consider vaccination of susceptibles at rate ν. The number vaccinated by time t is

$$\int_0^t \nu S(u)\, du. \tag{6.27}$$

At first approximation, the vaccination has a lifetime of τ_ω, after which it invariably vanes to zero. Biologically, τ_ω is the inverse of the waning rate or the mean period of protection. For values of $t > \tau_\omega$, the reversion to susceptibility at time t will be equal to the number vaccinated at the point in time $t - \tau_\omega$ (the principle of "first in, first out" applies here). That, of course, is $\nu S(t - \tau_\omega)$. Combining these yields the system of delay differential equations described as

$$\begin{aligned}
\frac{dS}{dt} &= \mu - \underbrace{\beta SI}_{\text{mass action}} \underbrace{-\nu S}_{\text{new vaccinations}} + \underbrace{\nu S(t - \tau_\omega)}_{\text{reversion}} - \mu S, \\
\frac{dI}{dt} &= \underbrace{\beta SI}_{\text{mass action}} - \underbrace{\gamma I}_{\text{recovery}} - \mu I,
\end{aligned} \tag{6.28}$$

$$\frac{dV}{dt} = \underbrace{\nu S}_{\text{new vaccinations}} - \underbrace{\nu S(t-\tau_\omega)}_{\text{reversion}} - \mu V,$$

$$\frac{dR}{dt} = \underbrace{\gamma I}_{\text{recoveries from } I} - \mu R.$$

Delay differential equations are more computationally expensive to solve than ODEs, but there exist accelerated optimizing solvers for DDEs; their use is discussed in Computational Note 6.2.

Computational Note 6.2 Solving delay differential equations computationally

A delay differential equation (DDE) differs from an ODE in that it has at least one term represented as a function of the system's state at a previous time. The term $\nu S(t-\tau_\omega)$ in Eq. (6.28), wherein the value of $\frac{dS}{dt}$ at time t, is a function of its value at time $t-\tau$. This makes integration significantly more challenging, and not quite feasible with much accuracy at all for an initial period (about twice the length of the delay). However, we can usefully integrate DDEs to elicit a system's long-term behavior.

JiTCDDE is an efficient solution for DDEs that is related to the JiTCODE project used in some other applications in this text (see, e.g., Computational Note 7.7) [204]. After importing `jitcdde` and explicitly importing the symbols `t` and `y`, we can specify our DDE:

```
f = [
    mu - beta * y(0, t) * y(1, t) - (nu + mu) * y(0, t) \
     + nu * y(0, t - tau),
    beta * y(0, t) * y(1, t) - (gamma + mu) * y(1, t),
    nu * (y(0, t) - y(0, t-tau)) - mu * y(2, t),
    gamma * y(1, t) - mu * y(3, t)
]
```

Unlike in previous instances where we defined differential equations as Python functions of y, t, and its parameters, JiTCDDE expects a more formal definition. This is because JiTCDDE performs a symbolic analysis of the function to interpret it and render it as optimized C code.

In JiTCDDE, the system state is represented as the vector `y`, which in this case is a vector of length 4, each element corresponding to S, I, V, and R. In addition, JiTCDDE can take a time parameter. By default, `y(2)` refers to the third (zero-indexed) element of the state vector, but in addition, we can specify a time relation. Thus `y(2, t - tau)` means "the value of the third element of the state vector at the time $t - \tau$". It is not strictly necessary to specify the time if it is t, although this is good practice to keep in mind which of the parameters are delay-contingent and which are not.

Next, we initialize the JiTCDDE solver by providing it with the system of differential equations we defined above, as well as the delay parameters. In this case, we have a single static delay parameter: we assume that τ days after vaccination, the vaccine's effect completely disappears in all cases. This is not strictly reflective of practical realities, which is why a waning function is often helpful. Nevertheless, it is a useful approximation of reality especially for short-lived vaccines, or where the pathogen's rate of mutation is high enough to render the previous season's immunity ineffective (as is the case with influenza vaccination).

Subsequently, we need to tell JiTCDDE about the past. For the first τ time, JiTCDDE will be asked to look into a past before we started integration, that is, a past that, strictly speaking, does not exist. JiTCDDE supports a number of ways this could be specified, but by far the simplest is "constant past":

```
DDE.constant_past([0.95, 0.05, 0, 0], time=tau)
```

Here, we essentially declare the starting parameters (much as we provided $S(0)$, $I(0)$, and so on to other solvers), and tell JiTCDDE that up until $t = \tau$, it should assume those values for each of the four quantities.

Finally, we will have to deal with the fact that our definition of the starting parameters results in a discontinuity in the case of constant past, essentially a straight line, up until integration begins, where it discontinuously assumes the first integrated value. The trivial solution is to initialize with `DDE.step_on_discontinuities()`, which adaptively integrates at discontinuities. We now have everything we need to run our DDE integrator with a simple iteration:

```
res = []

for time in np.arange(DDE.t, DDE.t + 30000, 1):
    res.append(DDE.integrate(time))
```

The results of a run of this DDE integrator are displayed as Fig. 6.4. JiTCDDE is an enormously powerful tool for exploring the long-term dynamics of systems, and this simple example can be expanded easily with a more complex waning function or varying delays. A caveat is that since JiTCDDE (along with the entire JiTCODE project) relies on symbolic optimization, many frequently used numeric functions (such as `np.pi` to obtain the value of π or `np.exp` to perform exponentiation to e), are not available. Instead, `symengine`, which is at the backend of JiTCODE, and is thus installed at the same time, must be imported, and the relevant symbols be explicitly sourced from `symengine` (e.g., for exponentiation, the symbol `exp` must be imported from `symengine`).

A notebook implementing the contents of this Computational Note is available on the book's companion Github repository in the folder `/ch06/waning_dde`.

Fig. 6.4 shows three critical effects of waning immunity:

1. Waning creates periodicities. As the disease process depletes the population of suitable hosts, it begins to decline. Waning replenishes the pool of susceptibles, allowing an epidemic to once again spread. This is, of course, a result of the phenomena explored in the beginning of Chapter 2.
2. Eventually, waning results in an equilibrium state, where the disease becomes endemic, so that for each recovery, there will be an infection. In Fig. 6.4, the system converges onto an equilibrium point at around $S = 0.0475$ and $I = 0.003$. At this point, the disease may persist indefinitely.
3. Where vaccination for a particular pathogen is rolled out to a large population in a single large campaign (as was the case for the SARS-CoV-2 vaccine), it is often helpful to keep the periodic dynamics of waning in mind. Scheduling booster or follow-up campaigns at the right time (preferably before, or around the time of, local minima) may often suppress the infection to levels that eliminate the pathogen's ability to spread. This is analogous to pulse vaccination (see Subsubsection 6.1.2.3) at the pulse frequency equal to the inverse of the typical waning period.

Models of waning immunity are helpful in understanding how the interactions between a pathogen's dynamics and human intervention result in new equilibria, and may be powerful tools for the planning of disease eradication.

Practice Note 6.4 DDEs and the terms of dynamics

DDEs are uniquely powerful tools to mathematically explore systems, whose right hand side at t depends on an earlier state. The tradeoff is that in the beginning, DDEs operate on an assumption of what the past was before the first moment of integration. Since this is rarely easy to specify, in the overwhelming majority of cases, DDEs are not very useful in the very beginning.

Eventually, dynamical systems converge to states that are no longer governed by their initial states (see Subsection 7.3.5 for an important qualification to this statement). The initial discontinuities of a DDE are eventually subsumed into the system's wider dynamics. As tools for exploring the long-term evolution of a system (past $2-4\,\tau$), DDEs are uniquely powerful. For predicting short-term effects, they should be handled with care, if at all.

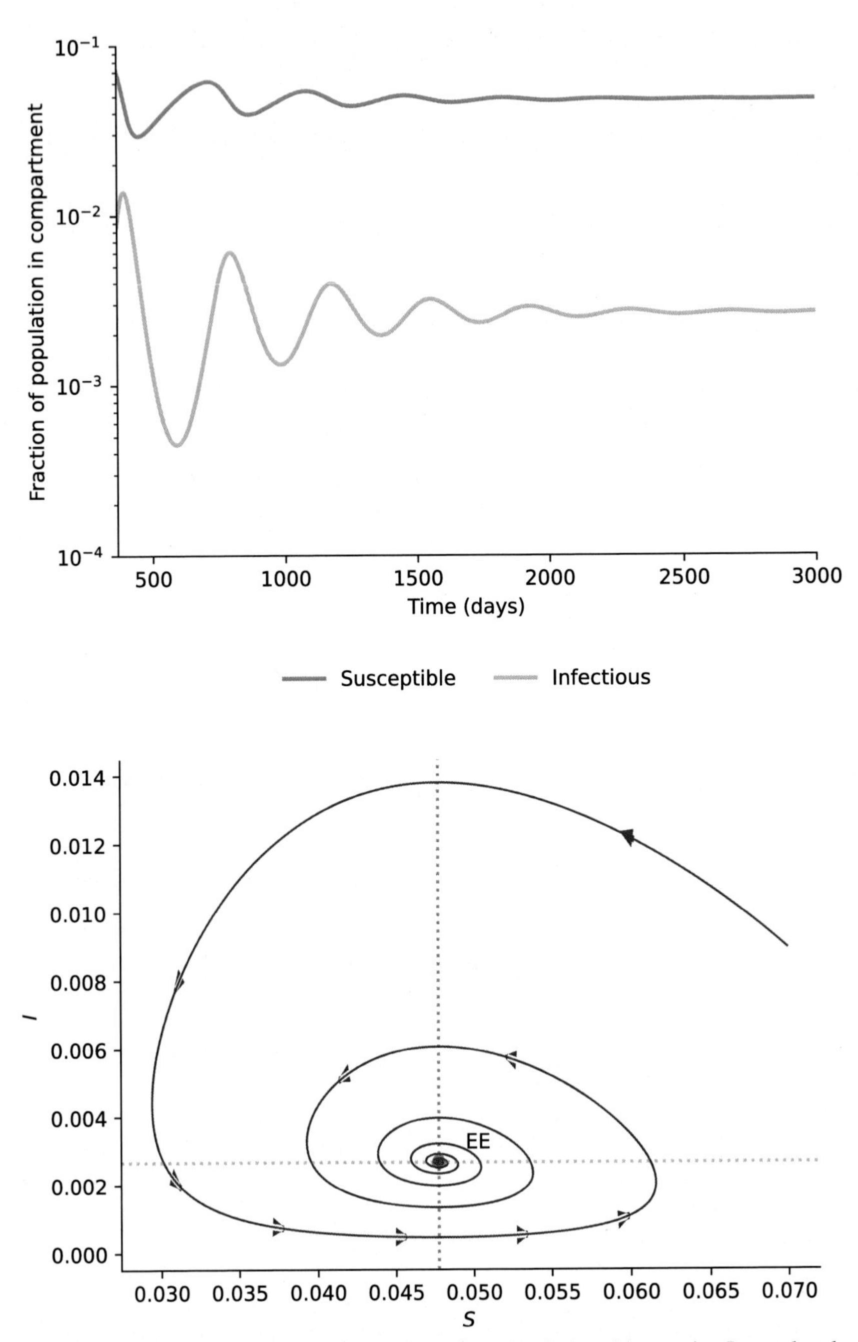

Figure 6.4 Periodicities induced by the waning of vaccine-induced immunity. Integral and phase portrait of a delay-differential system with $\tau_\omega = 180$, $\beta = 1.5$, $\gamma = \frac{1}{14}$, $\mu = 2 \times 10^{-5}$, and $\nu = 10^{-6}$ per day.

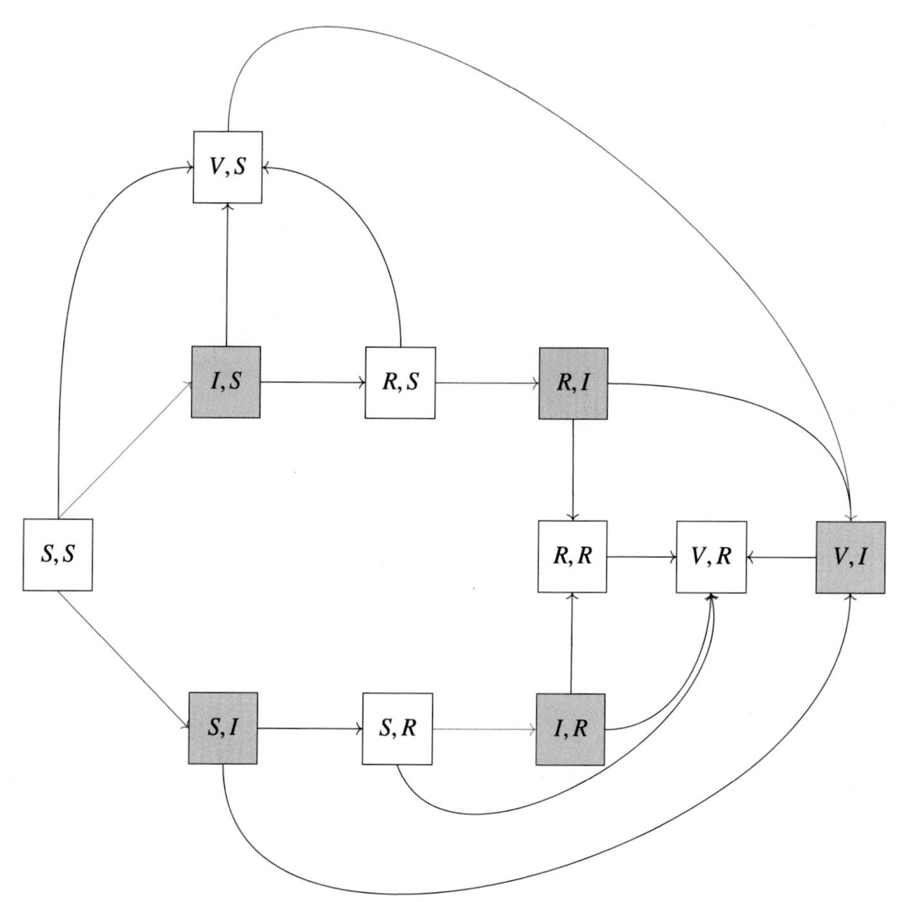

Figure 6.5 Two strains of a pathogen with a vaccine that only protects against one strain. This diagram shows the state space of two pathogens in the state representation, i.e., C_1, C_2 describes the compartment of individuals who are in C_1 with respect to pathogen 1 and C_2, with respect to pathogen 2. Red boxes denote the infectious subsystem of pathogen 1 and blue boxes that of the pathogen 2, respectively. Arrows that transition into infectious states are colored based on which infectious subsystem governs the mass action term.

6.2.4 Mutating out of immunity

In the vast majority of cases, in particular where there is no significant divergence between strains or epitopic subtypes of the pathogen, immunity is more or less constant. In creating an infectious disease model, one needs to be aware of the particular circumstances (major mutations, geographic dispersal, and separate evolution) that yield significantly (i.e., antigenically/epitopically) different variants.

Let us consider a pathogen with two strains in the presence of a vaccine protective against strain 1, but not against strain 2. We can analyze this as essentially a special case of two-pathogen competition with no coinfection and with no cross-immunity (since cross-immunity would mean that vaccine-induced immunity would be protective against both strains). Fig. 6.5 lays out such a scenario, where a vaccine protects against the first, but not the second, pathogen. The result of such a system, assuming a constant vaccination rate ν, can be described as the following system of differential equations (where strain 1 is susceptible to the vaccine, but strain 2 is not):

$$
\begin{aligned}
\frac{dN_{S,S}}{dt} &= N_{S,S}(-\underbrace{\beta_1 I_1 - \beta_2 I_2}_{\text{infections}} - \underbrace{\nu}_{\text{vaccinations}}),\\
\frac{dN_{V,S}}{dt} &= \underbrace{\nu(N_{S,S} + N_{I,S} + N_{R,S})}_{\text{new vaccinations}} - \underbrace{\beta_2 N_{V,S} I_2}_{\text{infections with resistant strain}},\\
\frac{dN_{I,S}}{dt} &= \underbrace{\beta_1 N_{S,S} I_1}_{\text{new infections}} - N_{I,S}(\underbrace{\gamma_1}_{\text{recoveries}} + \underbrace{\nu}_{\text{vaccinations}}),\\
\frac{dN_{R,S}}{dt} &= \underbrace{\gamma_1 N_{I,S}}_{\text{recoveries from 1}} - N_{R,S}(\underbrace{\beta_2 I_2}_{\text{infections with 2 from } R,S} + \underbrace{\nu}_{\text{vaccinations}}),\\
\frac{dN_{S,I}}{dt} &= \underbrace{\beta_2 N_{S,S} I_2}_{\text{infections with 2 from } S,S} - N_{S,I}(\underbrace{\gamma_2}_{\text{recoveries}} + \underbrace{\nu}_{\text{vaccinations}}),\\
\frac{dN_{S,R}}{dt} &= \underbrace{\gamma_2 N_{S,I}}_{\text{recoveries from } S,I} - N_{S,R}(\underbrace{\beta_1 I_1}_{\text{infections with 1}} + \underbrace{\nu}_{\text{vaccinations}}),\\
\frac{dN_{I,R}}{dt} &= \underbrace{\beta_1 N_{S,R} I_1}_{\text{infections with 1 from } S,R} - N_{I,R}(\underbrace{\nu}_{\text{vaccinations}} + \underbrace{\gamma_1}_{\text{recoveries}}),\\
\frac{dN_{R,I}}{dt} &= \underbrace{\beta_2 N_{R,S} I_2}_{\text{infections with 2 from } R,S} - N_{R,I}(\underbrace{\nu}_{\text{vaccinations}} + \underbrace{\gamma_2}_{\text{recoveries}}),\\
\frac{dN_{R,R}}{dt} &= \underbrace{\gamma_1 N_{I,R}}_{\text{recoveries from } I,R} + \underbrace{\gamma_2 N_{R,I}}_{\text{recoveries from } R,I} - \underbrace{\nu N_{R,R}}_{\text{vaccinations}},\\
\frac{dN_{V,R}}{dt} &= \underbrace{\nu(N_{R,R} + N_{V,I} + N_{S,R} + N_{S,I})}_{\text{vaccinations}},\\
\frac{dN_{V,I}}{dt} &= \underbrace{\beta_2 N_{V,S} I_2}_{\text{infections from } V,S} + \underbrace{\nu(N_{R,I} + N_{S,I})}_{\text{vaccinations}} - \underbrace{\gamma_2 N_{V,I}}_{\text{recoveries}}.
\end{aligned}
\tag{6.29}
$$

I_1 and I_2 describe the infectious subsystems of the pathogens, respectively; they are made up of $N_{I,S} + N_{I,R}$ on one hand and $N_{R_I} + N_{S,I} + N_{V,I}$ on the other. The susceptible subsystem in respect of pathogen 2, that is, the proportion of the population available for infection by pathogen 2, consists of $\sum N_{*,S}$, i.e., the sum of all states with state vectors, where the second element is S. This amounts to $N_{S,S} + N_{I,S} + N_{V,S} + N_{R,S}$. On the other hand, for pathogen 1, this system only includes $N_{S,S} + N_{S,I} + N_{S,R}$. Denoting the two subsystems as $\mathcal{S}_1$ and $\mathcal{S}_2$, respectively, we obtain the difference $N_{I,S} + N_{V,S} + N_{R,S} - N_{S,I} - N_{S,R}$, which we shall call ΔS_ν, or the difference between susceptibles between the two strains given the vaccination rate of ν. We can calculate it as

$$\Delta S_\nu = \beta_1 I_1 (N_{S,S} + N_{S,R}) - \beta_2 I_2 (N_{R,S} + N_{S,S}) + \nu (N_{S,S} + N_{S,R} + N_{S,I}). \quad (6.30)$$

We may be interested in the overall evolutionary effect of mutating out of protection. Kennedy and Read [205] correctly note that antimicrobial resistance is much more frequent than resistance to vaccination, which they attribute to two factors:

1. vaccines are prophylactic, and immunity is thus often preexistent by the time of pathogenic exposure, consequently there is no initial period before treatment during which the pathogen can establish a large enough seed population from which to mutate, and
2. vaccines target a wider range of epitopes, whereas antimicrobials typically target a single cellular function or enzyme.

A counterexample that strengthens the rule is, of course, Marek's disease (see Subsection 6.2.1).

In the example of Eq. (6.29), the mutated pathogen gains a fitness advantage in terms of being able to access a larger pool of susceptibles than the vaccine-susceptible strain. We denote this fitness advantage vis-a-vis vaccination as ΔF_ν^+, and calculate it as

$$\Delta F_\nu^+ = \frac{\sum N_{*,S}}{\nu \sum N_{S,*}}. \quad (6.31)$$

Since the denominator $\nu \sum N_{S,*}$ increases strictly monotonically with increasing values of ν, whereas $\sum N_{*,S}$ is not dependent on ν at all, ΔF_ν^+ increases at higher vaccination rates ν.

On the other hand, it is common to see a resistant pathogen pay a fitness cost in terms of reduced transmissibility, so that

$$\Delta F^- = \frac{\mathfrak{R}_{0,2}}{\mathfrak{R}_{0,1}} = \frac{\beta_2 \gamma_1}{\beta_1 \gamma_2}. \quad (6.32)$$

In this case, $\mathfrak{R}_0$ is analogous to the way competitive fitness studies use the number of offspring as an indicator of the fitness cost of resistance [206]. This gives us the overall fitness criterion

$$\Delta F_\nu = \frac{\sum N_{*,S}}{\nu \sum N_{S,*}} - \frac{\beta_2 \gamma_1}{\beta_1 \gamma_2}. \quad (6.33)$$

Consequently, the resistant strain will become dominant once

$$\nu \sum N_{S,*} > \frac{\beta_1 \gamma_2 \sum N_{*,S}}{\beta_2 \gamma_1}. \quad (6.34)$$

Since the system of ODEs we are considering is autonomous, the time at which this condition is met, given a value of ν and initial conditions, can be calculated numerically. This gives us a quantitative tool of immense value with which to analyze the evolutionary effect of vaccination on a pathogen, and where a vaccine-resistant strain exhibits other clinically relevant features, such modeling may inform considerations of vaccination policy.

6.2.5 Boosting

Boosting refers to the use of an additional exposure to an antigen to prolong or reinforce immunity. The classical model by Alexander et al. [207] conceives of boosters as conferring final immunity, so that the vaccinated compartment is essentially in an anteroom of immunity. This model is appropriate where immunity indeed becomes life-long after the booster vaccination, but fails to reflect cases where the booster prolongs temporary immunity, rather than conferring indefinite immunity. This was the case with the COVID-19 vaccine, where multiple boosters were required to prolong and increase immunity [208]. Most variants of the boosting problem, e.g., Carlsson et al. [209], adopt some form of "stacking" approach, where there are n compartments that each represent $1, 2, \ldots, n$ boosters. Such situations are perhaps more adequately modeled by a differential outcome model.

Let us assume that the duration of immunity is τ_e, that is, τ_e time after the last booster, an individual must receive another booster or revert to susceptibility. For simplicity's sake, we will assume that the vaccination rate for the entire process is constant (ν) and regardless of the number of boosters received, the likelihood to boost is p_b. Then, we obtain the following system of delay-differential equations:

$$
\begin{aligned}
\frac{dS}{dt} &= \mu - \beta SI - \underbrace{\nu S}_{\text{vaccinations}} + \underbrace{(1-p_b)\nu S(t-\tau_e)}_{\text{lapsed, unboosted vaccinees}} - \mu_s, \\
\frac{dI}{dt} &= \beta SI - \gamma I - \mu I, \\
\frac{dV}{dt} &= \nu S - \underbrace{(1-p_b)\nu S(t-\tau_e)}_{\text{lapsed, unboosted vaccinees}} - \mu V, \\
\frac{dR}{dt} &= \gamma I - \mu R.
\end{aligned}
\tag{6.35}
$$

Lifelong boosting is relatively rare, however. In most cases, a limited number of boosters establishes final immunity. For this case, we may adapt Eq. (6.35) with a given initial vaccianted compartment V, and vector p, which has the number of elements corresponding to booster stages, so that p_i is the likelihood of a person who has had the $n-1$th booster to receive the i-th booster (with the 0th booster being the initial vaccination and governed by ν). The $|p|$-th booster is the terminal booster, which confers terminal immunity.

$$
\begin{aligned}
\frac{dV}{dt} &= \nu S - \mu V - \mu S(t - \tau_e),\\
\frac{dB_i}{dt} &= \Big(\prod_{m=1}^{i} p_m\Big)\nu S(t - i\tau_e) - \mu_{B_i} - \Big(\prod_{n=1}^{i+1} p_n\Big)\nu S(t - (i+1)\tau_e),\\
\frac{dS}{dt} &= \mu - \beta S I - \nu S + \sum_{k=1}^{|p|}\Bigg(\Big(\prod_{m=1}^{k}(1 - p_m)\Big)\nu S(t - k\tau_e)\\
&\quad - \Big(\prod_{n=1}^{k+1}(1 - p_n)\Big)\nu S(t - (k+1)\tau_e)\Bigg).
\end{aligned}
\tag{6.36}
$$

6.3 Isolation and quarantine

Isolation and quarantine are the oldest methods in the arsenal of public health [210]. The word itself hints at its origins: a *quarantena* was the forty-day period ships had to spend in anchorage before entering Venice to avoid the transmission of infectious diseases [211]. Quarantines may be enforced or voluntary (self-quarantining): though the public perception of quarantine is principally a restrictive measure, voluntary quarantine has also long history. Boccaccio's *Decamerone* is one of the immortal legacies of self-quarantine, describing the story of ten affluent young people from Florence, who shelter in the seclusion of a villa outside the city from the 1348 outbreak of the Black Death [212,213].

Throughout history, both general quarantines and specific quarantines have been used to prevent the spread of disease. A general quarantine applies to a population on the whole, regardless of health status. For instance, several countries require companion animals (pets) to be quarantined—usually in their country of origin—for a given amount of time before entry [214]. A specific quarantine applies to people who exhibit the symptoms of a particular disease, or to people who are particularly vulnerable (reverse quarantine or shielding).

It is sometimes common in the literature of public health to see the quarantine refer to identified infectious cases and prophylactic medical isolation to exposed suspected cases. However, since from a quantitative perspective, this distinction is not necessarily useful, we shall use these terms interchangeably. Table 6.1 attempts to provide an overview of the most frequently used nonpharmacological interventions.

We can explain quarantines mathematically by reference to the mass action term (see Subsection 2.1.3). Transmission is a function of the product of a constant β, the proportion of infectious, and the proportion of susceptibles. Separating either some of the infectious or some of the susceptible (or, indeed, both) into a compartment that cannot communicate with the rest of the population decreases the mass action term:

Table 6.1 The most common forms of isolation, quarantine, and related nonpharmaceutical interventions.

	Applied to	Limitations
Isolation	Exposed individuals	Movement, interaction
Barrier isolation	Exposed or vulnerable individuals	Direct physical interaction
Quarantine	Symptomatic individuals	Movement, interaction
Reverse quarantine	Groups, up to entire populations, of generally susceptible individuals	Interaction, movement of individuals from outside the quarantine zone
Cordon sanitaire	Areas surrounding (and sometimes including) the source of an infection	Presence
Lockdown	Regions or wider populations, regardless of infectious status (exemptions of essential workers are common)	Movement, interaction, presence
Circuit breaker	Regions or wider populations, regardless of infectious status (exemptions of essential workers are common)	Movement, interaction
Social distancing (physical distancing)	Groups, up to entire populations, typically regardless of infectious status	High-risk interactions (proximity, congregation, physical contact)
Facility closures	All facilities of a certain type (e.g., schools, bars, restaurants) within an area	Operation
Travel limitations	Geographic areas, typically borders	Movement
Conditional entry	Individuals who do not meet certain conditions (e.g., vaccination status)	Presence, service use
Post-exposure quarantine	Exposed individuals	Movement, interaction
Shielding	Individuals at risk for infection or adverse clinical outcomes and their direct contacts	Movement, interaction with non-shielders

- by reducing the ability of some of the infectious individuals to come in contact with susceptible individuals (quarantine, isolation); or
- by reducing the likelihood of susceptible individuals to come in contact with infectious individuals (reverse quarantine, protective isolation, shielding).

In either case, the effect is to decrease the overall mass action term.

Despite this similarity, not all forms of quarantine are made equal. Computational modeling can help us understand how, why, and when quarantines work, and assist in decision-making in a public health emergency.

6.3.1 Case study: nonpharmaceutical interventions against Covid-19

SARS-CoV-2 and the viral syndrome it causes, COVID-19, made their world debut in the late days of 2019. Until the emergence of COVID-19 vaccines in late 2020, nonpharmacological interventions (NPIs) were the mainstays of the public health response. Despite its time-honored provenance, quarantine measures have been at the forefront of the initial measures aimed at combating the spread of SARS-CoV-2. The scale of these measures was truly unprecedented: by late 2021, almost all countries (with the notable exceptions of Sweden, Japan, South Korea, and some US states) have imposed a form of lockdown.

Lockdowns are highly effective responses to an epidemic threat, but also constitute significant limitations of individual freedoms of movement and the exercise of associated rights. Economically, lockdowns are costly and disproportionally affect certain disadvantaged groups, who may not have the financial wherewithal to sustain themselves over a period of enforced business closures.

A strategy for minimizing such costs, while achieving a comparable effect, is the "circuit breaker," first deployed in Singapore [215] and later adopted in the UK and other states [216]. The "circuit breaker" is a short period of lockdown triggered by an objective indicator, such as a rise in test-positivity ratios or a spike in healthcare capacity utilization. Circuit breakers can achieve favorable public health objectives at lower economic costs.

Practice Note 6.5 Health equity and NPIs

It is important to consider nonpharmacological interventions as what they are: fundamental (albeit justified) limitations of individual liberties. Connected to that are a range of economic and social interests, including serious concerns of equity and social solidarity. For instance, the impact of lockdowns is significantly stronger on hourly workers with little to no savings than it is to salaried employees, who may be able to work from home. Equally, public health measures can be abused as a method of "justified" exercise of prejudice: it is hard to consider the lifelong imprisonment of Mary Mallon (better known as Typhoid Mary) to be entirely separate from the prevailing anti-Irish sentiment at the time [74]. For this reason, it is crucial to see quarantines, lockdowns, and similar measures not in isolation but as a combined social, legal, and epidemiological intervention that needs to be justified in each of those three dimensions.

Socially, lockdowns are justified by appealing to the public's overriding benefit and the importance of public health. A cornerstone of this process is ensuring that the social justification is clearly laid out and communicated in clear, intelligible terms, avoiding both panic and downplaying the risks. Legally, lockdowns will often need to rely on specialized emergency legislation. Such legislation might have to comply with superordinate norms (e.g., constitutional legislation and civil liberties). This often requires lockdowns to be articulated in the least onerous way possible.

Finally, lockdowns must be epidemiologically sound. Gaining and maintaining public trust in the epidemiological and public health professions is paramount. For this reason, disease modelers must be prepared to communicate the risks of the pathogen, as well as the effects of the lockdown, taking into account wider social and socio-economic effects. Decision-makers in a crisis situation are often suffering from a bias of "tunnel vision," focusing on the most urgent problem at hand. Epidemiologists can assist in alleviating the detrimental effects of NPIs on social equity and the situation of disadvantaged populations by highlighting the effect of NPIs on society's most vulnerable and advocating for measures to minimize the detrimental impact.

6.3.2 General quarantine

A general quarantine is not selective to the class of individuals, and for this reason, everyone enters and exits the quarantine at the same time. The easiest way to characterize general quarantine is to conceive of it as a time-dependent depression of β by a discount factor χ (where $0 \leq \chi \leq 1$). For a τ_q-day quarantine starting on t_q,

$$\beta(t) = \begin{cases} \beta\chi \text{ if } t \in [t_q, t_q + \tau_q], \\ \beta \text{ otherwise.} \end{cases} \tag{6.37}$$

This model is somewhat akin to the time-varying β models that utilize a step function, e.g., term-time forcing (on which see Subsection 7.3.3), except that we are modeling a single, nonrecurrent event. A welcome convenience of this approach is that it does not introduce a new compartment. Though this would ordinarily make analytical solutions rather more elusive than a simple transfer function to another compartment, it makes numerical solutions almost trivially easy, as we shall see in Computational Note 6.3.

Computational Note 6.3 Modeling the effect of different quarantine regimes

A SIR model can be adapted to take account of a quarantine period rather easily. Since we define our derivative function as a Python function, any Python construct operating on its parameters is fair game. This includes conditionals:

```
def deriv_with_quarantine(t, y, beta, gamma, chi, tau_q, t_q):
    S, I, R = y
```

```
    if t_q < t < t_q + tau_q:
        beta_eff = beta * chi
    else:
        beta_eff = beta

    dSdt = -beta_eff * S * I
    dIdt = beta_eff * S * I - gamma * I
    dRdt = gamma * I

    return dSdt, dIdt, dRdt
```

It is often quite practical to integrate over a range of parameters. `solve_ivp` is particularly useful for this purpose, because it returns a lot of parameters in a single object. Different parameters often lead to the function being evaluated at different points in time, and for this reason, they need to be plotted with their own times of evaluation. For the derivative above, we can obtain the times of evaluation (the property `.t` of the result object) and the corresponding result vector (the property `.t`).

A notebook implementing the contents of this Computational Note is available on the book's companion Github repository in the folder `/ch06/quarantine`.

Practice Note 6.6 The ethics of quarantines

Quarantines restrict fundamental human freedoms we recognize as inherent rights of every human being. As such, they are *prima facie* infringements on human rights and civil liberties. The American Medical Association's Code of Medical Ethics Opinion 8.4 concerning the ethical use of quarantine and isolation lays down some fundamental rules that could be considered good practice for all of public health [217]:

- Quarantine and isolation must be scientifically and ethically sound.
- The least restrictive means that are sufficient to control disease must be used.
- Quarantine and isolation must be exercised equitably, without bias against classes or groups of patients.

The power to quarantine individuals is perhaps the most extensive power over another human being available to the state without a judicial process. Like the criminal justice system, it is empowered to deprive people of their liberty temporarily, yet it is not subject to many of the safeguards that protect a defendant in a criminal trial. Unequal or discriminatory application of quarantines, or, indeed, even the perception of it, as Desclaux et al. [218] note, may vitiate a quarantine's perception as a measure for the public good. Complex social and economic contexts, as were evidenced both during the West African EBOV outbreak and, later, during the COVID-19 crisis (see for example Hesselman et al. [219] for energy poverty, Crawford and Waldman [220] for period poverty, Ahmed et al. [221] for race, Phillips et al. [222] and Sachdeva et al. [223] for the LGBTQ+ community), factor into these perceptions.

The quarantine power may be enforceable by the state's policing powers, but this is not the case indefinitely. Quarantines require civil cooperation, and such cooperation is easily lost if the measures appear biased, inequitable, or calculated to affect some groups more harshly than others. A general quarantine is an emergent and extraordinary option, and wherever possible, alternative mechanisms, such as circuit breakers (see Subsection 6.3.4), should be considered. There will always be a "next pandemic," and errors of judgment in the present can jeopardize adherence in the future.

6.3.3 Quarantines and healthcare capacity

As Fig. 6.6 demonstrates, different quarantine regimes have a profound effect on the long-term dynamics of the disease. Importantly, longer quarantine times slow down the spread and reduce the maxima of infections. The consequence is that appropriate quarantines can reduce the demand on healthcare services and prevent overloading.

This is instructively demonstrated by a model with contingent mortality. Let μ_T be the base mortality rate of the disease with treatment and μ_U the untreated mortality rate, both denominated in terms of deaths per person per day. Let c, furthermore, be the carrying capacity of the healthcare system, expressed as new infections per day.

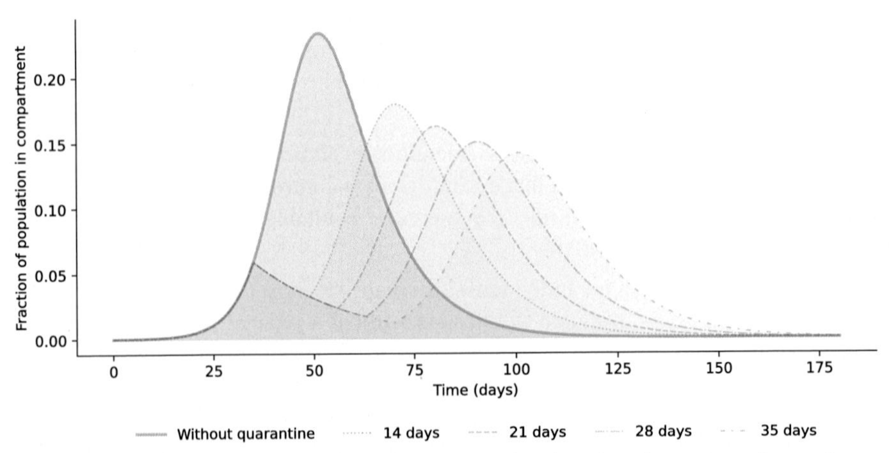

Figure 6.6 Results of a SIR model with quarantines of various lengths commencing on day 35 and a quarantine effectiveness of $\chi = 0.3$. The model was initialized with $\mathfrak{R}_0 = 2.5$, $\gamma = \frac{1}{8}$ and an initial infected population of 1×10^{-4}.

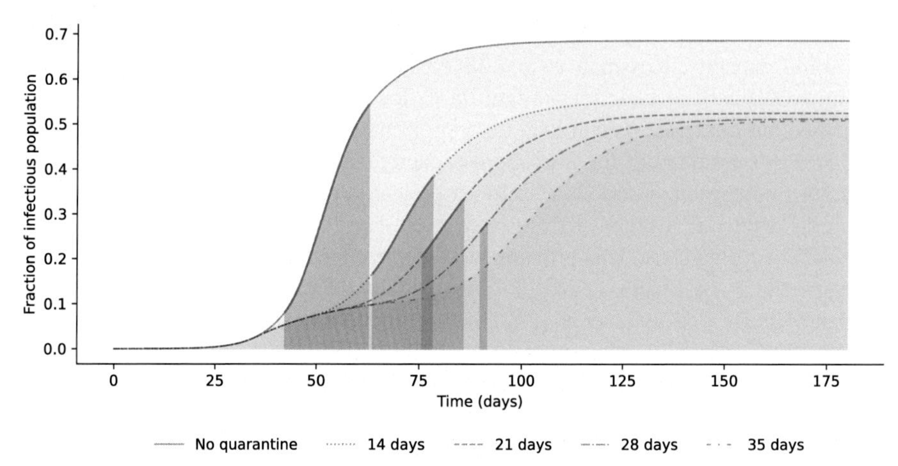

Figure 6.7 The effect of quarantines on capacity-dependent mortality regimes for a model identical to that in Fig. 6.6, with a carrying capacity $c = 0.15$, $\mu_T = 0.5$, and $\mu_U = 0.15$. Regimes where I exceeds c are marked and shaded in red.

A critical phenomenon occurs at $I = c$, at which point excess cases above c will receive lower level care, or no care at all. We may approximate this by separating treated and untreated mortality rates, where the first c patients at any given time experience the treated mortality rate μ_T and the remainder experiences the untreated mortality rate μ_U. This allows us to formulate overall illness-related mortality (i.e., ignoring natural mortality) as

$$\frac{dD}{dt} = (\underbrace{\mu_T \max(I, c)}_{\text{first } c \text{ patients}} + \underbrace{\mu_U \max(I - c, 0)}_{\text{overload}})I. \tag{6.38}$$

Fig. 6.7 highlights the effects of quarantines in preserving a healthcare system's ability to care for patients. Not only is the overall mortality lower due to the attenuated epidemic curve witnessed in Fig. 6.6, but the periods during which the healthcare system is beyond the critical point, and mortality occurs at the over-capacity regime are drastically shortened, until they are eliminated altogether in the case of the 35-day quarantine.

Practice Note 6.7 Mortality below, at, and above capacity

In much of the section above, we discussed mortality that sets in once healthcare capacity is exceeded. This is largely discontinuous: receiving care is largely a binary variable, splitting the population into two cohorts, whose clinical outcomes will be governed to a great extent by the difference between mortality with versus without treatment.

There is another and entirely distinct source of mortality that is related to hospital capacity. Rossman et al. [224] examined mortality rates of patients with COVID-19 in Israel depending on patient load, and found that under increased but moderate patient load (at approximately 60% of the Israeli Ministry of Health's estimate of the maximum capacity), mortality rates were 22.1% to 27.2% higher than in periods with lower patient loads. Strålin et al. [225] found similar results in Sweden with respect to 60-day all-cause mortality following COVID-19 diagnosis. This phenomenon has been observed in other infectious diseases (e.g., by Maciás et al. [226] in the context of dengue) and noninfectious phenomena (e.g., Kayiga et al. [227] in the context of obstetric care and Crandall et al. [228] in respect of trauma care).

This suggests that in addition to the critical phenomenon once $I = c$, there is a continuous subcritical phenomenon as I converges on the capacity constraint. As Rossman et al.'s study showed, this phenomenon appears at a rather disconcertingly early stage. Exploring the capacity envelope of the healthcare system in an epidemic should, if at all possible, be avoided. Even subcritical excess loads present a cost in terms of mortality, and recommendations on quarantine length should take this into account.

6.3.4 Circuit breakers

A circuit breaker is a form of generalized, short-term quarantine that is triggered by a sentinel variable, such as the number of new infections or hospitalized individuals. The term originates from a type of safety mechanism in stock exchanges, notably the New York Stock Exchange (NYSE), which automatically applies a trading halt when an indicator declines more than a certain preset value over a trading day. Circuit breakers were introduced during the COVID-19 pandemic in Singapore, and have been quite successful in attenuating the early dynamics of COVID-19 [229,230]. A key benefit of circuit breakers is that they are highly efficient: the effect of a circuit breaker policy is comparable, if not superior, to a longer general quarantine, without the economic and social costs of sustained quarantining.

Circuit breakers can be modeled most conveniently by using a time-dependent coefficient of transmission:

$$\beta(t) = \begin{cases} \beta_1 & \text{if circuit-breaker is active at } t, \\ \beta_0 & \text{otherwise.} \end{cases} \tag{6.39}$$

Circuit breakers are activated if a sentinel indicator reaches a particular threshold value. Where the key objective is preserving the healthcare capacity, the number of hospitalized cases is a good sentinel indicator, as it is quite easy to ascertain from hospital information systems. This was, indeed, the approach adopted in Singapore [229]. Though the example in Computational Note 6.4 uses the size of the infectious compartment as the sentinel variable, this figure is in practice much more difficult to ascertain.

Practice Note 6.8 Choosing a sentinel indicator

The circuit breaker strategy is only as good as its sentinel indicator. Good sentinel indicators meet the three As: ascertainability, availability and accuracy.

- Ascertainability means that the indicator can be calculated or at least very accurately estimated. The number of infectious individuals, for instance, is relatively hard to ascertain, especially if the pathogen is novel, its presentation is nonspecific and/or there is a nontrivial asymptomatic period.
- Availability means that the sentinel indicator should be accessible and updated frequently. Any indicator with a less than daily temporal resolution is generally not a suitable candidate. Hospital admissions data, perhaps the most widely available potential sentinel indicator in resource-rich settings, where electronic medical records (EMRs/EHRs) are common, are connected to a reporting system maintained by the public health authorities, and cases can be reported in near real time. In resource-constrained settings, including in the aftermath of natural disasters, communications may not be sufficient to support a circuit breaker strategy at all.
- Accuracy means that the indicator reasonably closely reflects reality. This can be rather challenging, especially in the context of a novel emerging pathogen. Often, a more sensitive but less specific indicator, such as the total number of presentations with any type of respiratory illness, might be preferable to a more specific but less sensitive metric, e.g., COVID-19 diagnoses.

Circuit breakers are powerful tools that limit the time spent in quarantine, while providing benefits comparable to a longer quarantine period. Quarantines have economic [231], social [232], and emotional [233,234] costs of which one must remain mindful. Circuit breakers have the potential to reduce the scale of these effects without losing the public health benefits of quarantines, but are strongly dependent on the availability of the right kind and quality of data to support it. It is not a "one size fits all" solution, and its relative success during COVID-19 must be weighed against its specific resource requirements.

Circuit breakers work best when their duration τ_q is relatively long compared to the mean infectious period. Singapore's two circuit breakers lasted a little less than two months each. Byrne et al. [235] estimated the median infectious period of COVID-19 as 6.5 to 9.5 days for asymptomatic cases and the median maximal infectious period as 18.1 days. Singapore's circuit breakers were thus approximately 3 times as long as the median maximal infectious period. The simulation in Computational Note 6.4 and Fig. 6.8 uses a quarantine period 1.75 times the mean infectious period, and shows a positive impact. The length of the quarantine period is bounded on the lower end by the maximal infectious period; any shorter and an infectious case may still transmit

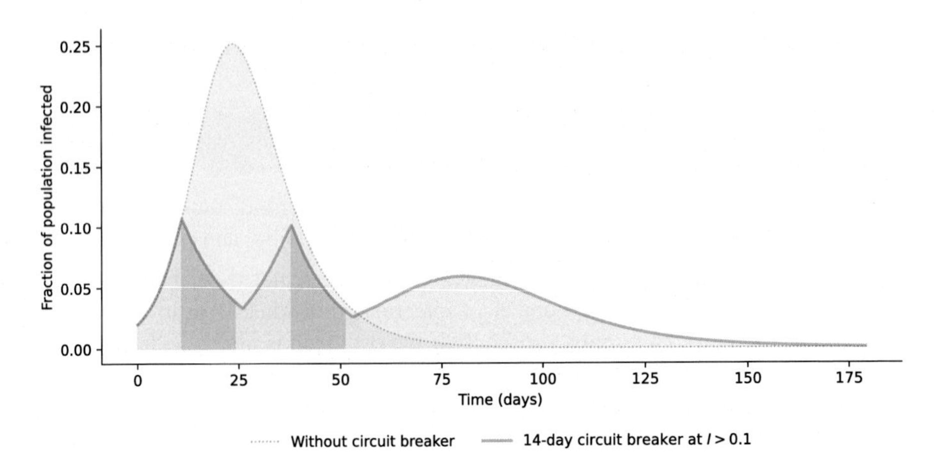

Figure 6.8 A model of a circuit breaker quarantine for a pathogen with $\mathfrak{R}_0 = 2.5$ and $\gamma = \frac{1}{8}$. The transmission coefficient is $\beta_0 = 0.3125$ outside the circuit breaker and $\beta_1 = 0.0625$ during the circuit breaker's activity. The circuit breaker is triggered at $I(t) = 0.1$ and lasts for 14 days.

disease when emerging from the circuit breaker quarantine. The minimum safe values of $\frac{\tau_q}{\tau}$ for circuit breakers have not been determined, but Singapore's example suggests that for a highly efficient (and restrictive) quarantine, a factor of 3 is appropriate. For lower effectiveness, higher time factors are likely to be necessary.

Computational Note 6.4 Iterative stateful evaluation

ODE integrators are powerful tools, but generally do not accommodate nonintegrated state variables too well. It is possible to model circuit breakers as delay differential equations with a conditional function for β, but in practice, computationally ascertaining the result is much easier by way of iterative evaluation.

Since a differential equation describes the state of change of a system, an initial value problem can be computationally solved by iterating over a span of time $t_0 \rightarrow t_n$ and in each step, applying the differential equation's terms to the previous step's results. We implement the circuit breaker through four main steps:

- an array that keeps track of whether the circuit breaker is active or not for each time step,
- a variable that retains the last time the circuit breaker was triggered,
- an expression that adjusts β, depending on whether the circuit breaker is in effect, and
- an expression that turns the circuit breaker on and off.

We shall take these in turn.

We initialize a two-dimensional array with a row for each day (given by the integration period t_span) and four columns, three for each compartment and one for the current state of the circuit breaker. The rationale for using an array rather than an expanding list is due to the memory benefits of changing values in an existing and predefined array over consistently reshaping the array, which would require the array to be re-mapped over memory.

We initialize the first row with the vector y_0, comprising $S(0)$, $I(0)$, and $R(0)$, and 0 to represent that we initialize the system with the circuit breaker off:

```
y = np.zeros(shape=(t_span, 4))

y[0, :] = [*y0, 0]
```

Thus $S(t)$, $I(t)$, and $R(t)$ correspond to y[t, 0], y[t, 1], and y[t, 2], respectively, whereas y[t, 3] gives the state of the circuit breaker. The destructuring assignment in [*y0, 0] is a quick and efficient way to create the initial state that is partly provided by an iterable (y0, for S, I and R) and partly by a concrete value (0, for the state of the circuit breaker).

Next, we initialize a variable to store the start of the circuit breaker:

```
circuit_breaker_start = None
```

The iterative execution will loop through each day, from 0 to t_span, and calculate the integral by adding the differential, evaluated at the previous time step, to the value at the previous time step:

```
for i in range(1, t_span):
    beta = beta_0 if y[i - 1, 3] == 0 else beta_1

    y[i, :3] = y[i - 1, :3] + (
        - beta * y[i - 1, 0] * y[i - 1, 1],
        beta * y[i - 1, 0] * y[i - 1, 1] - gamma * y[i - 1, 1],
        gamma * y[i - 1, 1]
    )
```

The step function for β is defined here, too: if the circuit breaker was off in the previous time step, $\beta = \beta_0$, otherwise $\beta = \beta_1$.

Finally, we need to perform some housekeeping in each step with respect to the state of the circuit breaker:

- If the circuit breaker is on and it has been more than τ_q days since the initiation of the circuit breaker, it is switched off.

- If the circuit breaker is off and the sentinel value (in this example, I) exceeds the threshold (defined as `threshold`), the circuit breaker is turned on. The time is noted by setting `circuit_breaker_start` equal to the current time.

All put together, the function reads as follows:

```
def model_with_circuit_breaker(t_span:int,
                               y0: tuple,
                               beta_0: float,
                               beta_1: float,
                               gamma: float,
                               tau_q:float,
                               threshold:float) -> np.array:

    y = np.zeros(shape=(t_span, 4))

    y[0, :] = [*y0, 0]

    circuit_breaker_start = None

    for i in range(1, t_span):
        beta = beta_0 if y[i - 1, 3]  == 0 else beta_1

        y[i, :3] = y[i - 1, :3] + (
            - beta * y[i - 1, 0] * y[i - 1, 1],
            beta * y[i - 1, 0] * y[i - 1, 1] - gamma * y[i - 1, 1],
            gamma * y[i - 1, 1]
        )

        if y[i - 1, 3] == 1:
            if i > circuit_breaker_start + tau_q:
                y[i, 3] = 0
            else:
                y[i, 3] = 1

        if y[i - 1, 3] == 0 and y[i, 1] > threshold:
            y[i, 3] = 1
            circuit_breaker_start = i

    return y
```

Fig. 6.8 shows the number of an infected population with a trigger value of $I \geq 0.1$, showing the way circuit breakers can flatten the epidemic curve, and thus preserve healthcare capacity.

> *A notebook implementing the contents of this Computational Note is available on the book's companion Github repository in the folder* `/ch06/circuit_ breaker`.

6.3.5 Quarantine of the infectious

As the name suggests, in the quarantine of the infectious, only individuals who are infectious are placed in quarantine. Often, this is not entirely accurate; typically, infectious individuals are selected by serological markers, and depending on the methodology of testing, an infectious individual may not immediately test positive. This is the case in particular where tests look for antibodies, which take a few days to emerge, rather than the antigen, which is typically present immediately, but may be harder to test for. A system of delay differential equations can characterize this dynamic as

$$
\begin{aligned}
\frac{dS}{dt} &= \mu - \beta SI - \mu S, \\
\frac{dI}{dt} &= \beta SI - \gamma I - \mu I - \int_{t-\tau}^{t-\tau_d} p_c I(u - \tau_d)\, du, \\
\frac{dQ}{dt} &= \int_{t-\tau}^{t-\tau_d} p_c I(u - \tau_d)\, du - \int_{t-(\tau+\tau_q)}^{t-(\tau_d+\tau_q} p_c I(v - (\tau_d + \tau_q))\, dv - \mu Q, \\
\frac{dR}{dt} &= \gamma I + \int_{t-(\tau+\tau_q)}^{t-(\tau_d+\tau_q)} p_c I(v - \tau_d - \tau_q)\, dv - \mu R,
\end{aligned}
\tag{6.40}
$$

where τ_q is the fixed duration of the quarantine, τ_d is the diagnostic or symptomatic delay (i.e., the time before an infectious individual becomes symptomatic and/or diagnosable, and thus amenable to capture by quarantine) and τ is, of course, the length of the entire infectious period. The capture of infectious individuals is described by the integral

$$
\int_{t-\tau}^{t-\tau_d} p_c I(u - \tau_d)\, du. \tag{6.41}
$$

In other words, at time t, quarantine can capture individuals who have contracted the infection no earlier than $t - \tau$ (since anyone who has obtained the infection earlier would by t be recovered) and no later than $t - \tau_d$ (since anyone who has obtained the infection later would not be symptomatic or have a detectable infection).

- Individuals are liable to capture throughout their symptomatic period, which is $\tau - \tau_d$ long. Their capture likelihood is the constant p_c.
- Quarantinees are released τ_q time after their quarantine. This is so regardless of their time of infection, or how far into the infection they were captured. Thus the number of individuals released on t equals the inflow into quarantine on $t - \tau_q$,

namely

$$\int_{t-(\tau+\tau_q)}^{t-(\tau_d+\tau_q)} p_c I(u-(\tau_d+\tau_q))\,du. \tag{6.42}$$

The benefit of the delay integro-differential formulation of the model is that it provides a stateful memory to specify quantities with respect to the past. Thus though a simple delay-differential formulation might give the quarantine system a "single bite of the cherry" (one chance to recognize and capture a symptomatic individual at τ_d days after the infection), this model of quarantine reflects the reality that individuals can in fact be captured throughout their infectious career. The computational approach outlined in Computational Note 6.2 can be applied to this model, *mutatis mutandis*.

The delay differential equation in (6.40) can be vastly simplified in many cases, most appropriately where quarantine is not of a fixed duration, but is rather contingent on biomarkers (testing negative) or the clinical course. Instead of using a delay, we simply use a transfer term that considers the mean rate of moving out of the quarantined compartment, which is the inverse of the mean time spent in quarantine, $\tau_{\overline{q}}$:

$$\begin{aligned}
\frac{dQ}{dt} &= \underbrace{p_c I}_{\text{capture}} - \underbrace{\frac{1}{\tau_{\overline{q}}} Q}_{\text{release}}, \\
\frac{dR}{dt} &= \underbrace{\gamma I}_{\text{recovered from } I} + \underbrace{\frac{1}{\tau_{\overline{q}}} Q}_{\text{released from } Q}.
\end{aligned} \tag{6.43}$$

The effect of quarantine on $\mathfrak{R}_t$ deserves mention. An analysis of the simplified model's next generation matrix (see Subsection 2.5.2) reveals that given a quarantine of identified infectious individuals with the capture probability p_c,

$$\mathfrak{R}_t = \frac{\beta}{\gamma + p_c}. \tag{6.44}$$

Solving the above for $\mathfrak{R}_t = 1$ reveals that the capture rate of a quarantine system must be larger than $\beta - \gamma$ if it is to control the infection. We may use this finding to understand the effect of lags. Let us consider a single patient, who experiences the illness for $\tau = \frac{1}{\gamma}$ days. On each of these, he stands a p_c chance of being captured and quarantined. For $\tau_d < \tau$, we can conceive of each of the days in τ_d to be a "missed opportunity" to detect him. Thus a delay of τ_d days reduces the effective value of p_c by $\frac{\tau_d}{\tau}$. This means that Eq. (6.44) now becomes

$$\mathfrak{R}_t = \frac{\beta}{\gamma(1 - \tau_d p_c)}. \tag{6.45}$$

The maximum value of capture is, of course, unity, representing perfect capture. Setting $p_c = 1$ and $\mathfrak{R}_t = 1$, we can solve for τ_d. This gives us an important finding: if

$\tau_d > \frac{\beta-\gamma}{\gamma}$ (under the assumption that $\frac{1}{\gamma} >> \tau_d$), then even perfect control of detected cases will be ineffective at curbing the disease, since the time to detection will suffice to provide enough secondary cases for the pathogen's survival.

Practice Note 6.9 Time and life

The consequence of this limitation of quarantine is that for diseases with a higher $\mathfrak{R}_0$ (in general), control through quarantine of the infectious becomes increasingly difficult. If we consider that the vast majority of infectious diseases rarely show initial symptoms for days (and even then, the symptoms may be non-specific), control through quarantine of the infectious becomes an increasingly tenuous proposition. For this reason, quarantine of the infectious is rarely a sufficient mainstay of public health response to an epidemic.

This is not to say quarantining the infectious is not an important measure. Together with general quarantine and other NPIs, it can fulfill an important role. Moreover, in many settings, infection is presumed rather than substantiated. This is the rationale of human and animal quarantines at ports of entry. This eliminates the potential risk of missing an infection for too long until it moves beyond control, but comes with serious economic and social costs. As in so many cases of public health, the best option is often a mix of judiciously applied options rather than a single policy.

6.3.6 Post-exposure quarantine

Post-exposure quarantine refers to quarantine models, where the quarantine process is triggered by an exposure event. Often, this is described not as quarantine but as asymptomatic medical isolation, but the essence is the same: individuals subject to post-exposure quarantine will be restricted to an isolation unit or their place of residence.

An interesting feature of this model is that one can be infectious and quarantined at the same time. Since this is not compatible with the fundamental properties of a compartmental model discussed in Chapter 2, we will consider quarantined status to supervene infection. We can do so safely if $\tau_q >> \tau$, as long as we assume that the quarantined do not take part in the infectious process, and therefore do not need to be considered in the mass action term. Given a latency period of τ_d days, the model may be specified as

$$\frac{dS}{dt} = -\underbrace{\beta SI}_{\text{mass action}},$$

$$\frac{dE}{dt} = \underbrace{\beta SI}_{\text{mass action}} - \underbrace{\tau_d^{-1}}_{\text{lapse of latency}},$$

$$\frac{dI}{dt} = \underbrace{(1-p_c)\tau_d^{-1}}_{\text{non-captured exposures}} - \underbrace{\gamma I}_{\text{recovery}}, \quad (6.46)$$

$$\frac{dQ}{dt} = \underbrace{p_c\tau_d^{-1}}_{\text{captured exposures}} - \underbrace{\tau_q^{-1}Q}_{\text{released from } Q},$$

$$\frac{dR}{dt} = \underbrace{\tau_q^{-1}}_{\text{released from } Q} + \underbrace{\gamma I}_{\text{recovered from } I}.$$

In this model, the exposed compartment's outflow ($\tau_d^{-1}E$) is partitioned into those who are captured (the fraction $^{\Delta}\nu$) and those who are not, and therefore go on to constitute the infectious compartment ($1 - p_c$).

6.3.7 *Shielding (quarantine of high-risk susceptibles)*

Shielding is a form of reverse quarantine, where a segment of the susceptible population is separated to protect them from infection. However, in addition to traditional self-quarantine, we have an intermediate group of "shielders," who are not at high risk but take special additional measures to avoid infection, because they are in frequent contact with vulnerable individuals. Thus we have three strata: the general population G, the shielders S, and the vulnerable V. The vulnerable are only in contact with shielders, whereas shielders are in contact with the vulnerable and the general population, albeit at a lower rate.

It may be useful to reutilize the WAIFW matrix concept from Eq. (3.1), for

$$\mathbf{b} = \begin{pmatrix} \beta_{G\to G} & \beta_{G\to S} & \beta_{G\to V} \\ \beta_{S\to G} & \beta_{S\to S} & \beta_{S\to V} \\ \beta_{V\to G} & \beta_{V\to S} & \beta_{V\to V} \end{pmatrix}. \quad (6.47)$$

Under the assumption of perfect shielding, there are no direct interactions between the general population and the vulnerable population, therefore $\beta_{G\to V}$ and $\beta_{V\to G}$ are both zero. We may describe the model as a classical SIR model with n subpopulations $\{G, S, V\}$, i.e.,

$$\frac{dS_i}{dt} = -S_i \sum_{j=1}^{n} \mathbf{b}_{i,j} I_j.$$

$$\frac{dI_i}{dt} = S_i \sum_{j=1}^{n} \mathbf{b}_{i,j} I_j - \gamma I_i, \quad (6.48)$$

$$\frac{dR_i}{dt} = \gamma I_i.$$

The overall impact of this is not only to reduce the mass action term βSI, but also to selectively protect a subset of susceptible individuals. The "cost" of shielding is the shielders' reduced ability to interact with others. Shielding is only effective if β_S is quite significantly lower than β_G [236]. Experience from the COVID-19 pandemic suggests that a $\frac{\beta_G}{\beta_S}$ of between 2.0 and 4.9 is most effective [236].

Practice Note 6.10 Stratified shielding

The example above approaches shielding as a monolithic concept, but that need not be so. Precision shielding or stratified shielding is an approach that balances the risk to the shielded individual with the extent of shielding [237]. The difficulty is that risk stratification is often quite individual, and creating a "permissible activities framework" for shielders depending on the shielded vulnerable person's individual degree of vulnerability is often quite complicated and not necessarily amenable to analytically sound assessment.

It is also important, however, to keep the economic aspects of shielding in mind. Shielding affects not only the vulnerable person but also the shielders. The ability of households to be able to sustain shielding over a prolonged period of time depends on economic and societal factors. Stratification may be imperfect, but even a rough stratification may reduce the overall economic burden on shielders, and thus, overall, benefit the vulnerable, too.

Temporal dynamics of epidemics

Epidemics in time

It is only in appearance that time is a river. It is rather a vast landscape and it is the eye of the beholder that moves.

Wilder, The Eighth Day, 1967 [238]

7.1 Equilibrium states and stability analysis

Equilibria—specifically, stable equilibria—are long-term destinies of dynamic systems. A system in a stable equilibrium will not subject to any external force or dislodge itself therefrom. As epidemiologists, we are interested in equilibria, because stable equilibria tell us when a system has attained stability or where it will, eventually, attain stability. Epidemics are "extraordinary events." The term *outbreak*, beloved of the popular media when commenting on epidemics, emphasizes that we are dealing with a phenomenon that goes counter to "business as usual." Stable equilibria are nothing more than mathematical descriptions of states, in which the system can settle again and attain a measure of normalcy.

A compartmental model is at equilibrium when the derivative of each compartment with respect to time is zero. For instance, the equilibrium of a SIR model is

$$\frac{dS}{dt} = \frac{dI}{dt} = \frac{dR}{dt} = 0. \tag{7.1}$$

We denote this state by $(S^\star, I^\star, R^\star)$. There are, in general, two possible equilibria:

1. the trivial equilibrium, which is more commonly known as the **disease-free equilibrium**, where $I^\star = 0$, and
2. the nontrivial or **endemic equilibrium**, in which $I^\star > 0$.

The stability of each of these, and their attraction, determines the long-term destiny of any infectious process. In much of this chapter, we will be concerned—directly or indirectly—with the way systems behave at, around, or towards these equilibria.

7.1.1 Case study: pandemics, epidemics, and endemics

The definition of a pandemic is relatively clear [239–241]. The distinction between epidemics and endemic processes, at least for us arriving at the matter from the perspective of quantitation, is much less so.

First, there is an unhelpful tendency to use the term "epidemic" to refer to any phenomenon that rapidly arises in time; from the more widely discussed obesity [242] and opioid epidemics to notions such as that of a selfie epidemic [243], which

Computational Modeling of Infectious Disease. https://doi.org/10.1016/B978-0-32-395389-4.00016-5

might perhaps sound closer to social phenomena like the medieval *Tanzwut* (dancing plague) [244,245] than actual epidemics. These share an important feature with actual epidemics: early near-exponential growth, but the resemblance ends there.

In mathematical and computational epidemiology, we are rather more careful with our words. Importantly, we distinguish between epidemics and endemics. The fundamental difference is that an endemic disease is at, or converging at, an equilibrium, whereas an epidemic is growing and moving away from an equilibrium state.

Definition 7.1 (Endemic vs. epidemic disease). A disease process is **endemic** if it is at an equilibrium ($\frac{dI}{dt} = \frac{dS}{dt} = 0$) and $I > 0$. A disease process can only be stable and endemic if a stable equilibrium $(S^\star, I^\star, R^\star)$ exists so that $\frac{dS}{dt} = \frac{dI}{dt} = \frac{dR}{dt} = 0$.

A disease process is an epidemic if $\frac{dI}{dt} > 0$ and $I > 0$. An epidemic is not an equilibrium state, and is therefore never stable.

The counterintuitive result of this is that a disease with a dozen cases might, in theory, be an epidemic, but a disease with hundreds of thousands cases a year might be "merely" endemic. The crucial point to recall is that endemicity versus epidemicity is not about absolute (or relative) numbers, but rather whether the state of the system occupies a stable endemic equilibrium point. It is perfectly possible, for instance, for a disease with relatively high $\mathfrak{R}_0$, to occupy an endemic equilibrium, where the infectious compartment houses thousands of individuals at a time, whereas for another disease, a hundred cases constitute no doubt an epidemic if the endemic equilibrium is at tens of cases. The popular understanding of epidemics as being primarily about numbers is misleading: more than anything, it is about rates and about the relationship to the equilibrium point.

7.1.2 Disease-free and endemic equilibria

A compartmental model may have two equilibria: the disease-free equilibrium and the endemic equilibrium.

Definition 7.2 (Disease-free equilibrium). The **disease-free equilibrium** (DFE) is an equilibrium state, where $I^\star = 0$. This equilibrium is always stable if $\mathfrak{R}_0 < 1$.

Definition 7.3 (Endemic equilibrium). The **endemic equilibrium** is an equilibrium state with a nonzero value of $I^\star$. This equilibrium is always unstable if $\mathfrak{R}_0 < 1$. It may be stable at $\mathfrak{R}_0 > 1$.

The two equilibria are, essentially, the alternative long-term destinies of pathogens: they become endemic or they disappear altogether (along with immunity to them). For this reason, we are deeply interested not only in where these equilibria occur, but also whether they are stable (i.e., whether the system, once it achieves that equilibrium, will remain in that state indefinitely unless influenced from the outside).

7.1.3 Identifying equilibria

A system is at an equilibrium when, as we have stated mathematically in Eq. (7.1), the derivative of all compartments in respect of time is zero. We thus identify possible equilibria by setting each of the terms of the model to zero, and solve for the variables. For a SIR model with equal births and deaths μ, we essentially obtain the following system of equations:

$$\begin{aligned} \mu - \beta S I - \mu S &= 0, \\ \beta S I - \gamma I - \mu I &= 0, \\ \gamma I - \mu R &= 0. \end{aligned} \tag{7.2}$$

This system has a trivial solution in $(1, 0, 0)$: if $S = 1$, then μ and μS will be equal. All other terms, which contain either I or R, will be zero. Consequently, we have an equilibrium at $(1, 0, 0)$, the disease-free equilibrium.

The endemic equilibrium can be found either symbolically or through some pen-and-paper mathematics. The second equation is a good place to start. Factoring out I yields,

$$I(\beta S - \gamma - \mu) = 0.$$

This is satisfied if either side of the product is zero.

- $I = 0$ is the disease-free equilibrium that we have already worked out.
- $\beta S - \gamma - \mu = 0$ can be rearranged so that $\beta S = \gamma + \mu$, and hence $S = \frac{\gamma+\mu}{\beta}$.

Notably, this is the inverse of $\mathfrak{R}_0$. Thus we have the value for $S^\star$ for the endemic equilibrium. Inserting this into the expression for I simplifies to $\frac{\mu(\mathfrak{R}_0-1)}{\beta}$. $R^\star$ is then obtained through $1 - S^\star - I^\star$. This gives the endemic equilibrium of a SIR model as

$$\begin{aligned} S^\star &= \mathfrak{R}_0^{-1}, \\ I^\star &= \frac{\mu(\mathfrak{R}_0 - 1)}{\beta}, \\ R^\star &= 1 - \mathfrak{R}_0^{-1} - \frac{\mu(\mathfrak{R}_0 - 1)}{\beta}. \end{aligned} \tag{7.3}$$

Computational Note 7.1 Symbolic identification of equilibria

SymPy is a very useful tool for the identification of equilibria, especially if we have a more complex system. The `solve` function solves a system of equations symbolically.

We initialize our symbols as

```
mu, beta, gamma, S, I, R = sympy.symbols("mu beta gamma S I R")
```

By definition, the `solve` function accepts either equations or zero-valued expressions. Since we are essentially setting each of the differentials to zero, this will be straightforward. The partition constraint $S+I+R=1$ can be expressed in the form $1-S-I-R$, or as an equality:

```
sympy.Equality(S + I + R, 1)
```

In an `Equality` object, the two arguments correspond to the left-hand side and the right-hand side of the equality, respectively. This gives us

```
sympy.solve((
    mu - beta * S * I - mu * S,
    beta * S * I - gamma * I - mu * I,
    gamma * I - mu * R,
    sympy.Equality(S + I + R, 1)
), [S, I, R])
```

This returns a list of tuples, which each represent a solution for the system of equations. Unfortunately, SymPy is not particularly strong at substituting subexpressions with variables, so the nontrivial endemic equilibrium will look somewhat different from what is documented in Eq. (7.3), although it is mathematically entirely equivalent of course.

A notebook implementing the contents of this Computational Note is available on the book's companion Github repository in the folder `/ch07/sir_ stability`.

7.1.4 Equilibrium stability analysis

We have now found equilibria, but we are interested in the stability of these equilibrium points. A system of ordinary differential equations at an equilibrium point may be unstable or asymptotically stable. To determine this, we ordinarily use the eigenvalues of the Jacobian matrix of the differential equation vis-a-vis each of its compartments.

Definition 7.4 (Jacobian matrix). For a system of n ordinary differential equations with n variables $x_1, x_2, ..., x_n$, let f_i be $\frac{dx_i}{dt}$ for all i in n. The **Jacobian matrix**, or Jacobian for short, is then defined as

$$\mathbf{J} = \begin{pmatrix} \frac{\partial f_1}{\partial x_1} & \cdots & \frac{\partial f_1}{\partial x_n} \\ \vdots & \ddots & \vdots \\ \frac{\partial f_n}{\partial x_1} & \cdots & \frac{\partial f_n}{\partial x_n} \end{pmatrix}. \tag{7.4}$$

The eigenvalues of the Jacobian evaluated at an equilibrium point $x^\star = x_1^\star, x_2^\star, ..., x_n^\star$ define the system's stability:

1. If all eigenvalues have negative real parts ($\text{Re}(\lambda_i) < 0$ for all i in n), the equilibrium is asymptotically stable.
2. If at least one eigenvalue has a positive real part ($\text{Re}(\lambda_i) > 0$ for at least one i in n), the equilibrium is unstable.
3. If all eigenvalues have zero or negative real parts, with at least one eigenvalue being zero, the Jacobian method does not work, and typically, a Lyapunov function needs to be constructed for equilibrium analysis.

Consider a simple SIR model, as described in Subsection 2.1.4, with a constant birth and death rate μ. The Jacobian of this system at the equilibrium point $(S^\star, I^\star, R^\star)$ is

$$\mathbf{J} = \begin{pmatrix} -\beta I^\star - \mu & -\beta S^\star & 0 \\ \beta I^\star & \beta S^\star - \mu - \gamma & 0 \\ 0 & \gamma & -\mu \end{pmatrix}. \tag{7.5}$$

If we substitute the disease-free equilibrium $(S^\star, I^\star, R^\star) = (1, 0, 0)$, the Jacobian assumes the following shape:

$$\mathbf{J}_{\text{DFE}} = \begin{pmatrix} -\beta - \mu & 0 & 0 \\ \beta & -\mu - \gamma & 0 \\ 0 & \gamma & -\mu \end{pmatrix}. \tag{7.6}$$

The eigenvalues of this Jacobian are $-\mu$ and $\beta - \mu - \gamma$. This equilibrium is stable if $-\mu < 0$ and $\beta - \mu - \gamma < 0$. That holds true if $\beta < \mu + \gamma$, or $\frac{\beta}{\mu+\gamma} < 1$. Since $\frac{\beta}{\mu+\gamma}$ is, of course, $\mathfrak{R}_0$, the disease-free equilibrium is obtained if and only if $\mathfrak{R}_0 < 1$, a result that should, in view of the previous, be no cause for any surprise.

If we evaluate the Jacobian at the endemic equilibrium, we get

$$\mathbf{J}_{\text{EE}} = \begin{pmatrix} -\mu(\mathfrak{R}_0 - 1) - \mu & -\gamma - \mu & 0 \\ \mu(\mathfrak{R}_0 - 1) & 0 & 0 \\ 0 & \gamma & -\mu \end{pmatrix}. \tag{7.7}$$

The eigenvalues associated with the endemic equilibrium are

$$\lambda_{2,3} = -\frac{\mu\mathfrak{R}_0}{2} \pm \frac{\sqrt{(\mu\mathfrak{R}_0)^2 - 4(\mu+\gamma)\mu(\mathfrak{R}_0 - 1)}}{2}. \tag{7.8}$$

Since $\frac{1}{\mu+\gamma}$ is the mean infectious period τ, and $\frac{1}{\mu(\mathfrak{R}_0-1)}$ is the mean age at sustaining infection (see Eq. (2.55)), the above may be simplified to

$$\lambda_{2,3} = -\frac{\mu\mathfrak{R}_0}{2} \pm \frac{\sqrt{(\mu\mathfrak{R}_0)^2 - \frac{4}{\tau A}}}{2}. \tag{7.9}$$

In line with Keeling and Rohani [39]'s approach, we can discount the $(\mu\mathfrak{R}_0)^2$ component, as this is typically a very small value; recall that the value of μ is the inverse

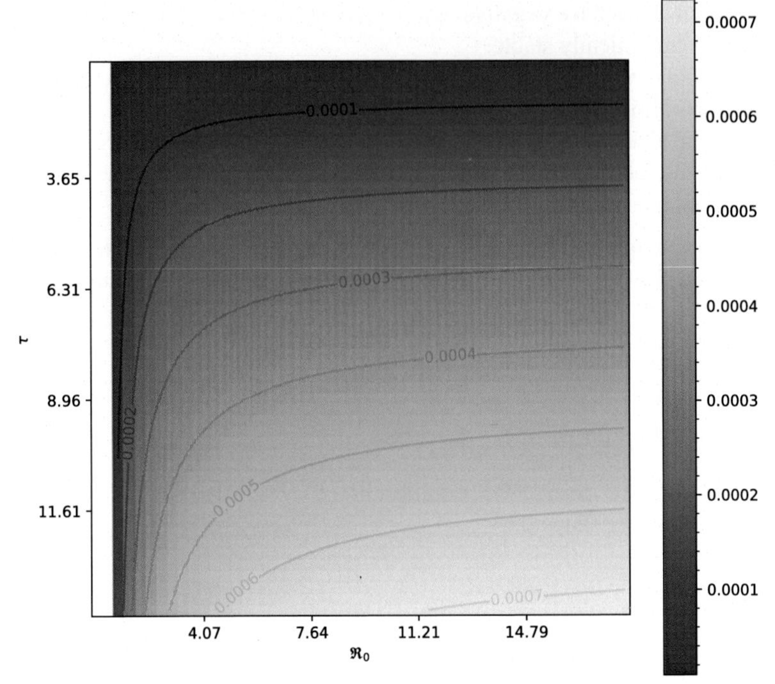

Figure 7.1 Size of the infectious compartment at the stable endemic equilibrium for a SIR model with $\mu = 0.02$ per annum as a function of τ and $\mathfrak{R}_0$. The white area denotes the space below $\mathfrak{R}_0 = 1$, where no stable endemic equilibrium can exist.

of the mean life expectancy of the population. This simplifies the equation to

$$\lambda_{2,3} \approx \frac{\mu \mathfrak{R}_0}{2} \pm \frac{i}{\sqrt{\tau A}}. \tag{7.10}$$

This expression is negative for all values of $\mathfrak{R}_0$ greater than one. Consequently, the endemic equilibria of SIR models are asymptotically stable if $\mathfrak{R}_0 > 1$. (See Fig. 7.1.)

Computational Note 7.2 Numerical equilibrium analysis

In practice, we are more often interested in the equilibria of a particular system with known parameters, rather than an abstract symbolic solution. For this reason, as long as we have a relatively good idea of the parameters of the system, we can obtain a solution and numerically evaluate it.

Consider the simple SIR system in Computational Note 7.1. We can obtain the values of the endemic equilibrium, given values of β, γ, and μ, by taking the nontrivial solution and substituting our known values.

```
def get_endemic_equilibrium(g, m, r0):
    return [solution.subs({"gamma": g,
                           "mu": m,
                           "beta": R0 * (g + m)})
                           for solution in solutions[1]]
```

With this function, we can calculate the endemic equilibrium for any arbitrary value of γ, μ, and $\mathfrak{R}_0$ (since $\mathfrak{R}_0 = \frac{\beta}{\gamma+\mu}$. However, we are also interested in whether the endemic equilibrium is going to be stable. For this, we once again construct the Jacobian:

```
jac = sympy.Matrix([
    mu - beta * S * I - mu * S,
    beta * S * I - gamma * I - mu * I,
    gamma * I - mu * R,
]).jacobian([S, I, R])
```

Recall that an equilibrium is asymptotically stable if and only if the real part of all of its eigenvalues are negative. We must, therefore, proceed in a few well-considered steps:

1. Substitute the equilibrium solution and the parameters for β, γ, and μ, into the Jacobian.
2. Obtain the eigenvalue. Since this is a purely numerical problem at this point, there are no benefits to using SymPy. SciPy's `linalg` subpackage provides the `eigvals` function to obtain the eigenvalues. To be able to do so, however, we will have to convert the substituted Jacobian into a NumPy `array` using `np.array(jac).astype(np.float64)`. The type casting into `float64` is necessary, because the `eigvals` function does not operate on object arrays.
3. We then take the real parts of each eigenvalue, using `.real` on the result. Since we are interested in whether all of the results are less than zero, we use `np.all(eigenvalues.real < 0)` to obtain a single truth value.

This would produce a tidy function that determines the endemic equilibrium, returns it, and returns a Boolean indicator of its stability:

```
def get_endemic_equilibrium(g: float,
                            m: float,
                            r0: float,
                            jac: sympy.Matrix) -> tuple:

    solutions = sympy.solve([
        mu - beta * S * I - mu * S,
        beta * S * I - gamma * I - mu * I,
        gamma * I - mu * R,
```

```
        sympy.Equality(S + I + R, 1)
    ], [S, I, R])

    endemic_sol = [solution.subs({"gamma": g,
                                  "mu": m,
                                  "beta": R0 * (g + m)})
                    for solution in solutions[1]]

    substituted_jacobian = jac.subs({"S": endemic_sol[0],
                                     "I": endemic_sol[1],
                                     "R": endemic_sol[2],
                                     "gamma" : g,
                                     "mu": m,
                                     "beta": R0 * (g + m)})

    eigenvalues
     = eigvals(np.array(substituted_jacobian).astype(np.float64))

    return *endemic_sol, np.all(eigenvalues.real < 0)
```

It makes sense to precalculate the Jacobian, and supply it to the function in case we wish to evaluate the function iteratively. Symbolic calculation of the Jacobian is relatively expensive, and since the symbolic result does not change by changing the parameters that govern it, we may calculate it once, then call it an arbitrary number of times.

The method discussed above can be applied to any model of any specification, although more complex models will be computationally more expensive. The iterative evaluation of equilibria, and their stability, at various points and for various parameters is a helpful tool in understanding how the parameters of an infectious disease govern its long-term destiny.

A notebook implementing the contents of this Computational Note is available on the book's companion Github repository in the folder `/ch07/sir_stability`.

7.1.5 Bifurcations and equilibria

As we have seen, the existence of equilibria and their stability depends on $\mathfrak{R}_0$. Bifurcation diagrams, such as that exhibited in Fig. 7.2 for a simple SIR model, show the critical point at $\mathfrak{R}_0 = 1$ in determining the equilibrium number of infecteds ($I^\star$):

- For $\mathfrak{R}_0 < 1$, the disease-free equilibrium (DFE) is stable at $I^\star = 0$.
- At $\mathfrak{R}_0 \geq 1$, the model experiences a transcritical bifurcation, and a new equilibrium emerges (the endemic equilibrium, EE). At this point, the DFE becomes unstable and the EE becomes stable. As $\mathfrak{R}_0$ increases, the equilibrium value of $I^\star$ increases.

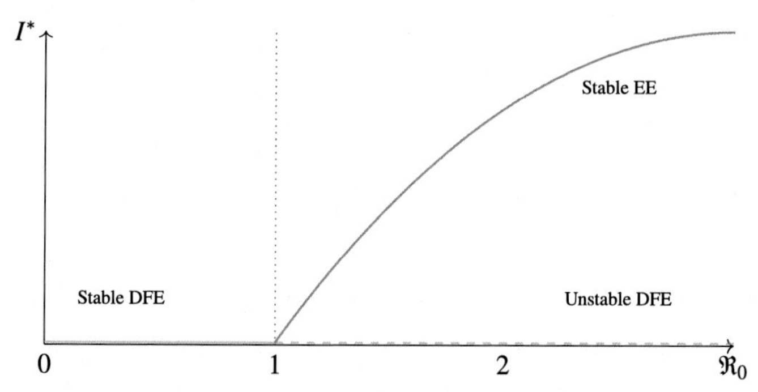

Figure 7.2 A compartmental model has a stable disease-free equilibrium (DFE) if $\mathfrak{R}_0 < 1$. Above $\mathfrak{R}_0 = 1$, the DFE is not stable, and the endemic equilibrium (EE) exists and is stable. This is an example of forward bifurcation.

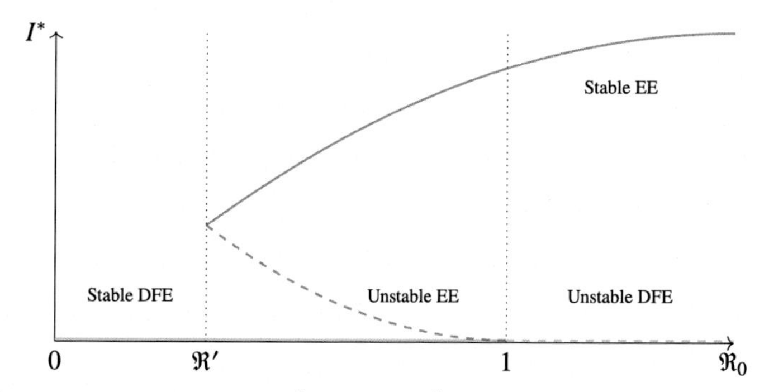

Figure 7.3 Backward bifurcation at $\mathfrak{R}'$. Between $\mathfrak{R}'$ and 1, a stable DFE (solid red line), a stable (upper) EE (solid blue line) and an unstable EE (dashed blue line) coexist. Consequently, only the necessary condition for the pathogen's elimination is met at $\mathfrak{R}_0 = 1$; the sufficient condition is only met at $\mathfrak{R}_0 = \mathfrak{R}'$.

This does not, however, hold universally. There exists an alternative pattern of bifurcation, known as backward (or subcritical) bifurcation, which is epidemiologically interesting for us, because, in such cases, reduction of a pathogen's $\mathfrak{R}_0$ might not be sufficient for the control of the disease. In backward bifurcation, the situation is the same for $\mathfrak{R}_0 > 1$, but quite significantly different below 1: for a part $[\mathfrak{R}', 1]$, three equilibria coexist: a stable DFE and two endemic equilibria, of which the upper equilibrium is stable and the lower equilibrium is unstable. This is illustrated by Fig. 7.3.

The most robust mathematical formulation of determining if backward bifurcation occurs is the center manifold theory proposed by Castillo-Chavez and Song [246], which allows the quantitative determination of whether a system would exhibit backward bifurcation. Qualitatively speaking, backward bifurcation occurs in cases where a pathogen has some way to persist in the population, such as

- waning immunity, whether vaccine-induced or post-infectious [247],
- imperfectly protective vaccines [247],
- reimportation of pathogenic populations [246], and

- vector populations [248].

The importance of backward bifurcation is particularly significant for diseases that can persist in a latent infectious state, such as tuberculosis [246]. It is worth noting, as Greenhalgh and Griffiths [249] has done, that more complex bifurcations can exist, with more than two subcritical endemic equilibria.

7.1.6 Equilibria of SEIR models

The disease-free equilibrium of a SEIR model as laid out in Eq. (2.43) is, of course $(1, 0, 0, 0)$. We are rather more interested in the endemic equilibrium, which is given, for an equal birth and death rate μ, as

$$\begin{aligned} S^\star &= \frac{(\mu+\gamma)(\mu+\sigma)}{\sigma\beta} = \mathfrak{R}_0^{-1}, \\ E^\star &= \frac{(\mathfrak{R}_0 - 1)\mu(\mu+\gamma)}{\sigma\beta}, \\ I^\star &= \frac{(\mathfrak{R}_0 - 1)\mu}{\beta}, \\ R^\star &= 1 - \mathfrak{R}_0^{-1} - \frac{(\mathfrak{R}_0 - 1)\mu(\mu+\gamma)}{\sigma\beta} - \frac{(\mathfrak{R}_0 - 1)\mu}{\beta}. \end{aligned} \tag{7.11}$$

The eigenvalues of the Jacobian for the SEIR model are somewhat challenging to identify. However, subject to some simplifications [250], it can be evaluated at the equilibrium point. The eigenvalues are negative if $\mathfrak{R}_0 > 1$. Thus similarly to SIR models (the equilibria of which we have discussed in Subsection 7.1.4), the endemic equilibrium of a SEIR model is stable at $\mathfrak{R}_0 > 1$. The inclusion of the exposed period means that perturbations to a SEIR model will however have a longer period of the decaying oscillations as the system converges on the endemic equilibrium.

Computational Note 7.3 Symbolic equilibrium analysis of SEIR models

The eigenvalues of complex models make it often quite difficult to discern information about stability. Thankfully, symbolic manipulation can help us quite considerably.

Given the SEIR model

$$\begin{aligned} \frac{dS}{dt} &= \mu + (\beta I + \mu)S, \\ \frac{dE}{dt} &= \beta SI - (\sigma+\mu)E, \\ \frac{dI}{dt} &= \sigma E - (\gamma+\mu)I, \end{aligned} \tag{7.12}$$

$$\frac{dR}{dt} = \gamma I - \mu R.$$

First, we obtain the equilibrium solutions of the system:

```
solutions = sympy.solve([
    mu - (beta * I + mu) * S,
    beta * S * I - (sigma + mu) * E,
    sigma * E - (gamma + mu) * I,
    gamma * I - mu * R,
    sympy.Equality(S + E + I + R, 1)
], [S, E, I, R])
```

This yields the rather unsurprising DFE solution of $(1, 0, 0, 0)$, which is stable. It also yields the solutions described in Eq. (7.11), in a somewhat unwieldier form. Thankfully, we will not have to directly engage with it. Next, we obtain the Jacobian of the system:

```
jac = sympy.Matrix([
    mu - (beta * I + mu) * S,
    beta * S * I - (sigma + mu) * E,
    sigma * E - (gamma + mu) * I,
    gamma * I - mu * R,
]).jacobian([S, E, I, R])
```

The Jacobian of the above is

$$\mathbf{J} = \begin{pmatrix} -I\beta - \mu & 0 & -S\beta & 0 \\ I\beta & -\mu - \sigma & S\beta & 0 \\ 0 & \sigma & -\gamma - \mu & 0 \\ 0 & 0 & \gamma & -\mu \end{pmatrix}. \tag{7.13}$$

We replace S and I with their equilibrium solutions from the endemic equilibrium (the second solution in the `solutions` object):

```
substituted_jacobian = jac.subs({"I": solutions[1][2],
                                 "S": solutions[1][0]})
```

This gives us the Jacobian

$$\mathbf{J} = \begin{pmatrix} -\mu - \frac{\mu(\beta\sigma - \gamma\mu - \gamma\sigma - \mu^2 - \mu\sigma)}{(\gamma+\mu)(\mu+\sigma)} & 0 & \frac{-(\gamma+\mu)(\mu+\sigma)}{\sigma} & 0 \\ \frac{\mu(\beta\sigma - \gamma\mu - \gamma\sigma - \mu^2 - \mu\sigma)}{(\gamma+\mu)(\mu+\sigma)} & -\mu - \sigma & \frac{(\gamma+\mu)(\mu+\sigma)}{\sigma} & 0 \\ 0 & \sigma & -\gamma - \mu & 0 \\ 0 & 0 & \gamma & -\mu \end{pmatrix}. \tag{7.14}$$

This has four eigenvalues, all of which are fairly complex. Fortunately, we do not need to evaluate or even see them. We are not interested in them, but rather only in whether, and how, they fulfill the Jacobian equilibrium criteria. We obtain a list of the real parts of the eigenvalues as

```
real_parts
 = [sympy.re(i) for i in substituted_jacobian.eigenvalues()]
```

The system is asymptotically stable if the real parts of all eigenvalues are negative. Of the four eigenvalues, the fourth eigenvalue has the real part $-\mu$, which means it is negative by default. At this point, we could theoretically solve for a neat analytical result, by setting each of the first three eigenvalues to zero, then solving for values of $\mathfrak{R}_0$, β, σ, and μ so that all eigenvalues have zero-valued real parts, and $\mathfrak{R}_0 > 1$. In practice, a symbolic analytical solution for this is not obtainable (as Keeling and Rohani [39] note, albeit arriving to the issue from a different angle). Symbolic computation is powerful, but often, highly complex dynamical systems are not amenable to analytical solutions. This is, then, a highly instructive failure: it shows that in many cases, a neat symbolic solution is not necessarily obtainable through computer algebra, and numerical stability analysis often proves much more suitable if values or approximations are known for some of the parameters.

A notebook implementing the contents of this Computational Note is available on the book's companion Github repository in the folder `/ch07/seir_ stability`.

7.1.7 Equilibria of SIS models

Models that do not engender immunity (or for which immunity upon recovery is short-lived) are best approximated by a SIS model (see Subsubsection 2.3.2.2). We obtain the endemic equilibrium of a SIS model by first expressing $\frac{dS}{dt}$ in Equation 2.39 in terms of I. Since there are only two compartments, the partition property (see Definition 2.4) implies that anyone not in I is in S, hence $S = 1 - I$. Consequently,

$$\frac{dI}{dt} = \beta I(1 - I) - \gamma I. \tag{7.15}$$

Factoring out β and replacing $\frac{\gamma}{\beta} = \mathfrak{R}_0^{-1}$, we get

$$\frac{dI}{dt} = \beta I(1 - \mathfrak{R}_0^{-1} - 1). \tag{7.16}$$

Setting Eq. (7.16) to zero and solving for I gives $I^\star = 1 - \frac{1}{\mathfrak{R}_0}$ and $S^\star = \frac{1}{\mathfrak{R}_0}$. The complementarity of these two results is, of course, the necessary corollary of $S^\star + I^\star = 1$. This equilibrium value is stable for $\mathfrak{R}_0 > 1$.

7.1.8 Equilibria of SIRS models

SIRS models differ from the traditional SIR model in that immunity wanes, but it is also distinct from the SIS model in that immunity exists in the first place. It is thus an intermediate of the two. Solving the SIRS model specified as

$$\begin{aligned}\frac{dS}{dt} &= \mu - (\beta I + \mu)S + \omega R,\\ \frac{dI}{dt} &= \beta S I - (\gamma + \mu)I,\\ \frac{dR}{dt} &= \gamma I - (\omega + \mu)R,\end{aligned} \tag{7.17}$$

gives us the trivial solution of the DFE $(1, 0, 0)$, and the endemic equilibrium at

$$\begin{aligned}S^\star &= \mathfrak{R}_0^{-1},\\ I^\star &= \frac{(\mu+\omega)(\beta-\gamma-\mu)}{\beta(\gamma+\mu+\omega)},\\ R^\star &= \frac{\gamma(\beta-\gamma-\mu)}{\beta(\gamma+\beta+\omega)}.\end{aligned} \tag{7.18}$$

This gives us the Jacobian at the endemic equilibrium as

$$\mathbf{J}_{\text{EE}} = \begin{pmatrix} -\mu - \frac{(\mu+\omega)(\beta-\gamma-\mu)}{\gamma+\mu+\omega} & -\gamma-\mu & \omega \\ \frac{(\mu+\omega)(\beta-\gamma-\mu)}{\gamma+\mu+\omega} & 0 & 0 \\ 0 & \gamma & -\mu-\omega \end{pmatrix}. \tag{7.19}$$

We may reason about the stability of this system using the Routh–Hurwitz criteria. Recall that the eigenvalues of $\mathbf{J}_{\text{EE}}$ would be the set of all values λ so that

$$(\mathbf{J}_{\text{EE}} - \lambda\mathbf{I})v = 0, \tag{7.20}$$

where $\mathbf{I}$ is the identity matrix, and v is the eigenvector scaled by λ so that $\mathbf{J}_{\text{EE}}v = \lambda v$.

We begin by taking the characteristic polynomial for $\mathbf{J}_{\text{EE}}$. For a 3×3 matrix, the characteristic polynomial is

$$-\lambda^3 + \text{Tr}(\mathbf{J}_{\text{EE}})\lambda^2 - \frac{1}{2}(\text{Tr}(\mathbf{J}_{\text{EE}})^2 - \text{Tr}(\mathbf{J}_{\text{EE}}^2))\lambda + \det(\mathbf{J}_{\text{EE}}). \tag{7.21}$$

The trace of $\mathbf{J}_{\text{EE}}$ is

$$\text{Tr}(\mathbf{J}_{\text{EE}}) = -2\mu - \omega - \frac{(\mu+\omega)(\beta-\gamma-\mu)}{\gamma+\mu+\omega}. \tag{7.22}$$

This gives us the coefficients of the characteristic polynomial:

$$\begin{aligned} a_1 &= \frac{\beta\mu + \beta\omega + \gamma\mu + \mu^2 + 2\mu\omega + \omega^2}{\gamma + \mu + \omega}, \\ a_2 &= \frac{\beta\gamma\mu + \beta\gamma\omega + 2\beta\mu^2 + 3\beta\mu\omega + \beta\omega^2 - \gamma^2\mu - \gamma^2\omega - 2\gamma\mu^2 - 3\gamma\mu\omega - \gamma\omega^2 - \mu^3 - \mu^2\omega}{\gamma + \mu + \omega}, \\ a_3 &= \beta\mu^2 - \beta\mu\omega - \gamma\mu^2 - \gamma\mu\omega - \mu^3 - \mu^2\omega. \end{aligned} \tag{7.23}$$

A system with a third-order characteristic polynomial is Hurwitz stable if

1. a_1, a_2, and a_3 are all positive, and
2. $a_1 a_2 > a_3$.

Though this looks rather daunting symbolically, it can be numerically ascertained with ease if the parameters are known. The benefit of a Routh–Hurwitz stability analysis is that unlike the method of equilibria from eigenvalues, it can be applied to problems that otherwise would not be solvable. Moreover, Routh–Hurwitz stability theory generalizes conveniently to arbitrary higher degrees, i.e., more complex models.

It is worth noting that the reversion to susceptibility results in small perturbations around the equilibrium. As Keeling and Rohani [39] note, this dynamic has been used to explain the longer-term periodicities of infectious disease, most famously by Grassly, Fraser, and Garnett [70]'s model of the decadal cycle of syphilis incidence. The oscillatory period is given by

$$P = \frac{4\pi}{4(\mathfrak{R}_0 - 1)(\gamma + \mu)(\omega + \mu) - (\omega + \mu - A^{-1})^2}. \tag{7.24}$$

A number of terms in this equation relate to biologically significant quantities. A, of course, is the mean age at sustaining infection, which is calculated by

$$\frac{\mu + \gamma + \omega}{(\mu + \omega)(\beta - \gamma - \mu)}.$$

Significantly, $I^\star = \frac{1}{\beta A}$—infections are at an endemic equilibrium point at a level given by the inverse of the transmission rate and the mean age of sustaining infection. The biological significance is not only that infections with a higher coefficient of transmission will result in a higher proportion of infectious individuals at the endemic equilibrium, but also that all things being equal, sustaining an infection at a younger age results in a higher disease burden. This, indeed, is the principal idea behind childhood vaccination: since $I^\star \propto \frac{1}{A}$, raising A will lower $I^\star$, and thus reduce the overall disease burden.

7.2 Seasonality and periodicity in infectious diseases

Epidemics take place in time. The dynamics that we have examined in previous chapters inherently assumed a temporal element. The fact that we use differential equations means we are more concerned, in general, with change over time, rather than particular values at particular points in time. However, this should not keep us from realizing the time-dependent dynamics of epidemics. For this reason, we often represent various epidemic parameters, such as the incidence of a particular disease, as a time series.

Definition 7.5 (Time series). A **time series** is an array of one or more variables indexed by a discrete unit of time. We refer to the time series $X = (X_t : t \in T)$, and to each X_t as an element or a data point in the time series. T is called the time domain of the time series. Typically, we treat t as zero-indexed, i.e., t_0 is the first value in a time series.

It is common to visualize time series of epidemic processes, particularly where a single outbreak is concerned, in a curve of the number of cases by the time of onset or time of confirmed diagnosis. This is frequently referred to as an epidemic curve.

Definition 7.6 (Epidemic curves). An **epidemic curve** (often called the **epidemiological curve** or **epi curve** for short) is a visual representation of incidence over time. It is a representation of an epidemic process in the time-amplitude space. Commonly, incidence is credited either with respect to the date of onset, i.e., each date shows the number of patients with onsets on a particular date, or the date of diagnosis/recognition.

An early finding of temporal analysis of epidemic dynamics relates to the symmetric nature of epidemic curves for outbreaks, known as Farr's law. The developments that have led to this finding—and with that, the first systematic study of the temporal dynamics of epidemics—are described in Subsection 7.2.1.

7.2.1 Case study: Farr's Law and smallpox in Britain

Dr. William Farr (1807–1883) was undoubtedly one of the heroic forerunners of the field that, over the centuries, grew to become modern epidemiology. His remarkable findings are the product not only of an agile mind, but also good fortune. When Britain embarked upon the first modern census, the United Kingdom Census of 1841, Farr obtained a position with the General Register Office, the organization newly established in 1837 to manage the registration of births, deaths, and marriages.

A physician who opened a (not very successful) practice in the 1830s, Farr's main interest was in statistics. This fit very well with the agenda of his supporters, the social reformer Edwin Chadwick and the Registrar General (the head of the General Register Office), Thomas Henry Lister, who saw the census as a way to illuminate health disparities in Britain, and thus bolster a socially progressive agenda. Farr used this support to create the first comprehensive tabulation of causes of death. As a side effect, he stumbled upon what became known as Farr's law. In an 1840 contribution to the Second Annual Report of the Registrar General of Births, Deaths, and Marriages in England, he comments on a smallpox epidemic, asserting that "the small-pox increases at an accelerated and then a retarded rate; that it declines first at a slightly accelerated, and at a rapidly accelerated, and lastly at a retarded rate, until the disease attains the minimum intensity, and remains stationary" [251].

Epidemics are processes in space *and* time. Their temporal course often takes a characteristic shape, which we can mathematically exploit. Farr's law suggests that case counts are typically normally distributed around a peak value, with epidemics disappearing as fast as they appear. Indeed, this is often a perplexing phenomenon to behold: a raging outbreak can disappear with a rapidity that belies its seriousness, as if a switch had been flipped. Modern computational approaches [252] and clinical experience both confirm this phenomenon. The 2013–2016 West African Ebola outbreak is a good example: the epidemic struck Liberia in mid-May 2014, peaked in October of the same year, and by June 2015, it was nearly completely gone, just as suddenly as it appeared.

This section deals with the use of time series methods to understand epidemic processes.

7.2.2 Seasonality and decomposition

A time series may exhibit various patterns that we can usefully analyze. We are, more than anything, interested in two types of patterns:

1. **recurrent relationships**, which includes cycles, periodicities, and other patterns in which peaks or troughs occur at relatively constant intervals, and
2. **trends**, which are typically aperiodic changes over a long time.

We may thus express a time series as the sum of a seasonal (recurrent) process S_t, a nonseasonal (trend) process T_t and a residual R_t:

$$X_t = S_t + T_t + R_t. \tag{7.25}$$

This is known as an additive composition. Alternatively, the composition may be multiplicative, i.e.,

$$X_t = S_t \times T_t \times R_t. \tag{7.26}$$

Time series decomposition is essentially the challenge of identifying S_t and T_t for all $t \in [t_{\text{min}}; t_{\text{max}}]$ so as to minimize $\sum_{i=t_{\text{min}}}^{t_{\text{max}}} R_i$. When decomposing time series, additive decomposition is appropriate, where we may safely assume that the seasonal component is equal every year. In all other cases, multiplicative decomposition is the better choice. For instance, the number of hospital admissions for frostbite and hypothermia is higher in the winter than in the summer, but how much higher it varies from year to year.

There are numerous mathematical methods to accomplish time series decomposition, with the most frequently used methods being X11 decomposition [253], the LOESS-based STL algorithm [254] and seasonal adjustment methods [255]. We shall

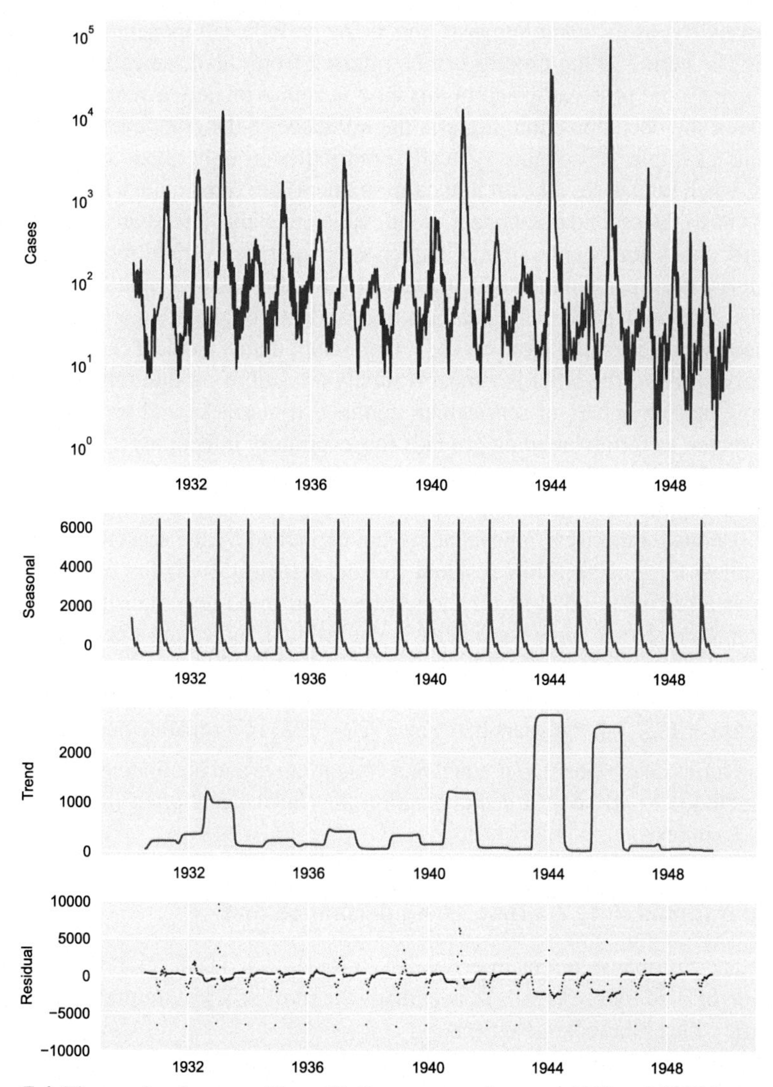

Figure 7.4 Time series decomposition of influenza cases between 1920 and 1952 into a seasonal and trend component using an additive model, with residuals displayed at the bottom. Based on data from the U.S. Nationally Notifiable Diseases Surveillance System (NNDSS), maintained by the CDC. Data accessed from the University of Pittsburgh's Project Tycho.

however be primarily concerned with the results, rather than the ways to arrive at it (since the latter is helpfully catered for by a large number of Python packages).

Fig. 7.4 shows a time series decomposition of influenza incidence in the United States between 1920 and 1952. Analyzing a decomposition plot is a mainstay of understanding a composite periodical process. The breakdown in Fig. 7.4 is the standard format of presenting the results of time series decompositions:

1. **Seasonality**: as the name suggests, this is the periodic component of decomposition. The period of the process can be inferred from the distance between peaks of the periodic process, which in this case is approximately a year. The absolute value of the decomposition suggests the influence of the component in question. In this example, seasonality typically contributes to only about 15,000 cases a year, when total cases (shown in the top segment) are often quite a bit larger. This leads us to the second component, trend, which explains interyear variability.
2. **Trend**: this component explains longer-term interyear variability. For instance, there is a clear peak in the years between 1928 and 1930 and smaller peaks in the 1940s. The Great Depression, malnutrition, and the congregation of large numbers of people in makeshift "Hoovervilles" with scant, if any, medical care explains the former, whereas the latter is almost certainly a result of prolific influenzal spread among large numbers of servicemen confined in barracks and transport vessels that served as fertile breeding grounds for respiratory infections.
3. **Residual**: it is common to plot the residual differently from the trend and seasonal components to make explicit the nature of the residual as the "error term." The residual comprises values that are not explained by the seasonal or the term variables. In general, a low residual (no more than 25% of the total value) indicates a relatively good fit for the seasonal and trend decompositions. It is not uncommmon to see some periodicity in the residual, indicating a periodic process that is absent or excessively present in some years. In that case, it may be worthwhile to look at time segments where the residual appears to be periodic (e.g., in the case of Fig. 7.4, the years between 1920–1928) in a separate decomposition.

A time series decomposition is equal parts quantitative and qualitative. It uses quantitative methods to arrive at a decomposition that however is often interpreted in a qualitative context.

Computational Note 7.4 Time series decomposition

There are multiple solutions to choose from for time series decomposition. The solution offered by `statsmodels` is perhaps the most efficient, returning each of the time series as a separate object.

`statsmodels` assumes that the input is a Pandas object with a `DateTimeIndex` or a similar time-index. If this is not the case, data points must be equally spaced, and the period between them must be supplied to the function using the `period` keyword.

First, we import `statsmodels`:

```
import statsmodels.api as sm
```

Next, we obtain our data from Project Tycho and import it our time series into a data frame, which, for convenience's sake, we shall call `df`. Project Tycho is a standardized interface to a wide range of disease surveillance data, including all of the National Notifiable Disease Surveillance System (NNDSS) in the United States. It can be accessed using an API key, which is how we will be obtaining the data. The project's website, tycho.pitt.edu, provides more information.

We begin by encoding our query parameters into a URL. It is a common practice to save API ketys as an environment variable, and obtain it using the `os.getenv()` function. This ensures we do not disclose or accidentally commit our API key to a repository. Because query items may contain strings that have spaces in them (e.g., to specify the United States as the geography of the query, we write `CountryName=UNITED STATES OF AMERICA`), we need to replace these with their "percent encoded" value to form a well-formed URL. The percent encoded value of a whitespace character is `%20`, so we replace all spaces in our initial query with `%20`:

```
flu_url = "https://www.tycho.pitt.edu/api/query?apikey="
        + os.getenv("TYCHO_APIKEY") + "&ConditionName
        =Influenza&CountryName=UNITED STATES OF AMERICA
         &PeriodStartDate>=1930-01-01
         &PeriodEndDate<=1950-01-01".replace(" ", "%20")
```

We now pass this URL on to Pandas, and perform some aggregations. In particular, we group by period end date and sum up values, and rename the columns to fit our specifications:

```
df=pd.read_csv(flu_url,
               low_memory=False)[["PeriodEndDate","CountValue"]].\
                                     groupby(["PeriodEndDate"]).\
                                     sum().\
                                     reset_index().\
                                     rename(columns={
                                        "PeriodEndDate":"Date",
                                        "CountValue":"Cases"})
```

In general, the best input for `statsmodels` is a datetime-indexed Pandas dataframe. In most cases, incidence data is of a common columnar format:

```
        Date        Cases
0       1919-11-01  344
1       1919-11-08  326
2       1919-11-15  312
3       1919-11-29  278
4       1919-12-06  293
...
```

To convert the `Date` column into a datetime index, we may do so using `pd.to_datetime(df.Date)`, then assign it as an index using `pd.set_index(df.Date)`.

We can confirm that we have a datetime-indexed data frame by calling the index object:

```
DatetimeIndex(['1919-11-08', '1919-11-29',
        ...
        '1941-02-01', '1941-02-01'],
        dtype='datetime64[ns]',name='Date',length=221574,freq=None)
```

We now go on to decompose this time series:

```
decomposition = sm.tsa.seasonal_decompose(df)
```

By default, `seasonal_decompose()` assumes an additive decomposition. The `model` parameter can be set to `multiplicative` for a multiplicative decomposition, e.g.,

```
decomposition = sm.tsa.seasonal_decompose(df,model="multiplicative")
```

The result is a `DecomposeResult` object. This comprises a number of attributes, each containing a component of the time series:

- `observed` contains the source time series,
- `seasonal` contains the seasonality component,
- `trend` contains the trend component, and
- `resid` contains the residuals.

We will discuss ways to represent the decomposition in a manner similar to Fig. 7.4 in Computational Note 7.5.

A notebook implementing the contents of this Computational Note is available on the book's companion Github repository in the folder `/ch07/ts_decomposition/`.

7.2.3 Perennial processes

A perennial plant is one which blooms every year (as opposed to an annual plant, which completes its entire life cycle in a single year). The revolution of the Earth around the Sun mediates a broad range of infectious disease parameters. For instance, cold weather in winter causes vasoconstriction, and thus a less efficient immune response, and therefore a greater liability to contract respiratory illnesses. Less obviously, the greater degree of social contacts during winter, occasioned by our very human aversion to being out in the cold, means that certain pathogens, typically respiratory viruses, can take hold of larger populations. Perenniality is the most typical form of periodic recurrence of human pathogens, and it therefore merits our special attention.

Fig. 7.4 shows the incidence of influenza in the United States between 1920 and 1952, decomposed into additive seasonal and trend components. This is an example of perennial phenomena with a trend. As the seasonal element shows, there is a strong perennial periodicity to influenza incidence.

Definition 7.7 (Perennial incidence). A perennially incident infectious disease is an infectious disease with a periodic incidence that exhibits a roughly 12-month periodicity.

Computational Note 7.5 Plotting time series decompositions

In Computational Note 7.4, we discussed the decomposition of a time series into trend, seasonal, and residual components. It is often helpful to plot these along a shared axis. Given a decomposition result object `decomposition`, we would use `GridSpec` to create a structured plot like Fig. 7.4:

```
fig = plt.figure(figsize=(20, 16))

gs = gridspec.GridSpec(4, 1, height_ratios=[3,1,1,1])
gs.update(hspace=0.2)

ax1 = plt.subplot(gs[0])
ax1.plot(decomposition.observed)
ax1.set_yscale("log")
ax1.set_ylabel("Cases")

ax2 = plt.subplot(gs[1], sharex=ax1)
ax2.plot(decomposition.seasonal)
ax2.set_ylabel("Seasonal")

ax3 = plt.subplot(gs[2], sharex=ax1)
ax3.plot(decomposition.trend)
ax3.set_ylabel("Trend")

ax4 = plt.subplot(gs[3], sharex=ax1)
ax4.scatter(decomposition.resid.index, decomposition.resid, s=0.05")
ax4.set_ylabel("Residual")

fig.align_labels()
```

The results of such a plot, for the time series decomposition in Computational Note 7.4, is displayed as Fig. 7.4.

A notebook implementing the contents of this Computational Note is available on the book's companion Github repository in the folder `/ch07/ts_decomposition`.

Of course, not all periodic peaks in disease are perennial. For instance, Grassly, Fraser, and Garnett [70] observed that the incidence of syphilis had an approximately 10-year periodicity, which is consonant with the time it takes for post-infectious immunity to wane to levels that once again permit an outbreak. While this finding is far from uncontroversial (see, e.g., Breban et al. [256] for a skeptical view on the alleged periodicity of syphilis infections), it is clear that annual cycles and longer multiyear processes, are crucial to our understanding of infectious disease dynamics.

7.2.4 Continuous wavelet transform (CWT)

One of the greatest ideas of mathematics is that all time series one is likely to encounter in nature can be described as a superposition of periodic functions. This means that given a signal in the time domain (i.e., the x-axis constituting time, the y-axis constituting the value of the signal), such as a time series, we can obtain its representation in the frequency domain, giving rise to Fourier analysis. In other words, we can transform a time-denominated series into information on the frequencies that make up the series. To us as epidemiologists, this matters primarily, because the frequencies we are discussing here constitute the periodicities of the infectious process. Thus by transforming a time series into its frequency domain representation, we can gain significant insights about the recurrent properties. The result we obtain is called the spectrum.

Definition 7.8 (Spectrum). The **spectrum** of a time series X_t is the representation of the time series in the frequency domain. The spectral value or **spectral power** of a frequency is the contribution of the periodic processes with that frequency to the overall time series.

The Fourier transform, a popular way of putting a time series into the frequency domain, analyzes time series using a complex exponential. Wavelet analysis is a significantly more powerful technique for the same purpose, using a wavelet as the analyzing function. This is relevant, because wavelets can be "scaled," allowing us some valuable insights into the rate at which values change.

Given a signal $X(t)$, such as a time series, the CWT at t_0 for the scale parameter s is given by

$$\mathrm{CWT}(s, t_0, f(t), \psi(t)) = \int_{-\infty}^{\infty} f(t)\frac{1}{s}\left(\psi\left(\frac{t-t_0}{s}\right)\right) dt. \tag{7.27}$$

CWT can be used to obtain a visual representation called a scalogram, which shows the power of each of a continuous range of frequencies at all points in time of a time series. This is commonly referred to as the signal in time-frequency space (as opposed to time-amplitude space, which is the epicurve we know and love). From an epidemiological perspective, scalograms are tools that help us identify the periodicity of an epidemic process and longer, multiyear dynamics.

The most frequently used approach to CWT analysis of time series originates with Torrence and Compo's 1998 paper, which applied CWT to elicit longer-term periodic dynamics of climate measurements [257]. The mathematics of wavelet transforms is complex, and I do not seek to repeat the work by Torrence and Compo [257] here. However, a brief overview is certainly appropriate.

A wavelet is, at its simplest, a short oscillating signal that begins at, and reverts to, zero, and is a function of a nondimensional time parameter, which we shall call η. We will, in particular, use a class of wavelets called Morlet wavelets, which are defined as

$$\psi_0(\eta) = \pi^{-\frac{1}{4}} e^{i\omega_0\eta} e^{-\frac{\eta^2}{2}}. \tag{7.28}$$

In this equation, ω_0 is the nondimensional frequency. Then, given a time series $X_t : x_1, x_2 \ldots x_N$, where each measurement is δ_t apart, the continuous wavelet transform of X_t is the convolution of X_t with the wavelet $\psi_0(\eta)$ scaled by the scale factor s and translated along the time index, t. Specifically,

$$W_t(s) = \sum_{i=0}^{N-1} x_i \overline{\psi_0\left(\frac{(i-t)\delta_t}{s}\right)}, \tag{7.29}$$

where the bar denotes the complex conjugate. $W_t(s)$ is a complex variable with a real part $\mathrm{Re}(W_t(s))$ and an imaginary part $\mathrm{Im}(W_t(s))$. The amplitude is the modulus of $W_t(s)$, i.e., $\sqrt{\mathrm{Re}(W_t(s))^2 + \mathrm{Im}(W_t(s))^2}$. The wavelet power spectrum is then $|W_t(s)|^2$.

We may conceive of this process as taking a signal (in this case, the weekly incidence data), and constructing a wavelet that is "dragged" across the signal. At each position, a similarity coefficient (similar, but not identical, to a correlation coefficient) is calculated. The process is then performed with the wavelet being scaled (stretched). At a sufficient granularity of scales, this eventually results in information about the magnitude (power) of periodicities over time, providing us with a view of what scale contributed to what degree to the overall signal. Since scale is inversely related to frequency, this grants us insight into the dominant periodicities of infectious processes.

To assist us in comparing the wavelet power spectra of different processes, it is helpful to perform some normalization. The expectation value of $|W_t(s)|^2$ for a white noise time series is the variance of the time series. Consequently, we divide the power by the variance σ^2 to normalize it. For the Morlet wavelet, the equivalent Fourier period is approximately equal to the inverse of the scale factor, whereas for other wavelets, it can be derived using the method described in the appendix of Meyers, Kelly, and O'Brien [258].

We can now plot the normalized power $\frac{|W_t(s)|^2}{\sigma^2}$ for each point of t and $\frac{1}{s}$, i.e., time and period. This is how we arrive at the scalograms (middle figures) in Fig. 7.5.

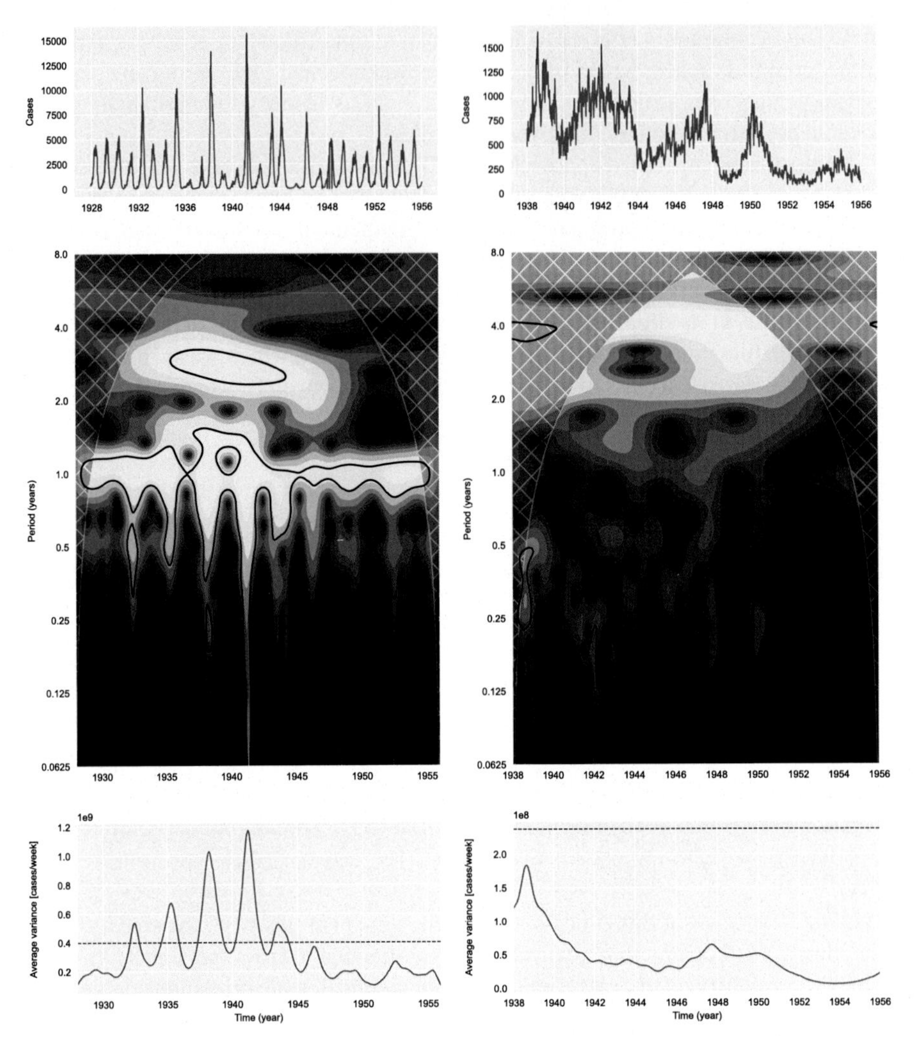

Figure 7.5 Continuous wavelet transform of the weekly incidence of measles (left) and pertussis (right): source data (top), scalogram (middle) and 3–36 month scale-averaged power (bottom). The scalogram is a visual representation of the intensity to which a signal of a particular frequency contributes to the overall value of a variable over time. In this case, it shows that while the incidence of measles has a strong dominant and statistically significant annual component, pertussis does not exhibit the same regularity and is largely influenced by 2–4 year trends.
Based on data from the U.S. Nationally Notifiable Diseases Surveillance System (NNDSS), maintained by the CDC. Data accessed from the University of Pittsburgh's Project Tycho.

Reading a scalogram is something of a learned skill, but may provide insights beyond simple Fourier plots due to its ability to represent changes over time. The scalogram is a plot of the relative power of different frequencies over time. Looking at it as a series of year-width columns, the scalogram shows the periodicities of the dominant periodic processes in that year. Typically, brighter colors indicate higher powers at the given frequency. It is common for a scalogram to show statistical significance as a black outline. In addition, there is typically a "cone of interest" marking on scalograms, which is often presented as a hatched or faded area. This is because towards the end or beginning of the timeframe, periodicities above a certain length are no longer reliably calculated.

Fig. 7.5's left-side scalogram shows a prominent band at the period of one year. This is not overly surprising; a brief perusal of the time series, plotted above, would conclude equally well that measles has an annual periodicity. However, less obvious are the high-intensity areas between 1935 and 1950 at 3–4 years, indicating that during these years, an additional longer-term dynamic, with a period of 3 to 4 years, was at play.

Scalograms are even more useful when the time series does not exhibit any obvious "first glance" periodicity. The time series of pertussis, shown on the right of Fig. 7.5, does not seem to have any clear periodicities. The scalogram, however, suggests that there is, in fact, a two-year and a four-year dynamic involved, albeit this does not rise to the level of statistical significance.

CWT and scalograms are powerful qualitative-quantitative tools to detect longer-term and time-varying periodicities that would escape most other methods of time series analysis. Though CWT is one of the more intricate and challenging techniques in this book, it rewards analysts with often quite exceptional insights.

Computational Note 7.6 Continuous wavelet transform

There are multiple packages for Python that offer continuous wavelet transform functionality, but of these, PyCWT is the most convenient and the most powerful at the same time. In this Computational Note, we will compare the continuous wavelet transforms of measles and pertussis between 1923 and 1956, obtained from Project Tycho.

First and foremost, we need to convert the `DateTimeIndex` to fractional years. It is a common notational convention in meteorology, where the CWT method was developed to analyze climate change, to describe dates as fractional years. To facilitate this conversion, we create a function for this:

```
def index_to_fractional_years(df: Union[pd.DataFrame, pd.Series])
    -> np.array:
    """
    Converts a datetime-indexed df's index to a numpy array of
    fractional years.
    """
    assert is instance(df.index, pd.DatetimeIndex)
```

```
    return df.index.map(lambda x: \
        (float(x.strftime("%j"))-1) / 366
        + float(x.strftime("%Y"))).to_numpy()
```

Next, we need to normalize our data by detrending. We do so by fitting a least squares first degree polynomial fit using the `np.polyfit()` function. This function returns the coefficients of the polynomial, which we then return to the `np.polyval()` function to evaluate the polynomial over the data. Finally, we normalize the data by dividing it by its standard deviation.

```
def normalize(data: pd.Series) -> (np.array, float):
    t = index_to_fractional_years(data)
    p = np.polyfit(t - t[0], data.values, 1)
    detrended_data = data - np.polyval(p, data)
    return detrended_data/detrended_data.std(), detrended_data.std()
```

The PyCWT package provides the `.cwt()` function to perform CWT analysis. For this to work, we will need to specify the "mother wavelet" and its possible scales. All wavelet filters are composed by varying the mother wavelet by way of two parameters: the number of "octaves" and the number of "suboctaves," sometimes referred to as "voices per octave." An octave increases the scale of the wavelet by doubling it (much as playing the same tone an octave higher corresponds to doubling its frequency). The number of suboctaves or "voices" refers to the number of equally-spaced parts that we subdivide each octave. Despite its name, the continuous wavelet transform is discontinuous in the sense that we need to evaluate it at particular values of the scaling factor s to obtain a computational answer. The granularity of this evaluation is determined by the number of suboctaves.

We need to provide, other than the data, a few starting parameters:

- `dt`, the frequency of measurements. Our measurements are monthly, but since months are of slightly different lengths, their fractional year equivalent is going to be different. We compensate for this by taking the mean of the differences of the fractional year index (`np.diff(index_to_fractional_years(data)).mean()`).
- `s0`, the starting scale. This is our assumption of the lowest possible value of whatever periodicity we expect to find. There is no harm in setting this at an unrealistically low level other than slightly increasing computational cost. For this reason, even though it is rare to find infectious diseases with a period under twelve months, we will set this to half a month.
- The number of octaves, `J`, is the upper limit on the largest resolvable scale. There is no real harm in setting this quite high, even beyond what we might

anticipate to be an epidemiologically realistic period. It is not entirely uncommon for epidemics, especially those that are closely connected to low-frequency climatic changes, to have oscillatory components with a period that numbers in the decades. We specify, in addition, the number of suboctaves per octave, `dj`.

- Finally, we need to specify the "mother wavelet," a wavelet that is then manipulated and shifted. Different wavelets are better at detecting different dynamic phenomena. The Morlet wavelet, discussed above, is specifically suitable to the analysis of signals, where the expected periodicity's frequency is likely to be mostly constant.

```
def cwt(data: pd.Series):
    mother = wavelet.Morlet(6)
    idx = index_to_fractional_years(data)
    dt = np.diff(idx).mean()
    s0 = 2 * dt
    dj = 1/52
    J = 8/dj
    alpha, _, _ = wavelet.ar1(data.values)

    normalized_data, stdev = normalize(data)

    wave, scales, freqs, coi, fft, fftfreqs
     = wavelet.cwt(normalized_data.to_numpy(),
                   dt,
                   dj,
                   s0,
                   J,
                   mother)
    ...
```

The `.cwt()` function returns the wavelet transform, including the Fourier spectra. One thing we are interested in particular is which scales and frequencies are statistically significant. For this, we will first need to obtain the power at each point, and also prepare our index for the values.

```
...
power = (np.abs(wave)) ** 2
period = 1/freqs
t = index_to_fractional_years(data)
...
```

We calculate the boundary of significance at $p < 0.05$ at each point of evaluation (i.e., each time unit and each scale and subscale):

```
...
signif, fft_theor
 = wavelet.significance(1.0, dt, scales, 0, alpha,
                        significance_level=0.95,
                        wavelet=mother)
sig95 = np.ones([1, len(normalized_data)]) * signif[:, None]
sig95 = power/sig95
...
```

Owing to the last line, `sig95` will be an array of the same size as the array of powers, with a value over unity for significant points. This is a practically quite sensible trick, as it allows us to plot the borders of the significant areas quite simply by drawing a contour where `sig95` takes a value of 1.

Finally, we calculate the average scale in the period between 0.25 and 3 years, and normalize the data against it:

```
...
sel = find((period >= 0.25) & (period < 3))
Cdelta = mother.cdelta
scale_avg
 = (scales * np.ones((len(normalized_data), 1))).transpose()
scale_avg
 = power / scale_avg  # As in Torrence and Compo (1998) eq. 24
scale_avg
 = stdev ** 2 * dj * dt / Cdelta * scale_avg[sel, :].sum(axis=0)
scale_avg_signif, tmp
 = wavelet.significance(stdev ** 2, dt, scales, 2, alpha,
                        significance_level=0.95,
                        dof=[scales[sel[0]],
                             scales[sel[-1]]],
                        wavelet=mother)

return t, dt, period, power, coi, sig95, scale_avg,
scale_avg_signif
```

The notebook on the book's companion repository details the plotting algorithm to obtain the scalogram. The `coi` output from the `cwt` function provides the "cone of influence," the cross-hatched area on the scalogram, within which edge effects become manifest. In short, as we move towards the start or the end of the time series, the data needs to be "extended" by padding it with zero-valued cells to be able to fit wavelets above a certain scale. This reduces amplitudes below their correct values, and CWT is not reliable in these edge areas. The `cwt` function therefore provides this zone, in which information becomes less reliable due to edge phenomena, and we represent this by cross-hatching.

A notebook implementing the contents of this Computational Note is available on the book's companion Github repository in the folder `/ch07/cwt/`.

7.3 Temporal forcing

Sometimes, there is a significant time-dependent exogenous process that affects a pathogen's spread. For instance, most pediatricians are quite familiar with the sudden upsurge of infectious illnesses and crowded waiting rooms around September, when the school term starts. Temporal forcing (or temporal modulation of the force of infection) describes the manner in which the temporal dynamics of infectious diseases are modulated by an exogenous variable.

7.3.1 Case study: summer, school, and poliomyelitis

Poliomyelitis is a disease caused by an enterovirus, which typically colonizes the gastrointestinal tract and causes mild disease resembling a rather typical gastrointestinal infection. In a small fraction of cases, estimated to be around one case in a hundred, the poliovirus affects the central nervous system, causing a serious syndrome characterized by meningitis, and in a very small number of cases, acute flaccid paralysis [259]. Though poliomyelitis was eradicated in the United States in the late 1970s [260], the global eradication of polio remains a major public health priority [261,262].

Many of the seasonal dynamics that drive the ebbs and flows of infectious disease relate to natural features. For instance, Martinez [263] categorized a wide range of infectious diseases by their prime drivers: vector seasonality, climatic features, non-climatic abiotic features (e.g., water salinity), and seasonal exposure differences. A specific subset of the last category, in the context of paediatric infectious disease, is the effect of school terms on disease transmission.

In countries with near-universal public primary and secondary education, few factors affect the school-age population's encounter rate with other individuals as much as the scheduling of the school term. Fig. 7.6 shows the incidence, per week, of poliomyelitis in the United States between 1945 and 1953, highlighting the 245th day of each year, around which school terms tend to start. It is not by accident that those weeks are the high water mark of incidence, preceded by a few weeks that are accounted for by earlier starts of school years and late summer activities (e.g., late summer camps).

The impact of seasonal dynamics is not limited to poliomyelitis, of course. Equally, it is important to keep in mind that these seasonal dynamics often mediate other fac-

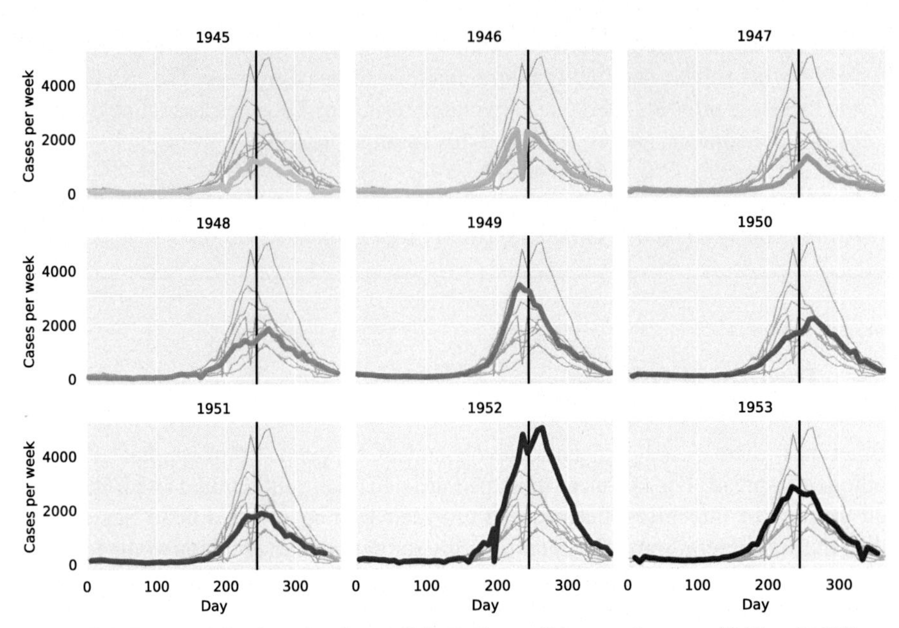

Figure 7.6 Temporal forcing of poliomyelitis. Poliomyelitis cases between 1945 and 1953 tend to peak around the start of the school year. Day 245 (corresponding to September 1/September 2) is marked with a black vertical line.
Based on data from the U.S. Nationally Notifiable Diseases Surveillance System (NNDSS), maintained by the CDC. Data accessed from the University of Pittsburgh's Project Tycho.

tors. As Dowell [264] notes, the summer-autumnal seasonality of many diseases in the Northern hemisphere is typically reversed in the Southern hemisphere, suggesting that the seasonality or time of the year in fact mediates the effect of temperature both on disease transmission and on human behavior. Sultan et al. [265], meanwhile, described how outbreaks of meningococcal meningitis in the Sahel correlate to periods of dry weather and intense winds that compromise the upper respiratory mucosa, thereby allowing pathogenic entry. These examples illustrate an important fact about seasonal dynamics that mediate seasonal climatic variabilities: the interaction between climate and individuals depends on the place and the population.

7.3.2 Sinusoidal forcing

The general model of temporal forcing is to replace the constant β with a time-dependent variable $\beta(t)$. This time-dependent variable is composed of a baseline transmission rate β_0 and a time-varying forcing component consisting of the amplitude of forcing, β_1, and the angular frequency of forcing, ω [39].

$$\beta(t) = \beta_0(\overbrace{1}^{\text{base } \beta} + \overbrace{\beta_1 \cos(\omega t)}^{\text{time-varying } \beta}). \tag{7.30}$$

For oscillations with an annual period—as the vast majority of oscillations are—$\omega = 2\pi$, and thus Eq. (7.30) becomes

$$\beta(t) = \beta_0(\overbrace{1}^{\text{base } \beta} + \overbrace{\beta_1 \cos(2\pi t)}^{\text{time-varying } \beta}). \tag{7.31}$$

At annual periodicity, in the winter, i.e. $2\pi t$ is close to either zero or one. This may be suitable to model the classical winter infections (e.g., influenza) on the northern hemisphere, but would need to be adapted for the southern hemisphere or for pathogens that are more abundant in the summer. For these instances, the cos function may be inverted by reversing the sign of the forcing term, i.e.,

$$\beta(t) = \beta_0(\overbrace{1}^{\text{base } \beta} - \overbrace{\beta_1 \cos(2\pi t)}^{\text{time-varying } \beta}). \tag{7.32}$$

For finer adjustments, it is possible to add a phase term ψ:

$$\beta(t) = \beta_0(\overbrace{1}^{\text{base } \beta} + \overbrace{\beta_1 \cos(\omega t + \underbrace{\psi}_{\text{phase}})}^{\text{time-varying } \beta}). \tag{7.33}$$

This is particularly useful for pathogens that occupy the spring ($\psi = -\pi$) and autumn ($\psi = \pi$) parts of the epidemic calendar, such as varicella and polio, respectively [263].

7.3.3 Term-time forcing

The sinusoidal model of forcing is a good approximation for processes that have a periodicity that largely depends on the season. For human populations, however, a significant factor is whether schools are in session at a particular time (which is also a relatively good proxy for when occupancy at workplaces is lower or higher). This is often accomplished using the Term function.

Definition 7.9 (Term function). The term function $\text{Term}(t)$ takes the value 1 if t falls within a school term and -1 if it does not.

School terms vary, but a summary of approximate dates for England is given in Table 7.1. This is a good starting point, although models should preferably use a termtime calendar from the population being considered.

The time-dependent transmission rate with the term-time function would be formulated as

$$\beta(t) = \beta_0(1 + \beta_1)^{\text{Term}(t)}, \tag{7.34}$$

where $\text{Term}(t)$ is the term-time function

$$\text{Term}(t) = \begin{cases} 1 & \text{if } t \text{ is in term time} \\ 0 & \text{otherwise} \end{cases}. \tag{7.35}$$

Table 7.1 Major school holidays in England, after Keeling, Rohani, and Grenfell [266].

Holiday	Days	Dates
Winter	356–6	21 December–06 January
Easter	100–115	10 April–25 April
Summer	200–251	19 July–08 September
Autumn half-term	300–307	27 October–03 November

Table 7.2 β_0 and β_1 parameters of three common infectious diseases in England and Wales, after Keeling, Rohani, and Grenfell [266].

Pathogen	β_0	β_1
Measles	1.175	0.25
Pertussis	0.664	0.25
Rubella	0.311	0.60

Fig. 7.6 is an example of both sinusoidal forcing and term-time forcing, which can coexist. The plot shows the incidence of poliomyelitis in the United States between 1945 and 1953. The peaks are centered around, and sometimes slightly to the right of, the start of September, when schools return from summer holidays. The combination of relatively warm late summer weather and the congregation of young susceptible individuals in close proximity creates a fertile breeding ground for illnesses, including transmission from still-infectious asymptomatic cases to susceptible individuals.

The basic reproduction number $\mathfrak{R}_0$ for a term-time forced model is essentially a weighted average of β_0 and β_1 divided by γ:

$$\mathfrak{R}_0 = \frac{1}{\gamma + \mu} \int_{t_{\min}}^{t_{\max}} \frac{\beta_0 (1 + \beta_1)^{\text{Term}(u)}}{t_{\max} - t_{\min}} \, du. \tag{7.36}$$

The term-time model can be adapted to other seasonal phenomena, e.g., for tourist seasons or work holidays. The variable β_0 and β_1 coefficients for some common pathogens is laid out in Table 7.2.

Practice Note 7.1 School terms: not just for kids

It comes quite naturally to an infectious disease modeler to consider the impact of term times on a school-age population, but in fact, term times are quite applicable to other populations as well. Term-time mediates workplace attendance behaviors, so that most people take their holidays out of term-time. Factory shutdowns often coincide with major school holidays in the region.

In specific populations and specific settings, however, the inverse is true. There are certain encounters that take place more often outside of term-time. Long distance travel, for instance, increases outside term-time [267,268]. This vastly enhances the speed by which a novel pathogen can spread all over the world.

In addition, certain specific settings, such as long-term care facilities and nursing homes, experience a higher volume of visitors during the holidays, which may expose their patients to pathogens from the outside world. Halvorsrud and Örstavik [269] describe a particularly unpleasant instance of rotavirus-associated gastroenteritis that made its way to a care facility for elderly patients during the holiday season. It is therefore always important to keep in mind that all are not created equal when it comes to the effects of seasonal fluctuations; what may increase the encounter rate in one population might, in another, drive it down.

7.3.4 Non-β forcing

β forcing assumes that the time-varying parameter is the transmission coefficient. However, a number of other factors may change, most importantly, as it relates to the population. In particular where domestic animals bred at quantity are concerned (e.g., poultry, rabbits), births tend to occur in pulses. Fig. 7.7 exemplifies this tendency. In this case, the fluctuations in the birth rate create complex long-term dynamics. Davis et al. [270] observed, for instance, that cases of bubonic plague fluctuate in accordance with fluctuations of its intermediate host in the region, the great gerbil (*Rhombomys opimus*). Periodic oscillations in the number of hosts and vectors, which in turn is related, by way of the Lotka–Volterra model, to seasonal fluctuations in the number of their natural predators, create "downstream" oscillations in the number of infections on the host plane.

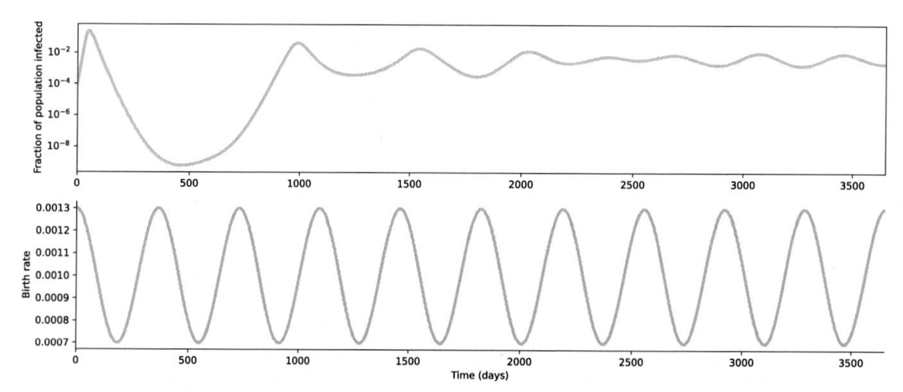

Figure 7.7 A birth-pulsed SIR system with a base birth rate of $\mu = 0.001$ and a cosine birth pulsing model with a multiplier of $\mu_1 = 0.3$.

A special case of non-β forcing is where there is a seasonal intervention, such as pulse vaccination (see Subsubsection 6.1.2.3). This effectively varies the pool of susceptibles in a pulsed manner. A concern with such approaches is that at and above certain values, a system may develop long-term nonlinearities, so that relatively small differences in the intervention (e.g., the time-varying aspect of the vaccination rate) might result in large differences with respect to long-term outcomes. For this reason, a detailed sensitivity analysis for such interventions, assisted by modeling of the effects of differences in phase, will often prove helpful. The method described in Computational Note 7.7 is a useful tool for the quantification of chaotic behavior.

7.3.5 *Bifurcations and chaos*

It is our ordinary expectation, gained from every-day life, that small changes to the parameters of a system will result in comparably small differences. A steak cooked to 140 degrees Fahrenheit might meet with a food inspector's disapproval, but it is not fundamentally different from a steak cooked to a safe 145 degrees Fahrenheit. Forced dynamical systems follow this rule, too; up to a point. As forcing increases, it becomes increasingly difficult to predict where the system will settle. The consequence is that above a certain critical point, extremely small differences in the forcing parameter will result in very different long-term outcomes [271]. The emergence of chaotic phenomena means that in the sufficiently long run, above a certain level of forcing, the relationship between the forcing parameter and the long-term equilibrium (as measured by the number of infected individuals measured at one-year intervals after running the model for several years) becomes quite discontinuous [272,273].

The analytical expansion of most epidemic models is often quite complex, and though an interesting problem from the mathematical standpoint, it is arguably of lower practical utility. Practically, however, understanding the nonlinear dynamics of small perturbations to an epidemic system at a sufficiently long time range helps us in reasoning about the accuracy of our predictions and identify when those might be affected by a strong nonlinear dependence on a parameter or initial condition. Computation affords us the ability to simulate outcomes (such as the proportion in the infectious compartment) at an arbitrary range of values for our initial parameter of choice, and thus numerically analyze a system's likelihood of nonlinear response.

Computational Note 7.7 Discrete Lyapunov exponents to estimate chaos

Discrete Lyapunov exponents are a quantitative way to identify when the evolution of a deterministic system takes on chaotic features. Recall that we understand chaotic behavior as the point where, past a critical point, minor changes in a parameter result in major fluctuations in an outcome variable, and these fluctuations are largely quite discontinuous. The Lyapunov exponent allows us to quantify this transition.

Since we expect to solve our problems computationally, a discrete time model is perfectly adequate for our purposes. Let us consider $f_{t,p}$ a function over discrete time with the discrete parameter p. If we evaluate this function at a fairly long time scale, we would expect the difference between $f_{t,p}$ and $f_{t,p+\epsilon}$, where $\epsilon \to 0$, to be quite small. Chaotic systems, however, are characterized by a strong dependence on the value of the discrete parameter, so that even the minute change to p that is represented by ϵ causes a very significant difference. We can thus analyze whether any value of p will result in chaotic behavior by looking at the differences between $f_{t,p}$ and $f_{t,p+\epsilon}$ at sufficiently large values of t. The maximal discrete Lyapunov characteristic exponent (MDLCE) accomplishes this by essentially taking the limit of the mean of the logarithm of the absolute value of the difference between each value of p and the next.

$$\lambda(f_p) = \lim_{t\to\infty} \lim_{\epsilon\to 0} \frac{1}{t} \ln \frac{|f_{t,p+\epsilon} - f_{t,p}|}{|\epsilon|}. \tag{7.37}$$

The idea is that as systems descend into chaotic behavior, the difference between the long term limits $\lim_{t\to\infty} f_{t,p}$ and $\lim_{t\to\infty} f_{t,p+\epsilon}$ increases for sufficiently large values of t, and the Lyapunov exponent allows us to express this in a single figure for each p, where values of $\lambda(f_p) > 1$ suggest chaotic long-term behavior.

In practice, computational approaches do not favor infinities, so we will often have to make do with bounded but long periods of time. A conventional way to calculate a bifurcation map and the corresponding Lyapunov exponents (see, e.g., Keeling and Rohani [39]) is to integrate the system of ODEs for a long time for different values, then sample the compartment of interest (typically, I) at regular intervals, e.g., every year for the last ten years, as our approximation for $t \to \infty$. We will do so for a fairly large number of values, which is performance-intensive. For this reason, we will have to use a more performant ODE solver.

JiTCODE is a highly performant ODE solver that has two remarkable features [204]. First, it precompiles code into C, which speeds up execution. Second, it comes with an automated functionality for determining the Lyapunov exponent.

We begin by importing `jitcode`, as well as the symbols for `sin` and `pi` from `symengine`, the symbolic computation tool that backs `jitcode`.

```
from jitcode import jitcode_lyap, jitcode, y, t
import symengine
from symengine import sin, pi
```

The `y` symbol we imported is JiTCODE's abstraction for a vector of the left-hand side quantities in the differential equation. Since we prefer to call these by their name, we redefine `y(0)` and `y(1)` to the variables `S` and `I`:

```
S, I = [y(i) for i in range(2)]
```

Next, we define our constants:

```
beta_0 = 1.3
gamma = 1/13
mu = 1/(70 * 365)

I_0 = 1e-3
S_0 = 1 - I_0
```

These are the constants we do not envisage to change. β_1, on the other hand, will. For this reason, we need to represent it symbolically, by declaring a `symengine` symbol:

```
beta_1 = symengine.Symbol("beta_1")
```

We may now finally define our model. Unlike `odeint` and `solve_ivp`, JiTCODE does not require the differential equation to be defined as a function. Among others, it can be defined as a simple Python dictionary:

```
time_dependent_sir = {
    S: mu - beta_0 * (1 + beta_1 * sin(2 * pi * t / 365)) * S * I
        - mu * S,
    I: beta_0 * (1 + beta_1 * sin(2 * pi * t / 365)) * S * I
        - mu * I - gamma * I
}
```

Next, we define the `ODE` object with the `jitcode_lyap` function, supplying it with the model. Since we are only interested in the maximum Lyapunov coefficient, we set the `n_lyap` parameter to `1`. We define `beta_1` as a control parameter, which means that the ODE will be capable of parameterization with various values of β_1:

```
ODE = jitcode_lyap(time_dependent_sir, control_pars=[beta_1],
                   n_lyap=1)
```

Next, we set the initial state of the ODE and set the integrator:

```
ODE.set_initial_value(initial_state, time=0.0)
ODE.set_integrator("dopri5")
```

JiTCODE supports a large number of integrators, including the explicit Runge–Kutta method of order 5(4) that is used by default by the `solve_ivp` function call of previous examples. `dopri5` is a fast Dormand-Prince integrator of order 5. Once the integrator is set, C code is generated for the function.

Finally, we need to define the space over which we shall evaluate β_1:

```
beta_1_range = np.linspace(0, 0.5, resolution)
```

The `resolution` variable is the number of different, linearly spaced, values of β_1 in the range of 0 to 0.5 with which the differential equation will be parameterized. A finer resolution (higher `resolution` variable) does result in a more granular picture of long-term behavior, but may be computationally vastly more expensive.

We also create two objects to hold our results:

- `bifurcation_map`, which holds our ten years' worth of samples (10 samples) for each point in the β_1 space, and
- `lyapunov_map`, which holds the estimated Lyapunov exponent for each value of the β_1 space.

We can initialize these as zero-valued arrays:

```
bifurcation_map = np.zeros(shape=(10, resolution))
lyapunov_map = np.zeros(shape=(resolution))
```

We are finally ready to write the main loop that will iterate over the entire β_1 space, and sample ten years after day 5000 on the same day of the year. We specify a burn-in period of 1000 days: typically, systems do not attain their stable long-term periodicities for quite some time, and this period reduces unnecessarily storing values we are not going to sample. We take the Lyapunov exponent for a β_1 value to be the mean Lyapunov exponent after day 1730, once again to reduce susceptibility to initial oscillations.

```
for b1idx, i in enumerate(beta_1_range):
    ODE.set_initial_value([S_0, I_0], time=0.0)
    ODE.set_parameters(i)

    ys, lyaps = [], []

    for T in np.arange(0, 7500, 1):
        y, lyap, _ = ODE.integrate(T)
        if T > 1000:
            ys.append(y[1])
            lyaps.append(lyap)

    for idx, val in enumerate(ys[4000::365]):
        bifurcation_map[idx, b1idx] = val
```

```
    lyapunov_map[b1idx] = np.mean(lyaps[730:])
```

Note lines 2 and 3, in which we set the ODE solver to its initial value. In JiTCODE, ODE objects are stateful, that is, they retain a memory of their past execution. Consequently, if the ODE object has been run once, it will be on day 7500, with the respective quantities as its current state with respect to S and I. The `.set_initial_value()` method call resets these values to the initial conditions.

Our results are, at the end of iterative integration, stored in the two objects we initialized above. These can, with some patience, be plotted to reveal the bifurcation behavior and the Lyapunov exponent's corresponding values, as we indeed do in Fig. 7.8.

A notebook implementing the contents of this Computational Note is available on the book's companion Github repository in the folder `/ch07/discrete_lyapunov/`.

The practical meaning of our computational exploration of a simple seasonally forced SIR system's chaotic dynamics is that such systems are capable of exhibiting highly nonlinear dependence on initial parameters after a certain point. As Fig. 7.8

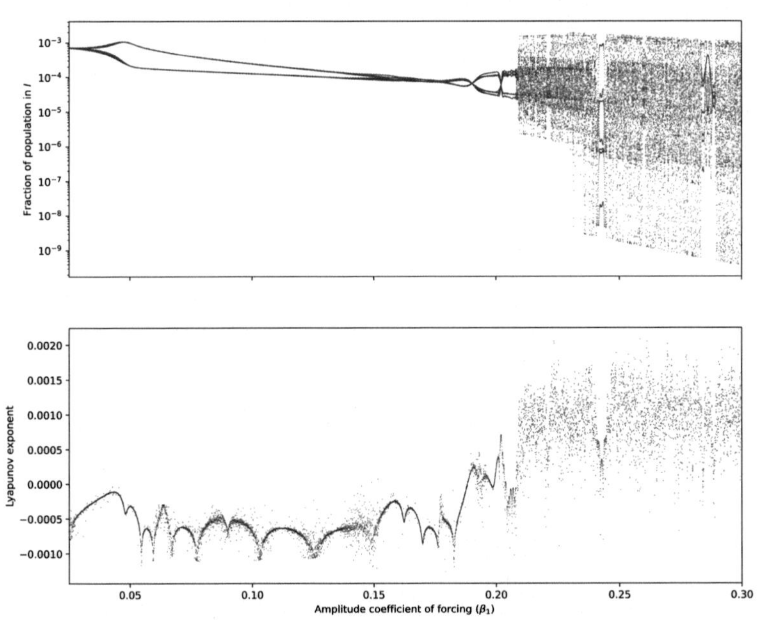

Figure 7.8 Bifurcation map of the temporally forced SIR model described in Computational Note 7.7 and its Lyapunov exponents. The model was initialized with $\beta_0 = 1.3$, $\gamma = \frac{1}{13}$, $\mu = \frac{1}{40}$ *per annum* and a starting infectious population fraction of 10^{-3}.

shows, our initially perfectly well-behaved system of linear equations experiences a bifurcation at around $\beta_1 \approx 0.04$, and becomes entirely chaotic by $\beta_1 \approx 0.21$. By $\beta_1 \approx 0.3$, minuscule variations of β_1 can result in a range of outcomes six orders of magnitude apart, the difference between approximately 3.3 million cases *per annum* in the United States and fewer than four cases in the same period.

Practice Note 7.2 Chaos in practice

There is, after a certain point, an inherent instability in dynamical systems, after which an incremental change might result in large differences. The consequence is that it is quite difficult to reason about these problems after a certain level. Assuming that β_1 is estimated to be 0.25, with a 95% confidence interval of 0.24–0.26, the model explored in Computational Note 7.7 admits to values six orders of magnitude apart, a significant difference. The long-term destiny of dynamical systems can be immensely complex to predict, estimate, and influence once strong nonlinear dependence on an initial parameter enters the scene.

The consequence is that just as the weather forecast for tomorrow is more likely to be accurate than for next week, forecasting the long-term evolution of epidemic processes might be fraught with difficulty [274]. The analogy is not accidental: much of modern chaos theory, which is aimed at a mathematical exploration of what happens once a system slips into complex nonlinearity, is intrinsically connected to modern efforts to understand weather phenomena [275].

Another nonnegligible aspect is where the strong nonlinear dependence on a parameter is in respect to the parameters of a human intervention, such as seasonal pulse vaccination (see Subsection 6.1.2.3) [276]. In these cases, more is not necessarily better, and minuscule changes in the pulse rate or the phase of the pulse vaccination with respect to the natural ebbs and flows of the disease can result in hard-to-predict long-term nonlinearities. Simulation can help by identifying ranges and patterns of stability ("safe spots"), in which the effect of such phenomena is likely to be lower.

A Lyapunov exponent analysis of a long-term epidemic model or interventional approach can be seen as a sort of sensitivity analysis: given the uncertainty of each parameter, what is the range in which, at the time horizon at which we intend to operate, we expect to find a given fraction of our values? Simulation of a system's evolution with values from our parameter's confidence interval is a useful, and regrettably underused, tool in evaluating, interpreting, and communicating long-term models of complex dynamics.

Spatial dynamics of epidemics
Epidemics in discrete and continuous space

That thing we call a place is the intersection of many changing forces passing through, whirling around, mixing, dissolving, and exploding in a fixed location. To write about a place is to acknowledge that phenomena often treated separately—ecology, democracy, culture, storytelling, urban design, individual life histories and collective endeavors—coexist. They coexist geographically, spatially, in place, and to understand a place is to engage with braided narratives and sui generis explorations.

Solnit, The Encyclopedia of Trouble and Spaciousness, 2014 [277]

8.1 Spatial lattice models

The simplest models of spatial interactions discretize space, which is inherently continuous, into a matrix. Under the assumption that space essentially consists of equal-sized square patches, we can represent space as an $m \times n$ matrix. This is a convenient (albeit rather inaccurate) formalism. The convenience that flows from the patch formalism is that we can conceive of patches as individual, spatially confined subpopulations that interact with a relatively limited neighborhood. Lattice models are easy to build as matrices, and for that reason, they are also efficient to calculate. This commends them as the optimal entry point into spatial modeling.

8.1.1 Case study: the game of life (and death)

Some games are deadly serious. The Game of Life ("Life" for short, and unrelated to its boardgame namesake) is one of the simplest games ever devised; it does not even need a player. Devised by John Horton Conway in 1970, Life is a cellular automaton; a game evolving, deterministically, from a starting state over a discrete-space cellular grid. It is perhaps somewhat of a misnomer to call Life a "game," in that few people could be said to have derived considerable entertainment value from it, but it is a serious model of evolutionary processes. Life is played on an infinite board with horizontally and vertically equal cells (or, more simply put: an imaginary infinite sheet of graph paper), each of which is "dead" or "alive." Life has only four simple rules:

1. A live cell with fewer than two live neighbors dies of loneliness.
2. A live cell with more than three live neighbors dies of overcrowding.
3. A live cell with two to three live neighbors lives on to the next generation.
4. A dead cell with three live neighbors springs to life.

Life is an example of a "cellular automaton": "cellular" because it operates over a cellular grid, and "automaton" because its initial state and its rules exhaustively define

Computational Modeling of Infectious Disease. https://doi.org/10.1016/B978-0-32-395389-4.00017-7

it. It is, of course, an almost risibly reductionistic view of natural biological processes, yet it is a surprisingly apt first approximation. For us as analysts of infectious disease, Conway's Game of Life is a model of the proximity-dependent spread of infectious diseases.

A cellular automaton is a simple example of a discrete space model. Each cell has a predefined size, and a defined state. In Life, these states are "dead" or "alive," but in an infectious disease model, they might be any of the compartments of a compartmental model. Consider a neighborhood laid out along a strict rectangular grid of households.

1. A household that does not have at least two infectious neighbors does not get infected.
2. A household with over three live neighbors does not get infected; either they self-quarantine or stay away from their neighbors.
3. A household with two infected neighbors continues trading germs.
4. A household that is not infected but is surrounded by three infected neighbors becomes infected.

Though this presupposes a particular infectious model (specifically, the original rules of Life correspond to an SI model with no immunity), cellular automata are a remarkably simple first approximation for a discrete space model of infectious disease. The most famous cellular automaton model of epidemics is the forest fire model. The forest fire model has four simple rules:

1. A burning (infected) cell eventually turns into an empty (recovered) cell.
2. A cell will catch fire (become infected) with a probability that depends on the number of burning cells in its neighborhood multiplied by a constant.
3. An empty (recovered) cell eventually turns into a susceptible cell.

It is clear that this corresponds closely to a SIR model with waning immunity. Rule 1 describes the waning process, which we describe in the differential formulation as the γI term. Rule 2 describes the mass action, with β being the rate of infection. Finally, Rule 3 represents the waning process, where the quantity ωR is transferred from the removed to the susceptible compartment. The forest fire model shows how easily the differential equation models can be translated into lattice space.

It also highlights an important fact, as we shall see in extensive detail later on: much depends on what we consider one's neighborhood.

8.1.2 Simple spatial lattices

A spatial lattice is an abstraction of continuous space into a mathematically convenient structure, namely a two-dimensional Cartesian grid. This discretization of space offers ways to explore the interaction between spatial units with the simplicity of a discrete model.

Conveniently, we can represent a discrete-time spatial lattice model as a tensor. This is a useful abstraction for three reasons:

1. Tensors, even large ones, are relatively inexpensive computationally and their computational representations often support the mathematical methods used to

extract the values we might feasibly need. For instance, the population at time t_1 of the entire compartment w is the sum of the population in all cells at time t_1 and compartment w. Because these selection operations can be made highly efficiently, we can simulate large metapopulations at low computational cost.

2. Many of the dynamics of a pathogenic process in space, including the effect of neighboring cells, can be described using relatively trivial linear algebra and implemented with high computational efficiency (e.g., through convolution kernels).
3. Finally, one of the motivations behind such models is to reflect the fact that real spaces are made up of smaller subunits that have their characteristic differences. Thus the same pathogen may have a different value for β between cells, or cells may have different populations. An infection may start in one locality and spread through space. Spatial lattice models can accommodate specifications of model parameters and initial conditions as matrices of, e.g., random values that follow an empirically observed distribution.

We can represent space and states as the $m \times n \times c \times |t|$ tensor $\mathbf{T}$ describing the entirety of the system at time t, where c is the number of compartments, so that $\mathbf{T}_{u,v,w,t}$ is the value in the w-th compartment for the coordinates u, v.

Definition 8.1 (Spatial lattice). A spatial lattice is a model of discrete space, in which space is represented by equal-sized, equidistantly distributed points called cells or patches. A cell may have a single real value or be vector-valued, which is the case where a cell may comprise different values in different states (compartments).

Computational Note 8.1 Simple spatial lattices

For relatively simple instances of spatial lattices, where there are only a few compartments, it may be feasible to simply maintain the individual state (compartment) lattices as separate variables, along with the parameter matrix. In the following example, we use a constant value of $\gamma = 1/6$ and spatially varying values of $I(0)$ and β. (See Fig. 8.1.)

First, we initialize our β matrix, which comprises the spatially unique value of β for each cell. We fill this with values drawn from a normal distribution so that $\beta_{u,v} \sim \text{Normal}(0.7, 0.005)$. The `scipy.stats` module provides a large number of commonly used statistical distributions, and the method `.rvs()` obtains samples from these. The `.rvs()` method allows for an integer-valued `size` parameter, which specifies the number of samples returned. Thus to obtain 100 random values drawn from the above-specified normal distribution, we write `stats.norm(0.7, 0.005).rvs(size=100)`. Since this is a single vector, we shall use the `.reshape()` method, which reshapes a Numpy array into a specified shape. We therefore initialize our β matrix as follows:

```
beta_matrix = stats.norm(0.7, 0.005).rvs(size=100).reshape(10, 10)
```

Next, we initialize the $I(0)$ matrix much in the same way, with $I(0)_{u,v} \sim \text{Normal}(10^{-5}, 10^{-7})$:

```
I_0_matrix = stats.norm(1e-5, 1e-7).rvs(size=100).reshape(10, 10)
```

Since for all u, v, $S(0)_{u,v} = 1 - I(0)_{u,v}$, we define the matrix of initial susceptibles as

```
S_0_matrix = np.ones(shape=(10,10)) - I_0_matrix
```

Given that we want to also be able to track the results of each evaluation (i.e., each day), much as we have when using ODE solvers, we create lists that are initialized with the day zero value:

```
I_arr, S_arr = [I_0_matrix], [S_0_matrix]
```

Finally, we can write the difference equation forms of our SIR differential equation. Note that since we are interested in the sizes of the compartments, we must adapt our equations to the difference equation form

$$\begin{aligned} S(t) &= S(t-1) - \beta S(t-1)I(t-1), \\ I(t) &= I(t-1) + \beta S(t-1)I(t-1) - \gamma I(t-1). \end{aligned} \tag{8.1}$$

Helpfully, in Numpy, multiplications of arrays are Hadamard multiplications (i.e., element-wise multiplication), rather than proper matrix multiplication. We may therefore write for the above difference equations a recursive form that uses the previous calculation's results by indexing at `-1` from the lists we created to store individual results:

```
for i in range(max_iterations):
    S_arr.append(S_arr[-1] - beta_matrix * S_arr[-1] * I_arr[-1])
    I_arr.append(I_arr[-1] + beta_matrix * S_arr[-1] * \
                            I_arr[-1] - gamma * I_arr[-1])
```

Ran for a set value of `max_iterations`, we obtain the results of a simple spatial lattice model. Each of the arrays are indexed first by day, then as a Numpy array's row × column indexing. Thus we obtain the fraction of susceptibles in the 16th row of the 7th column on the 5th day by

```
S_arr[4][15][6]
```

since in Python lists are 0-indexed. It is also possible to merge the row and column indices for a Numpy array (e.g., `S_arr[4][15,6]`), but a Numpy array

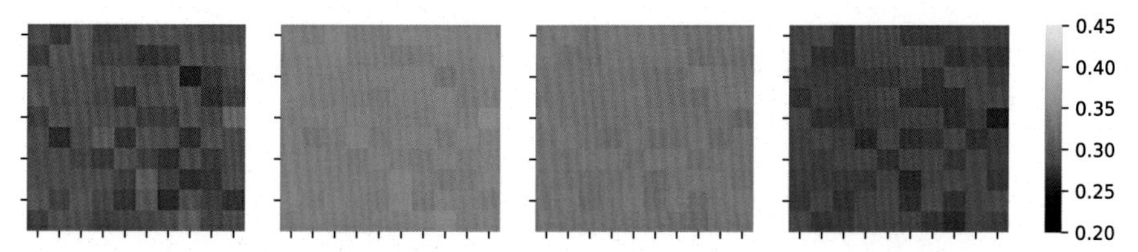

Figure 8.1 Snapshots of the spatial lattice of the infectious compartment from Computational Note 8.1 at days 26, 28, 30, and 32 for a pathogen with $\gamma = \frac{1}{6}$, $\beta \sim \text{Normal}(0.7, 0.005)$ and $I(0) \sim \text{Normal}(10^{-5}, 10^{-7})$.

inside a list is an object of its own. Therefore attempting to access the same element as `S_arr[4, 15, 6]` will not work. Multidimensional arrays, which we will explore in the next Computational Note, allows such indexing with ease.

A notebook implementing the contents of this Computational Note is available on the book's companion Github repository in the folder `/ch08/spatial_lattice`.

Let us consider a single point at u, v at time t, and denote the vector $\mathbf{T}_{u,v,*,t}$ as $\vec{\theta}$. Then, considering only the first two elements (since the recovered compartment follows necessarily from the previous two),

$$\frac{d\vec{\theta}}{dt} = \vec{\theta} \odot \begin{pmatrix} -\beta\vec{\theta}_2 \\ \beta\vec{\theta}_1 - \gamma \end{pmatrix}, \tag{8.2}$$

where $\odot$ is the Hadamard product. The benefit of formulating a spatial model in this way is that it provides certain mathematical conveniences. First, the number of individuals in the w-th state at time t is given by the double sum

$$N_w = \sum_{i=1}^{m} \sum_{j=1}^{n} \mathbf{T}_{i,j,w,t}, \tag{8.3}$$

and the total population in cell u, v is

$$N_{u,v} = \sum_{k=1}^{w} \mathbf{T}_{u,v,w,t}. \tag{8.4}$$

Finally, the temporal evolution of the infectious process can now be represented as

$$\begin{aligned}
\mathbf{T}_{*,*,S,t+1} &= \mathbf{T}_{*,*,S,t} - \frac{\beta\mathbf{T}_{*,*,S,t}}{N_t}\mathbf{T}_{*,*,I,t}, \\
\mathbf{T}_{*,*,I,t+1} &= \mathbf{T}_{*,*,I,t} + \frac{\beta\mathbf{T}_{*,*,S,t}}{N_t}\mathbf{T}_{*,*,I,t} - \gamma\mathbf{T}_{*,*,I,t}, \\
\mathbf{T}_{*,*,R,t+1} &= \gamma\mathbf{T}_{*,*,I,t},
\end{aligned} \tag{8.5}$$

where division and multiplication of matrices are element-wise.

Computational Note 8.2 Indexing and manipulation of multi-dimensional arrays

In Computational Note 8.1, we stored individual snapshots of the S and I states as arrays of 10 × 10 NumPy `array` objects, so that each moment in discrete time was represented by a 10 × 10 matrix. What, however, if we are interested in the time series of the proportion in the I or S compartment for each of the cells, so as to obtain a time series as we did when using integrating ODE solvers to calculate compartment sizes over time? For this, we would have to turn our list of 2-dimensional arrays into 3-dimensional arrays.

The `numpy.dstack()` function "stacks" each frame after each other, with the third dimension being the list index:

```
I_fractions = np.dstack(I_arr)
S_fractions = np.dstack(S_arr)
```

We can now access the proportion of susceptibles in $\{u, v\}$ at time t using `S_fractions[u, v, t]`. More generally, we can obtain the time series for the proportion of susceptibles in $\{u, v\}$ at all times using `S_fractions[u, v, :]`. Ordinarily, the `:` character is known as the slice operator, which obtains elements of an indexable array between the indices each side of the operator (e.g., `[3:5]` means "the subset of the array between the indices 3 and 5"). Where one side of the slice operator is left empty, it is taken to mean the boundary of the set, i.e., the lowest-indexed element for the left and the highest-indexed element for the right (thus `[3:]` means "the subset of the array from index 3 onward").

Since each cell is associated with a time series, yielding a 10 × 10 matrix of time series, we might benefit from transforming this into a single time series of the means of each frame.

```
I_means = [I_fractions[:, :, i].mean()
           for i in range(I_fractions.shape[2])]
S_means = [S_fractions[:, :, i].mean()
           for i in range(S_fractions.shape[2])]
```

This list comprehension iterates over the number of entries in the third dimension of `I_fractions` and `S_fractions`, which we obtain by way of the `.shape` attribute. For each unit in time, the entire frame is obtained by indexing the array with `[:, :, i]`, where `i` is the iterator of the array comprehension.

At the same time, taking the mean only obviates one of the benefits of spatial modeling, which is our ability to understand the spatial variation. Recall that in the Computational Note 8.1, we have initialized our model with slight differences in $I(0)$ and β between the cells of the lattice. Consequently, our results are going to vary. In practice, the use of models of spatial heterogeneity is to

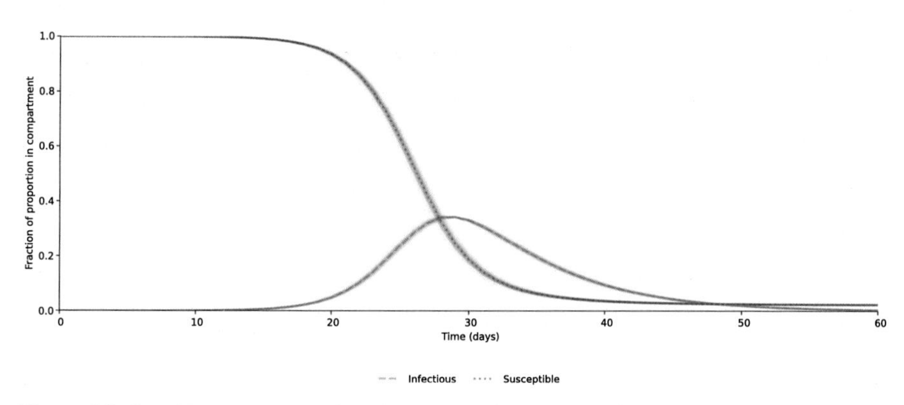

Figure 8.2 S and I compartment sizes from the lattice model in Computational Note 8.1.

understand the impact that these differences make, and for this reason, we want to know not only the mean but also the range of possible values. In this sense, we can consider our 10 × 10 lattice as 100 disconnected simulations (for now, this will change when we begin to consider spatial interactions), initialized with slightly different parameters of $I(0)$ and β. We can use `.reshape` to obtain a matrix, where each row corresponds to a cell and each column corresponds to a point in time:

```
reshaped_I_fractions = I_fractions.reshape(100, 101)
```

This reshapes the 10 × 10 × 101 array into a 100 × 101 array (since we have 100 iterations following $I(0)$ and $S(0)$), where each row represents a cell and each column a point in time.

Fig. 8.2 shows a plot of the compartment sizes for the S and I compartments for each day, together with the means of each day. It is evident that our lattice-based estimation without spatial interaction essentially converges to a SIR model with stochastically determined parameters for β and $I(0)$. This is unsurprising, since we have effectively performed a manual stepwise integration of the governing system of ODEs for each cell. In Subsection 8.1.3, we will begin to explore models that include spatial interaction as well.

A notebook implementing the contents of this Computational Note is available on the book's companion Github repository in the folder `/ch07/spatial_lattice`.

8.1.3 Spatial interaction in lattices

In the previous model, cells did not exert an effect on their surroundings. The simple lattice model in Computational Note 8.1 is no different from running the same model with slightly different parameters a hundred times in sequence.

Spatial interaction refers to the effect exerted on a cell of a lattice by cells other than itself. The cells that are capable of exerting such an effect are called the cell's neighborhood. This notion of neighborhood is somewhat vague, because from an epidemiological perspective, it matters more what a neighborhood does than how it is defined. In general, for a lattice, there exist two common definitions of neighborhood:

- the von Neumann neighborhood $\mathcal{N}_N(u, v)$, which comprises the cells $\{u-1, v-1\}$, $\{u-1, v+1\}$, $\{u+1, v-1\}$, and $\{u+1, v+1\}$ (which is equivalent to all cells that have a Manhattan distance of one from u, v), and
- the Moore neighborhood $\mathcal{N}_M(u, v)$, which comprises all cells i, j, where $u-1 \leq i \leq u+1$, and $v-1 \leq j \leq v+1$ and $i, j \neq u, v$ (which is equivalent to all cells with a Chebyshev distance or L_∞ metric of one).

Computational Note 8.3 Kernel neighborhoods

In theory, a simple way of obtaining the neighborhood of a cell $\{u, v\}$ would be to index the array for each of $\{u-1, v-1\}$, $\{u-1, v+1\}$, $\{u+1, v-1\}$, and $\{u+1, v+1\}$, looping over each cell. This is possible, but computationally rather expensive and more than a little inelegant. A more elegant way of calculating neighborhoods makes use of convolution kernels (also known as convolution matrices).

Consider the convolution of an $m \times n$ matrix $\mathbf{M}$ with the following kernel:

$$\mathbf{w} = \begin{pmatrix} 0 & 1 & 0 \\ 1 & 1 & 1 \\ 0 & 1 & 0 \end{pmatrix}. \tag{8.6}$$

The convolution of $\mathbf{M}$ with $\mathbf{w}$ is the matrix $\mathbf{M} * \mathbf{w}$ so that for any u, v, where $1 < u \leq m-1$ and $1 < v \leq n-1$,

$$(\mathbf{M} \circ \mathbf{w})_{u,v} = \mathbf{M}_{u-1,v-1}\mathbf{w}_{1,1} + \mathbf{M}_{u-1,v}\mathbf{w}_{1,2} + \cdots + \mathbf{M}_{u+1,v+1}\mathbf{w}_{3,3}. \tag{8.7}$$

More generally, for any kernel of size $p \times q$, we may write

$$(\mathbf{M} \circ \mathbf{w})_{u,v} = \sum_{r=-\infty}^{\infty} \sum_{s=-\infty}^{\infty} \mathbf{M}_{u-r,v-s}\mathbf{w}_{r,s}. \tag{8.8}$$

The convolution of the matrix with the kernel is essentially equivalent to "moving" the kernel across the matrix so that the kernel is centered around each element in the matrix once. The result of the convolution is the elementwise sum of the elementwise (Hadamard) product of the kernel and the subset of the matrix it covers, which then becomes the value corresponding to the element around which the kernel is centered. Thus convolving a matrix with the kernel

$$\mathbf{w_N} = \begin{pmatrix} 0 & 1 & 0 \\ 1 & 1 & 1 \\ 0 & 1 & 0 \end{pmatrix} \tag{8.9}$$

assigns to any u, v, where $1 < u \leq m - 1$ and $1 < v \leq n - 1$ of the input matrix the sum of its von Neumann neighborhood (including itself). The kernel

$$\mathbf{w_M} = \begin{pmatrix} 1 & 1 & 1 \\ 1 & 1 & 1 \\ 1 & 1 & 1 \end{pmatrix} \tag{8.10}$$

is the corresponding 3×3 matrix for the Moore neighborhood.

One issue we have to resolve is to deal with the edges. For a 3×3 kernel, centering around, say, $\{1, 1\}$, means part of the kernel will fall outside the matrix. There are three commonly used solutions for this problem (other than simply discarding the edge):

- toroidal wrapping, in which we consider the edges of the lattice to be joined to their opposite, and
- half-sample symmetric sampling, in which the input matrix is extended by reflection across the centre of the last pixel, and
- full-sample symmetric sampling, in which the input matrix is extended by reflection across the edge of the matrix.

In cellular automata, which are discussed in Subsection 8.1.1, it is quite common to use toroidal wrapping. A drawback is that this ignores the effect of spatial boundaries, which can act like reflective surfaces and create reflection interference patterns from epidemic waves. For this reason, if a spatial lattice is used to approximate real boundaries, one of the symmetric sampling methods is likely to be a better reflection of reality.

In `scipy`, convolution of multidimensional arrays is provided by the `ndimage.convolve()` function. This function takes three arguments: the matrix, the kernel to convolve with, and the chosen method of extending the matrix over the edges, supplied as the `mode` keyword. Toroidal wrapping corresponds to the `‘‘wrap’’` mode, whereas half- and full-sample symmetric sampling is provided by `‘‘reflect’’` and `‘‘mirror,’’`, respectively.

Thus we can obtain the von Neumann neighborhood of a matrix `M` using

```
ndimage.convolve(M, np.array([[0, 1, 0],
                              [1, 1, 1],
                              [0, 1, 0]]), mode="reflect")
```

> *A notebook implementing the contents of this Computational Note is available on the book's companion Github repository in the folder* `/ch08/kernel_neighbourhoods`.

8.1.4 Influence

Influence reflects interactions between adjoining populations that does not result in changes to the populations themselves. The phenomenon of influence is reflective of two fundamental facts. First, in reality, space is continuous, and the boundaries we establish through the discretization of space into a lattice are quite arbitrary. Second, nearby and connected populations interact in fundamental ways. As our initial simplistic model of spatial interaction discussed in Subsection 8.1.1 has shown, the dynamics of cells is affected by their direct neighborhoods. This holds true for more involved simulations, too, as we will see in this chapter.

Let us consider a cell u, v of a spatial lattice model, and its neighborhood, $\mathcal{N}$. A susceptible individual in u, v may contract the infection from another individual in u, v or any other individual in any cell in $\mathcal{N}$. Consequently, the mass action term for u, v term will be

$$\beta_{u,v} S_{u,v} \sum_{n \in N} I_n. \tag{8.11}$$

The SIR model with influence is therefore governed by the system of differential equations (assuming γ is spatially homogeneous),

$$\begin{aligned}
\frac{dS_{u,v}}{dt} &= -\beta_{u,v} S_{u,v} \sum_{n \in N_{u,v}} I_n, \\
\frac{dI_{u,v}}{dt} &= \beta_{u,v} S_{u,v} \sum_{n \in N_{u,v}} I_n - \gamma I_{u,v}, \\
\frac{dR_{u,v}}{dt} &= \gamma I_{u,v}.
\end{aligned} \tag{8.12}$$

Given a kernel $\mathbf{w}$ that describes the neighborhood, we can characterize the differentials of the matrices $\mathbf{S}$, $\mathbf{I}$, and $\mathbf{R}$ with respect to time. In practice, it is often sensible to implement such models with respect to population sizes, rather than merely population fractions. This is because there is often quite significant spatial heterogeneity in populations, which affects the mass action term. Thus a better approximation is to use the actual numbers of susceptible, infectious, and recovered individuals, and divide it by the sum of the neighborhood population. Since convolution is both associative and distributive, this would amount to the convolution of $\frac{\mathcal{I}}{\mathcal{S}+\mathcal{I}+\mathcal{R}}$ with $\mathbf{w}$:

$$\begin{aligned}
\frac{d\mathcal{S}}{dt} &= -\beta \odot \mathcal{S} \odot \left(\frac{\mathcal{I}}{\mathcal{S}+\mathcal{I}+\mathcal{R}} \circ \mathbf{w}\right),\\
\frac{d\mathcal{I}}{dt} &= \beta \odot \mathbf{S} \odot \left(\frac{\mathcal{I}}{\mathcal{S}+\mathcal{I}+\mathcal{R}} \circ \mathbf{w}\right) - \gamma \mathcal{I},\\
\frac{d\mathcal{R}}{dt} &= \gamma \mathcal{I},
\end{aligned} \tag{8.13}$$

where $\odot$ denotes the Hadamard product operator.

Computational Note 8.4 A neighborhood model of influence

In this Computational Note, we are expanding the simple spatial lattice model we introduced in Computational Note 8.1 to include a neighborhood model of influence by applying the kernel convolution discussed in Computational Note 8.3. Recall that in the previous example, we have initialized the spatially varying parameters as being drawn from a random distribution. In general, the random diffusion process that takes place in such a distribution is not very enlightening. For this reason, we are going to artificially create an area of higher initial number of infected individuals, resembling the focal source of an outbreak. In addition, we are going to use absolute populations rather than proportions for the compartments.

We thus begin by initializing a population matrix, which assigns to each cell $[u, v]$ a population distributed according to Normal$(10^3, 3 \times 10^2)$.

```
population_matrix = stats.norm(1e3, 3e2).rvs(size=xsize * ysize).\
reshape(xsize, ysize)
```

After generating the proportion matrices, just as we did in Computational Note 8.1, we create a square-shaped "spot" (the infectious focus) centered around $\frac{1}{3}$ of the height and width of the lattice, with the spot's width being $\frac{1}{5}$ of the lattice's width:

```
spot = (floor(xsize/3), floor(ysize/3))
width = floor(xsize/10)

I_0_matrix[spot[0]-width:spot[0]+width, spot[1]-width:spot[1]+width]
 = stats.norm(5e-2, 1e-3).rvs(size=(2 * width)**2)
   .reshape(2*width, 2*width)
S_0_matrix = np.ones(shape=(xsize, ysize)) - I_0_matrix
```

By taking the elementwise product of these matrices with the population matrix, we obtain the size of each compartment in each cell:

```
S_0_matrix *= population_matrix
I_0_matrix *= population_matrix
```

and we once again initialize arrays for the compartments, including the recovered compartment, which we initialize using `np.zeros_like()` to create a matrix of zeroes with the same dimensions as `S_0_matrix`.

We are now ready to construct the iterative function, in which we convolve a 3×3 Moore neighborhood kernel over each compartment's previous iteration to obtain the number of infecteds in the neighborhood, as well as to obtain the total number of individuals in the neighborhood:

```
for i in range(max_iterations):
    I_prev_neighborhood = ndimage.convolve(I_arr[-1],
            np.array([[1, 1, 1],
                      [1, 1, 1],
                      [1, 1, 1]]),
            mode="reflect")

    S_prev_neighborhood = ndimage.convolve(S_arr[-1],
            np.array([[1, 1, 1],
                      [1, 1, 1],
                      [1, 1, 1]]),
            mode="reflect")

    R_prev_neighborhood = ndimage.convolve(R_arr[-1],
            np.array([[1, 1, 1],
                      [1, 1, 1],
                      [1, 1, 1]]),
            mode="reflect")

    N_prev = I_prev_neighborhood + S_prev_neighborhood
             + R_prev_neighborhood

    S_arr.append(S_arr[-1] - beta_matrix * S_arr[-1]/N_prev
                 * I_prev_neighborhood)
    I_arr.append(I_arr[-1] + beta_matrix * S_arr[-1]/N_prev
                 * I_prev_neighborhood  \
                 - gamma * I_arr[-1])
    R_arr.append(R_arr[-1] + gamma * I_arr[-1])
```

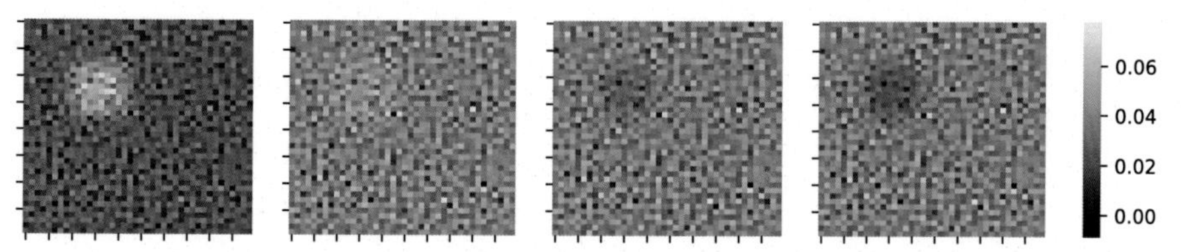

Figure 8.3 Snapshots of the infectious compartment of the spatial influence lattice model described in Computational Note 8.4 at days 7, 10, 13, and 16. The model is initialized with an initial spot of infectiousness, with the mean of the infectious proportion 5 times higher in this area. As time progresses, the disease exhausts the available supply of susceptible individuals, and at 16 days, the initial hotspot has a lower proportion of infectious individuals compared to the rest of the lattice.

Note that the 3×3 array of ones can be initialized in a perhaps more elegant manner as `np.ones(shape=(3, 3))`. This is sometimes not customary when specifying kernels, as it is often useful to see the shape of the kernel.

Fig. 8.3 shows a heat map of a 40×40 lattice on days 7, 10, 13, and 16. Models like this are particularly enlightening in illustrating the emergence of spatially differentiated epidemics. Because the kernel is relatively small, there is relatively little diffusion, so the initial hotspot behaves as an almost isolated zone with its own dynamics, ahead of the epidemic curve by starting out with a higher infected population. Larger kernels mean a wider net of influence and expand the range of interactions.

A notebook implementing the contents of this Computational Note is available on the book's companion Github repository in the folder `/ch08/filtering`.

The kernel is a descriptor of spatial dynamics of interaction; it is, in a sense, a mathematical abstraction for what we believe the spatial relationships to be:

- The size of the kernel describes our understanding of the spatial effect or "permeability" of the system, i.e., the maximum distance of mass action.
- The homogeneity of the kernel describes our expectation of how the population of one cell will affect others as a function of distance. A homogeneous kernel, such as we used in Computational Note 8.4, assumes that the effect is going to be the same with respect to every neighbor. This becomes more important for larger kernels. For instance, the kernel

$$\begin{pmatrix} 1 & 1 & 1 & 1 & 1 \\ 1 & 3 & 3 & 3 & 1 \\ 1 & 3 & 5 & 3 & 1 \\ 1 & 3 & 3 & 3 & 1 \\ 1 & 1 & 1 & 1 & 1 \end{pmatrix}$$

weights cells closer to the centre of the kernel stronger than more distant cells.

- The isotropy of the kernel reflects whether the kernel is biased towards directionality. The kernel in the previous point is isotropic, as its weights are the same regardless of direction. On the other hand, the kernel

$$\begin{pmatrix} 1 & 3 & 4 & 3 & 1 \\ 1 & 3 & 5 & 3 & 1 \\ 1 & 3 & 5 & 3 & 1 \\ 1 & 3 & 5 & 3 & 1 \\ 1 & 3 & 4 & 3 & 1 \end{pmatrix}$$

  is anisotropic, since its weighting along the vertical axis differs from that along the horizontal axis.

Thus, in designing a kernel, we essentially describe spatial interactions as a matrix that governs the way the mass action term of a cell in a lattice is affected by its surrounding, including how that surrounding is shaped or defined, with the size of the kernel, as well as its zero-valued elements, circumscribing the boundaries of mass action.

8.1.5 Neighborhood and quorum sensing

In previous examples, we used kernels to create a weighted sum of a neighborhood. However, sometimes, that is not the governing factor.

- A lattice in which each cell represents the catchment area of a medical facility might be governed by the overflow principle: the cell $[u, v]$ may only be affected if one or more of its neighbors exceed the threshold number of cases that a cell can manage. In that case, patients may be transferred or treated in facilities corresponding to another cell, and begin to affect the mass action term there.
- In other cases, similar to a redistribution problem, a cell is governed by the highest- or lowest-valued cell in its neighborhood. In this case, maximum or minimum pooling might be the key determining factor of mass action. Maximum-pooling dependent mass action is typical for a kernel that represents geographies that are typically rather separate, but transfer infectious patients to neighboring geographies if reaching capacity (but unlike in a previous case, this is determined by the single cell of highest need, rather than the aggregate of all cells exceeding a threshold value). Minimum-pooling dependent mass action occurs much less frequently, and characteristically describes a scenario of "stepping stones," in which a population stays in contact with only its least affected neighbor so as to balance risk reduction with maintaining supply chains and critical services.
- Finally, a cell might be subject to "peer pressure," adopting variable parameters of dynamics that most closely approximate its neighbors' positions (i.e., minimizing the sum of a distance metric vis-a-vis all of its neighbors). This is a commonly used model in opinion dynamics, but might apply to certain secondary factors in public health. For instance, the effectiveness of a mask mandate or a lockdown order is vastly decreased if adjoining spatial units do not follow the same policy, thus obviating the economic rationale behind such public health measures.

In such cases, the dynamics for $[u, v]$ might be governed by extrema or other values that are harder to estimate using kernels. In such cases, generic filtering may be useful to apply a function, such as determining extrema, to a sliding window similar to the way a kernel is applied in the convolution operation.

Computational Note 8.5 Minimum-filtered spatial lattice

A minimum-filtered spatial lattice replicates social behaviors of avoiding areas of high disease prevalence. The perception of risk modulates human behavior, and disease-avoidant behavior is one of the responses to risk. Minimum filtering means that the mass action term is primarily governed by the least-affected cell in the neighborhood. This simulates the tendency to seek out the "safest" available spot to engage in necessary human interactions.

`scipy.ndimage` provides the `generic_filter()` function to implement such filters. `generic_filter()` takes an array, a function and a `footprint` argument, and applies the function directly to a sliding window, as specified by the `footprint`. Just like convolution, by default, the filter is centered on the cell.

Our approach will be to use a 3x3 kernel, representing the neighborhood of interaction. Since such interactions are typically better approximated by a circular kernel, we will use a von Neumann neighborhood. Similarly to the process in Computational Note 8.4, we integrate stepwise, except we adjust the mass action term to be governed by the number of infectious individuals in the cell with the lowest number of such individuals.

```
for i in range(max_iterations):
    kernel = np.array([[0, 1, 0],
                       [1, 1, 1],
                       [0, 1, 0]])

    I_min_neighborhood
     = ndimage.generic_filter(I_arr[-1],
                              np.min, footprint=kernel)

    S_arr.append(S_arr[-1] - beta_matrix * S_arr[-1]
                 /population_matrix \
                 * I_min_neighborhood)
    I_arr.append(I_arr[-1] + beta_matrix * S_arr[-1]
                 /population_matrix \
                 * I_min_neighborhood - gamma * I_arr[-1])
    R_arr.append(R_arr[-1] + gamma * I_arr[-1])
```

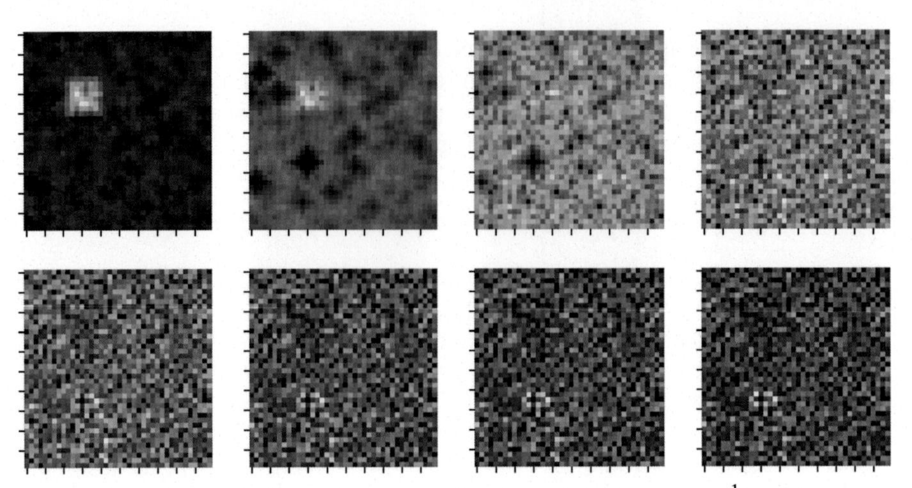

Figure 8.4 Evolution of a 40 × 40 minimum-filtered spatial lattice for $\gamma = \frac{1}{6}$ and $\beta \sim$ Normal(1.3, 0.05) at 3-day intervals beginning on day 2.

Fig. 8.4 shows the evolution of the spatial lattice. It is interesting to note the effect of minimum filtering: the preinitialized "hot" patch of higher initial infections diffuses into the general epidemic process quite rapidly, whereas "cold" spots of lower initial infections create quite enduring phenomena of delayed dynamics.

A notebook implementing the contents of this Computational Note is available on the book's companion Github repository in the folder `/ch08/filtering`.

8.2 Computational geospatial infectious disease analysis

In this section, we shall use computational methods to explore properties of space and the way connections are created over space, e.g., via street grids, to analyze the domain over which infectious disease is spread. At the same time, space is not only the domain of pathogenic spread but also the domain of response, both by the population affected (e.g., seeking medical attention) and by public health authorities (e.g., by providing vaccinations or testing services).

8.2.1 Case study: space to (stop) spread

The small town of Gunnison, Colorado, lies at the bottom of the valley carved by the Gunnison River into the Rocky Mountains. It is now crossed by the Colorado stretch of U.S. Highway 50, but in 1918 the town was mainly supplied by train and two at best mediocre roads. When the 1918–19 influenza pandemic reached Colorado as an unwelcome stowaway on a train carrying servicemen from Montana to Boulder, the town of Gunnison took decisive action. As the November 1, 1918, edition of the Gunnison News-Champion documents, a Dr. Rockefeller from the nearby town of Crested Butte was "given entire charge of both towns and county to enforce a quarantine against all the world" [278]. He instituted a strict reverse quarantine regime that almost entirely isolated Gunnison from the rest of the world. Gunnison became one of the few communities that largely escaped the ravages of the influenza pandemic, at least in the beginning [279]. In an instructive example of the limited human patience for the social, psychological, and economic disruption of quarantine, adherence eventually waned, and the front page of the Gunnison News-Champion's March 14, 1919, issue reports that the influenza pandemic got to Gunnison, too [280]. Nevertheless, Gunnison had a very lucky escape, of a population of over 6900 (including the county), there were only a few cases and a single death.

In Markel et al. [279]'s study of six communities that escaped the devastation wrought by the influenza pandemic, the enormous impact of spatial factors emerges when considering the location, accessibility, and interaction of each:

- Yerba Buena, CA, was a military base on an island, which could be easily isolated.
- Gunnison, CO, had a railway line and two main roads. It could be effectively isolated by blocking a small number of points of ingress.
- Princeton University, NJ, was situated on a campus that was a modest distance away from the town of the same name, and could equally be confined.
- The Western Pennsylvania Institute for the Blind in Pittsburgh, PA was located in a residential area, but due to the widespread absence of accessibility to the visually impaired residents at the time, contacts with the outside world were infrequent.
- Trudeau Sanatorium, was located on the edge of Saranac Lake, NY, itself a fairly isolated community at the time wedged between the McKenzie and High Peaks Wilderness areas.
- Fletcher, VT, was (and remains) a fairly small town at the intersection of two single carriageways.
- Bryn Mawr College in Bryn Mawr, PA, occupied a contiguous area between four roads, making quarantines relatively easy. As a college, it was relatively self-sufficient and could sequester its students and provide for them.

In the case of at least four of the six outliers in Markel et al. [279]'s study, geography played a role: some communities were located in naturally isolated places (Fletcher, VT, Yerba Buena, CA, and the Trudeau Sanatorium in Saranac Lake, NY), and some had a small number of links to the outside world that could be easily blocked (Gunnison, CO, and Fletcher, VT). The networks that connect us and create the life-giving interactions that together are "society" are also the networks through which epidemics propagate. Computational models can shed light on these connections, explain how spatial connectivities affect epidemic processes and how modulating these connectivities (e.g., through quarantine checkpoints and barriers) may contribute to the control of infectious disease in populations.

Practice Note 8.1 Only as good as your maps

We may think of maps and mapping as an objective process, but that would be an illusion. What gets mapped, and more importantly, what does not, is a product of various social, economic, and political phenomena. Quite apart from border disputes and contentious sovereignty, mapping also reflects political priorities. Creating the survey data that can be used in maps is expensive, and large-scale mapping endeavors are typically the preserve of states, whose ability to deliver that data often depends on resources that compete with other governmental priorities. This is true especially in resource-constrained settings.

A spatial analysis is only as good as the mapping data it relies on, and often, where data is (and is not) available reflects complex social and socio-economic dynamics. Thus we must be conscious of the biases sometimes inherent in mapping space, and ensure that we do not compound the disadvantages a group or a region might already be facing that led to being neglected in acquiring geospatial data in the first place by uncritical reliance on such map products. Exploratory analyses, e.g., the distribution of valid or meaningful data points versus the distribution of geographies or geometries, can be useful guides in discerning inequalities in mapping.

8.2.2 *Spatial autocorrelation*

In a 1970 article on modeling urban growth [281], American geographer Waldo Tobler expressed what has become known as Tobler's law or the first law of geography: everything is related to everything else, but nearby things are more related than things far apart. Spatial autocorrelation is the mathematical measurement of this relationship. Just as temporal autocorrelation is a metric of how the present correlates to the past a certain span of time away (the temporal lag of autocorrelation), spatial autocorrelation expresses how, if at all, what is here correlates to what is a certain distance away (known as the spatial lag).

Of the indices of spatial autocorrelation, by far the most popular is Moran's I. At the heart of Moran's I lies the concept of spatial weights, which is quite similar to the notion of a neighborhood as we have been using it in, e.g., Subsection 8.1.5. The spatial weights matrix w is a hollow matrix that determines whether two areas can be considered to be "neighboring" for the purposes of the spatial lag (where a neighbor corresponds to a spatial lag of 1, a neighbor's neighbor to a spatial lag of 2 and so on). There are two major approaches to creating spatial weights, and they differ in what they regard as the criterion of neighborhood.

- **Contiguity-based** weights consider two areas to be neighbors if they meet a contiguity definition.

   - Rook condition: under the rook condition, two areas are contiguous if their boundaries share an edge. New Mexico is a rook neighbor of Colorado and Arizona, as they share border edges.
   - Queen condition: under the queen condition, two areas are contiguous if their boundaries share a vertex. By definition, all rook neighbors are also queen neighbors, but the reverse is not true. New Mexico is a queen neighbor of Utah: they do not share a border edge, but they share a border vertex.
- **Distance-based** weights typically look at the distance between the centroids of two areas. By far the most popular is the k-nearest neighbor weighting, where, as the name suggests, an area's neighborhood is made up by a number of nearest areas. This is regardless of contiguity, in fact, it is possible to have noncontiguous nearest neighbors and contiguous areas that fall outside that definition.

For n spatial units with a weights matrix w, Moran's I with respect to the variable x is defined as

$$\frac{n}{\sum_{i=1}^{n}\sum_{j=1}^{n} w_{i,j}} \frac{\sum_{i=1}^{n}\sum_{j=1}^{n} w_{i,j}(x_i - \overline{x})(x_j - \overline{x})}{\sum_{i=1}^{n}(x_i - \overline{x})^2}. \tag{8.14}$$

We are overall less interested in the global Moran's I, which provides information about the entire area. Rather, we are usually interested in localized spatial autocorrelation, which describe the relationship between an area and its neighborhood:

- HH ("hot spots"): high values in the neighborhood are associated with high level in the index area, too.
- LL ("cold spots"): low values in the neighborhood are associated with low values in the index area, too.
- HL (positive outliers): low values in the neighborhood are associated with high values in the index area.
- LH (negative outliers): high values in the neighborhood are associated with low values in the index area.

p-values are obtainable using Monte Carlo simulation, which allows us to separate areas with statistically significant spatial autocorrelation. From an epidemiological standpoint, when performed over case counts or statistically significant local autocorrelations can be meaningful in four ways:

- HH ("hot spots"): hot spots indicate that an area follows its neighborhood with respect to rises in infectious dynamics. This is often the case for suburbs and bedroom communities, which follow peaks in infections in the adjoining urban area.
- LL ("cold spots"): cold spots occur where areas share features that predispose towards a lower number of infections. This is often the case where areas are under shared control with respect to quarantine measures, or share an isolating factor (e.g., multiple communities on a quarantining island could be expected to exhibit an LL pattern).
- HL (positive outliers): often, positive outliers indicate "entry communities" or "encounter communities." An entry community is an area where, as the name suggests, the pathogen enters a susceptible population (e.g., towns with significant

tourism/visitors, airports, sea ports), whereas an encounter community is a place that aggregates large numbers of individuals who would otherwise be separated. Positive outliers also indicate that the less affected neighborhood is at risk of rapid spread.

- LH (negative outliers): negative outliers are often communities worth paying attention to, such as Gunnison, Colorado during the 1918–19 influenza pandemic (see Subsection 8.2.1), these communities may have adopted policies that are highly effective. At the same time, a negative outlier is under constant pressure and at risk of infection from its more affected neighborhood. A negative outlier in case counts is typically a strong indication for reverse quarantine.

The respective local autocorrelations can guide evidence-based public health response: for instance, breaking up clusters of hot spots through quarantines and physical separation may reduce the coupled dynamics that allows the infection to spill over one area to its neighbors. Outliers are particularly useful for guiding quarantine response, with HL (positive outliers) indicating the need for quarantining the affected areas, whereas LH (negative outliers) suggesting a protective reverse quarantine

Practice Note 8.2 Projections

The Earth is spheroidal, and maps are flat. The consequence is that we need to engage in some mathematical finesse to translate one to the other. That process is called a projection, and how it is performed plays a significant role in the visual outputs of geospatial analyses at a wider scale.

For instance, in common Mercator projections seen in every schoolhouse across the world, Greenland appears to be almost as large, at least in latitude, as the whole of Africa. In reality, Greenland is less than a third the size of Africa in terms of latitude. In the same projection, Ellesmere Island (Umingmak Nuna) in Canada appears the same size as Australia, despite the latter being almost forty times larger.

The consequence of projections that are most accurate around the equator and increasingly distorting towards the poles is that much of what is termed the "developing world" appears significantly smaller than they actually are. As humans, we are evolutionarily geared to direct our attention to larger objects. This risks ignoring serious concerns that affect large populations being obscured by the mathematics of projections.

When selecting a projection at a large scale, it pays to be mindful of these differences. Selecting an appropriate projection with the right central meridian and generally eschewing super-scale maps (such as world maps) ensures that one's visual output does not lead data consumers into an incorrect analysis of respective scales.

Computational Note 8.6 Spatial autocorrelation of COVID-19

We begin our investigation of spatial autocorrelation of COVID-19 cases in the United States by obtaining the number of cases, and filtering it to the most recent day.

```
url = "https://raw.githubusercontent.com/nytimes/\
        covid-19-data/master/us-counties-recent.csv"
county_data = pd.read_csv(url,
                          usecols=["date", "fips", "state", "cases"],
                          dtype={"date": str, "fips": str, "state":
                                 str, "cases": int})

county_data = county_data[county_data.date
 == county_data.date.unique().max()]
```

Together with census data on population, this allows us to calculate the incidence normalized to unit population (i.e., cases per 100,000 population). We merge this information into a single GeoPandas `GeoDataFrame` called `gdf`.

Next, we create the spatial weights matrix between the counties using PySAL. In this case, we obtain a $k = 8$ k-nearest neighbor weighting:

```
w = weights.distance.KNN.from_dataframe(gdf, k=8)
```

Obtaining Moran's I is then as easy as calling the relevant function on `gdf` and supplying the weights as an argument:

```
morans_i = esda.moran.Moran_Local(gdf["cases_per_100k"], w)
```

While Moran's I is interesting in and of itself, our curiosity stretches further. First, we want to know what the direction or nature of the correlation is. A good first approximation is a "galaxy plot," displayed as the middle plot in Fig. 8.5. The galaxy plot is a density estimate heatmap overlaid over a scatterplot of each county's cases per unit population against the spatially lagged cases, with both standardized against the neighborhood mean. This allows for a neat separation into HH, HL, LL, and LH, with each occupying one quadrant (in clockwise order). It is typical in epidemics to have somewhat more hot spots than cold spots, as the asymmetry of the galaxy plot's upper right versus lower left quadrants exemplifies.

The `Moran_Local()` method of the `esda` package we used in the previous step to obtain a local Moran's I (typically called a Moran's I_s) conveniently provides us with a simulated p-value, accessible as the `p_sim` property off the object. This helps us filter nonsignificant spatial autocorrelation, as the bottom right plot in Fig. 8.6 exemplifies.

A notebook implementing the contents of this Computational Note is available on the book's companion Github repository in the folder /ch08/morans_ i.

8.2.3 Infectious disease and the urban space

The history of epidemics in human populations has always been closely connected to cities. From the Great Plague of Athens (430 BC) to the COVID-19 pandemic, cities have played a unique role in the lives of epidemics that affect human populations. The increased population density provides the pathogen with a vastly increased likelihood that during a given infectious period, an infected individual will make contact with a susceptible individual. Overcrowding and urban poverty also directly affect other epidemic processes, e.g., attracting vectors who thrive on the byproducts of urban human existence.

By far more important than mere population density is, however, the role of urban space as a space of interaction. The epidemiological correlate of population density is the fact that the number of susceptible individuals reached by an infectious individual (assuming, for argument's sake, a single infectious individual and an entirely susceptible population) is much higher than in a rural or otherwise more dispersed setting, as is, of course, the number of resulting infections. A corollary of this is that epidemic burnout is much less likely in an urban setting, as a study by Dalziel et al. [282] of influenza in cities in the United States has shown, influenza outbreaks burn out relatively swiftly in smaller cities while they find a more than sufficient supply of susceptible hosts in larger urban areas.

COVID-19 has reawakened interest in incorporating the consideration of epidemics and epidemic resiliency into urban planning and development [283]. The notion of pandemic-resilient urban planning is not new, in fact, epidemic processes and their avoidance has been among the driving forces of modern urban planning. As Gouveia and Kanai [284] writes, the COVID-19 pandemic has led to the threads of public health on one hand and urban planning on the other, which have been joined in a seamless braid until the emergence of vaccines and antibiotics (which provided aspatial control), to be re-uniting once again. Urban planning has, for instance, looked to the spatial segmentation of individual movement from the lens of the concept of "fifteen-minute cities," where every resident can reach all facilities required for their sustenance—work, play, healthcare, and so on—within fifteen minutes, as introduced by [285]. The fifteen-minute city is urban planning's answer to the hardships created by movement restrictions during COVID-19: in a fifteen-minute city, moving outside the fifteen-minute isochrone from one's place of residence would not ordinarily be necessary. This would reduce an individual's scale of mixing, and thus, possible exposure, to those within the thirty-minute isochrone of the individual's home. Though fifteen-minute cities have been described as pointing the way towards the

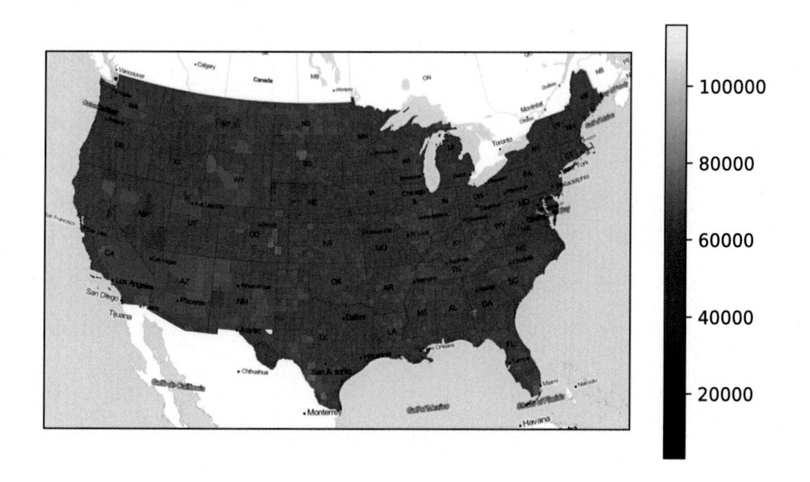

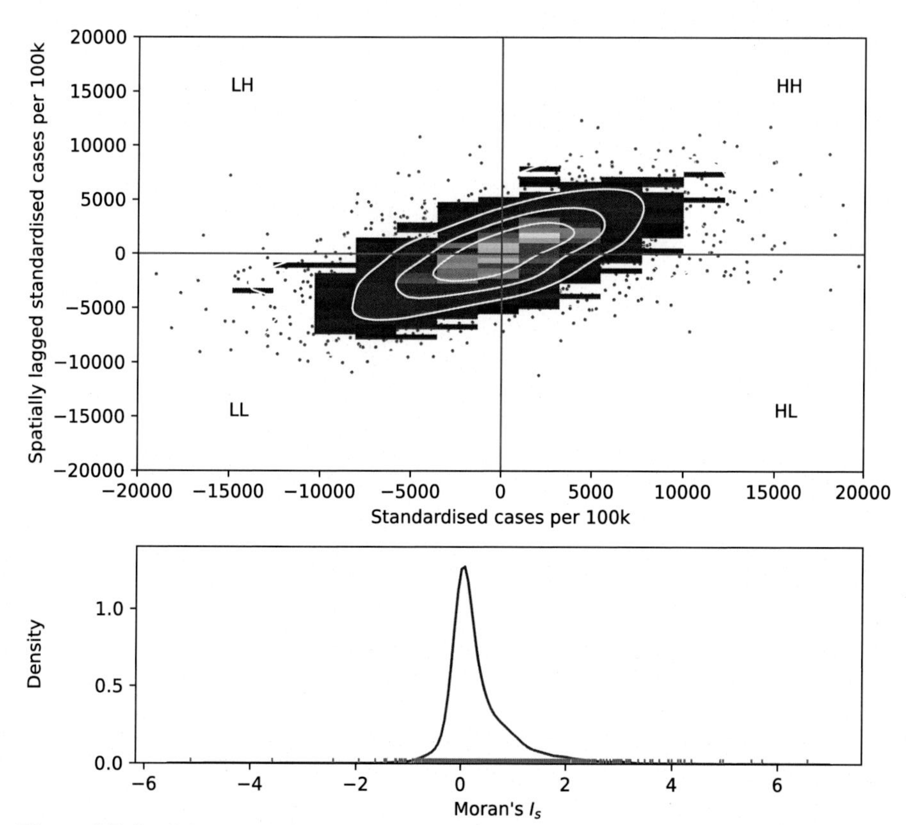

Figure 8.5 Spatial autocorrelation distribution of COVID-19 cases per 100,000 population by 5 May 2022.

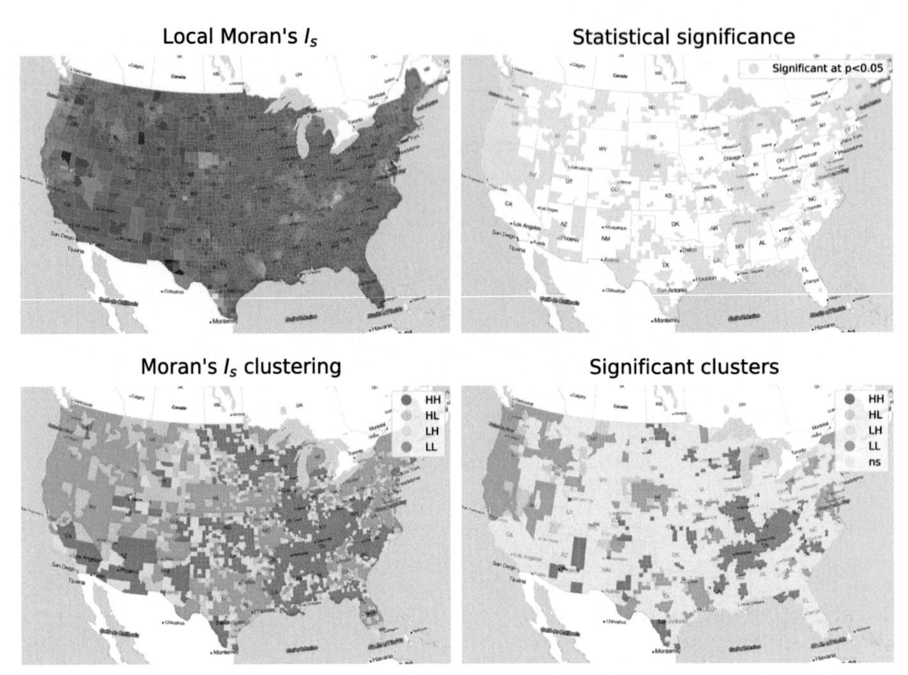

Figure 8.6 Choropleth maps of Moran's I for COVID-19 in the United States. Upper left: choropleth of I value. Upper right: counties with statistically significant values of I at $p < 0.05$. Lower left: raw quadrant assignment. Lower right: quadrant assignment filtered for statistical significance (ns: not significant at $p < 0.05$).

post-pandemic and pandemic-resilient city [286,287], urban planning paradigms in many places are ill-aligned with the goals, or even the possibility, of a fifteen-minute city.

Practice Note 8.3 Utopias and dystopias in the Smart City

The emergence of the Internet of things and new technologies that allow for an unprecedented volume of data to be collected, aggregated, and analyzed has unsurprisingly come to shape cities. Smart cities are thus not merely a static canvas that sets the scene for urban life, but a thing with a mind—or at least an algorithm—of its own, capable of synthesizing an operating picture from a wide variety of sensors (motion, traffic, occupancy, facility use, individual characteristics, and so on) and respond to optimize for a particular goal (e.g., reducing congestion and waiting times by intelligent traffic controls). It is unsurprising that the ability to direct, govern, and limit flows in an urban setting has gained attention as a possible avenue to respond to infectious disease outbreaks [288,289].

However, the line that separates a smart city from an Orwellian surveillance society is uncomfortably thin [290]. Ubiquitous surveillance is an intrusion into civil liberties, even if it is for an avowedly beneficial purpose. It must be narrowly construed and justified by strict reference to the objectives it seeks to accomplish. Though smart cities may open the door to a better data-driven pandemic response, they also open the door for concerning violations of privacy. Like with all responses to an epidemic, the rights of the individual must not be ignored in the pursuance of the welfare of society.

Despite the best efforts of urban planners, the tight links that connect cities as places of encounter and interaction to the spread of infectious diseases are unlikely to be easy to supervene through mere urban planning. Building a flood-resistant city is a relatively simple spatial challenge, because it involves only the space, and the interaction of a single extraneous force (water levels) with that space. In an urban pandemic, individuals move between locations, creating a much more complex problem that is much less amenable to a simple solution by way of urban planning. Often, the prevention of urban poverty, alleviation of unsanitary living standards, and elimination of unsafe living conditions (e.g., overcrowding) plays a much greater role in diminishing the potential of urban environments to contribute to epidemics.

Computational Note 8.7 Modeling the pandemic that never was

The world came uncomfortably close to a pandemic in late 1989, when, largely overshadowed by the sweeping political changes and the end of the Cold War, cynomolgus monkeys (crab-eating macaques, *Macaca fascicularis*) at a quarantine facility in Reston, Virginia, began to succumb with rather frightening rapidity to an outbreak of a haemorrhagic fever.

Located just a stone's throw outside the Beltway and within minutes from Dulles International Airport, the Reston monkey house could have been the ground zero of a devastating pandemic of proportions likely on par with the Black Death. In the end, the pathogen decimating the monkeys, now called Reston ebolavirus (RESTV), turned out not to be pathogenic in human populations [291]. The chilling story of humanity's narrow escape is documented by Jahrling et al. [292], but is made all the more real when we consider the spatial dimensions of the event. Fig. 8.7 depicts an isochrone map centered around the monkey house. An isochrone map (from the Greek iso, meaning "same," and chronos, denoting "time") marks points that are temporally equidistant, and are a powerful tool both for analyzing and responding to infectious disease.

Though we might not be able to recreate an isochrone map based on contemporary road conditions, we can create a fair approximation with present-day data. Creating an isochrone map requires us to map out the distances between

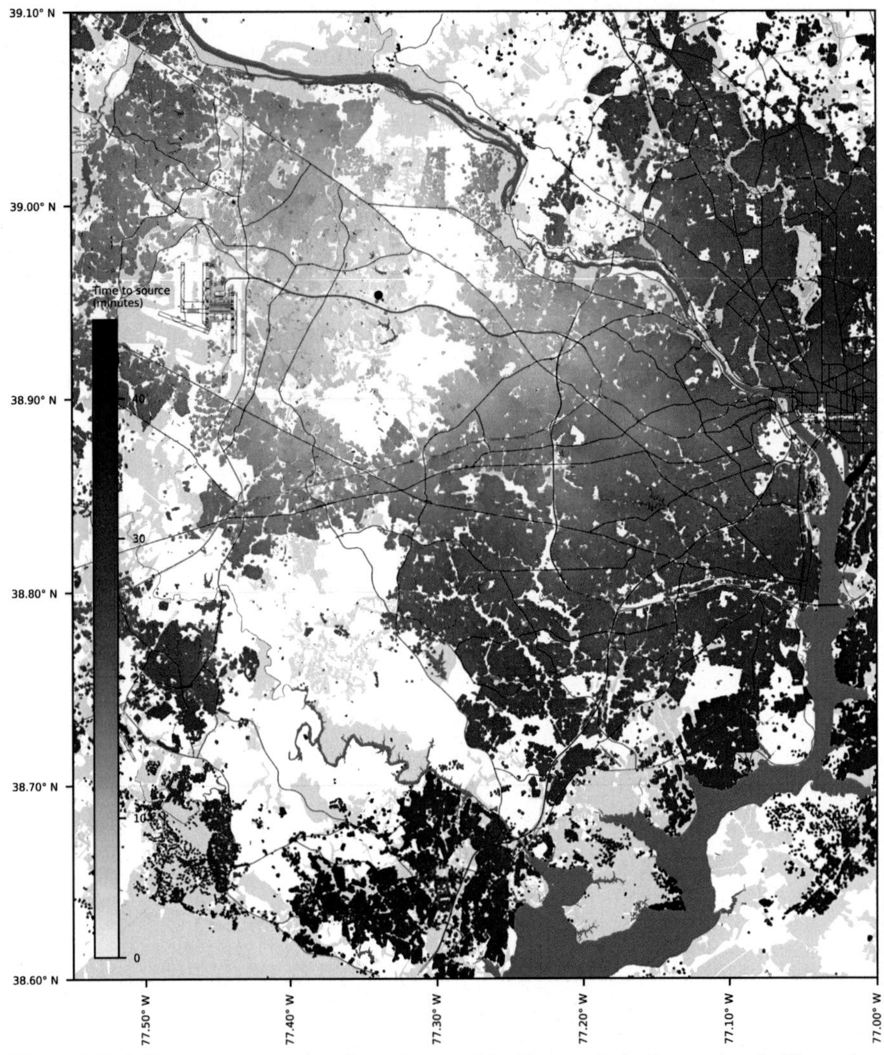

Figure 8.7 A time propagation (isochrone) map of the Reston ebolavirus outbreak, November 1989. The black spot is the location of the Hazleton Labs monkey house at 1946 Isaac Newton Square W, Reston, Virginia. Note the easy access to two major airports (Dulles International in the West and Reagan National in the east) and to central Washington, D.C. The distances calculated reflect road conditions as of 2022. Times are given in minutes.

the source location and a range of points on a street network and integrate the maximum allowable speeds over the distances. This is computationally highly intensive, but in an outbreak setting, such maps are indispensable tools to help us understand spatial risk over time.

For this calculation, we will be using an extract from OpenStreetMap via `osmnx` [293,294]:

```
source_address = "1946 Isaac Newton Square W, Reston, Virginia"
sourcey, sourcex = osmnx.geocode(source_address)
```

From this, we generate a NetworkX graph for a 25,000 m box around the focal point:

```
graph = osmnx.graph_from_point(center_point=(sourcey, sourcex),
                               dist=25_000,
                               dist_type="bbox",
                               network_type="drive")

graph = osmnx.distance.add_edge_lengths(graph)
```

Keep in mind that our "instinctive" ordering of parameters on the Cartesian grid (x, y) is the inverse of the conventional order of geographic coordinates, given in the format of latitude first, then longitude. Latitude corresponds to the y value, whereas longitude corresponds to the x value on a properly projected map.

Now, we can obtain the separate edge and node data frames using the `graphs_to_gdfs()` method of `osmnx`.

```
nodes, edges = osmnx.graph_to_gdfs(graph, nodes=True, edges=True,
                                   node_geometry=True)
```

Notice how the geometry contains distance (in metres), but we are interested in time. We will therefore have to translate distance into time based on the speed limit on a given road. Fortunately, `osmnx` provides two helpful methods to attach speeds and times to a graph:

```
graph = osmnx.speed.add_edge_speeds(graph, fallback=35*1.609)
graph = osmnx.speed.add_edge_travel_times(graph, precision=2)
```

The first of these adds edge speeds in km/h. As fallback, we specify 35mph, the most common urban speed in Virginia, and multiply it by 1.609, the conversion constant from miles to kilometers.

Next, we obtain the node on the graph closest to the geocoded source address:

```
monkeyhouse, distance_to_monkeyhouse
 = osmnx.distance.nearest_nodes(graph,
```

```
                                    X=sourcex,
                                    Y=sourcey,
                                    return_dist=True)
```

The distance to an index location is a useful indicator of graph quality. Ordinarily, distances less than 75 metres in an urban setting are acceptable. In our case, the distance to the monkey house is not all that relevant, but where graph simplification and point clustering has been used to simplify a complex graph, the index location's distance from the nearest node is often the moment of truth.

Next, we merge nodes to their time-denominated distance from the monkey house:

```
nodes = nodes.merge(pd.DataFrame.from_dict(
                                    travel_times,
                                    orient="index",
                                    columns=["time"]),
                         right_index=True,
                         left_on="osmid")
```

This could, in theory, be good enough, but we would like to associate individual buildings with a distance. For this, we shall obtain the geometries of all buildings within a bounding box:

```
bbox = (39.114, 38.74, -76.85, -77.70)

buildings = osmnx.geometries_from_bbox(*bbox,tags={"building":True})
```

We then perform a nearest-element spatial join on the `buildings` `GeoDataFrame` against the `nodes` `GeoDataFrame`. A nearest spatial join obtains a join with the nearest geometry from the target (which is ordinarily defined as the geometry that minimizes the distance between a polygon's centroid and a point or a nonpoint object's centroid). In this case, all nodes (road network intersection) are points and all building footprints will be polygons. The latter, we ensure by filtering out all point-type objects for buildings:

```
buildings = buildings[buildings.geom_type != "Point"]
```

Before we can perform a spatial join, we need to project both geometries to a projected coordinate reference system (CRS). To obtain an accurate spatial join, we need to project the `GeoDataFrames` into a suitable CRS. For Virginia, the North American Datum (NAD) 1983 Lambert equal area projection is best for our uses; this is an equal area projection, so distances will be accurately represented (within Virginia, about 1 m accurate), and though angles would be slightly distorted, this is not something we particularly care about here. In the EPSG system of identifying projections, the NAD83 Virginia Lambert CRS is known as EPSG:3968. We therefore write

```
buildings.geometry = buildings.geometry.to_crs(3968)
nodes.geometry = nodes.geometry.to_crs(3968)
```

Now, we can perform the spatial join:

```
buildings = buildings.sjoin_nearest(nodes,
                                    how="left")
```

With this, we have obtained an association between each building and the destination, so that each building is associated with the time it takes from the monkey house to reach the node nearest to the building. Finally, we reproject the geometries to the Web Mercator (EPSG:4326) CRS, which is convenient for plotting as it is the original CRS of web cartographic data:

```
buildings.geometry = buildings.geometry.to_crs(4326)
nodes.geometry = nodes.geometry.to_crs(4326)
```

To plot a building-level isochrone, we can quite simply use the `.plot()` method of `buildings`, providing `''time''` as the first argument, i.e., the argument that will be mapped against the color scale. Fig. 8.7 shows a building-level isochrone.

Two caveats deserve mention when it comes to isochrone maps.

- In this example, we have used the maximum speed limit as a proxy for actual movement speed. In reality, traffic does not always move at the speed limit, in fact, in densely populated urban areas notorious for traffic congestion, as the area of concern in our example would certainly be, it may be helpful to incorporate traffic data. A number of real-time traffic density APIs are available, many of which report the mean speed of traffic at any given time over a stretch of road. There are both commercial solutions and municipal open data initiatives (e.g., both in New York City and by the New York State Department of Transportation) to provide this data, which can be usefully incorporated into realistic models of traffic flows.
- This map associates the time-denominated distance with each building in the area under consideration. This does not mean that a real outbreak would reach all these points in that little time. This is a map of the minimum possible time for a pathogen to reach any one point. Traffic flows, traffic volume, and who make up that particular traffic (infected *vs.* uninfected) influence real-world epidemic propagation. Isochrones nevertheless shed light on the worst-case

scenario, and helpfully illuminate the extreme vulnerability of tightly connected urban areas to rapid pathogenic spread.

It is worth noting that urban road networks can get very big, and computation of distances can get computationally quite expensive. In many cases, as the graph gets bigger, `igraph` is superior to NetworkX in terms of speed. Computational Note 8.8 uses `igraph` in a similar shortest-path application, and the approach used in that example could be applied to the construction of isochrone maps alike.

A notebook implementing the contents of this Computational Note is available on the book's companion Github repository in the folder `/ch08/reston_virus`.

8.2.4 Access heterogeneities in space

Access to healthcare facilities governs a remarkable amount of factors in the human epidemiology of infectious diseases.

- In an outbreak setting, access often determines whether healthcare services can be sought out in the first place. Better access to healthcare services and diagnosis can assist in making informed decisions about quarantine and treatment, and in some cases, access to healthcare can facilitate isolation of symptomatic patients in the early stages of an outbreak to reduce further cases.
- In the longer run, as well as for endemic disease, early initiation of treatment that may reduce the transmission coefficient can make a significant difference in longer-term infectious dynamics. It is well-documented, for instance, that distance from the clinic is a significant factor in treatment nonadherence for TB [295–297], which in turn increases the likelihood not only of individual progression but also spread within the community.
- In emergent circumstances, distance to the nearest care facility is a direct negative correlate of survival [298,299]. Though this is best documented for cardiac arrest, there is evidence that the association holds for other diseases, and in fact even for no specific disease at all [300]. Where a pathogen may lead to a need for emergency medical attention, as is not uncommon with respiratory pathogens that can lead to ARDS, patients who have to undertake a long journey while *in extremis* fare significantly worse.
- Finally, the location of facilities for testing [301,302] and vaccination [303] often determine the success (or failure) of testing and vaccination initiatives. In an emergency, access to these services may well determine not only individual welfare but also the communal level of disease transmission.

For these reasons, we are critically interested in the ability of individuals to access appropriate healthcare. This is not only a matter of equity and welfare, but also has clear ramifications on epidemic processes. The analysis of spatial access parameters is thus a valuable tool in the management of an epidemic, since a lack of appropriate facilities within the feasible access radius of patients affects the spatial and overall quantitative dynamics of an epidemic.

Computational Note 8.8 Access and distance

While the spread of a pathogen is generally conditioned by the speed of movement, as we have seen in Computational Example 8.7, access is generally conditioned by distance. Without motor transportation, for instance, there is a maximum distance that an individual can travel within a reasonable time. Equally, the cost of getting from one point to another is a function of distance. Therefore when we assess facility access, the primary indicator is distance rather than time.

In this Computational Note, we will estimate the distance of residential households in Oxford, England, from their nearest hospital. We use osmnx [293,294] to obtain both the street graph of the District, and the location of residential neighborhoods and hospitals:

```
location = "Oxford, England"

graph = osmnx.graph_from_place(location, network_type="drive")
hospitals = osmnx.geometries.geometries_from_place(location,
            tags={"amenity": "hospital"})[["geometry"]].sort_index()
residential = osmnx.geometries.geometries_from_place(location,
              tags={"landuse": "residential"})
                    [["geometry"]].sort_index()
```

In OpenStreetMap, "geometries from place" are typically defined as "footprints" or areal geometries, i.e., polygons. Since polygons are sections of two-dimensional space, they are generally not very meaningful when we seek to relate them to a graph that ultimately determines distances. For this reason, we must convert the polygons to a point (by taking its centroid, which for a GeoPandas `GeoDataFrame` is as simple as calling the `.centroid` property of a geometry column), then finding the nearest point on the graph:

```
hospitals["nearest_node"] = osmnx.distance.nearest_nodes(graph,
                        hospitals.geometry.centroid.x,
                        hospitals.geometry.centroid.y)
residential["nearest_node"] = osmnx.distance.nearest_nodes(graph,
                    residential.geometry.centroid.x,
                    residential.geometry.centroid.y)
```

In this way, we have correlated the footprints with a point on the graph, without losing information about either.

To be able to accommodate a rather large data set, we will import igraph as `ig`, and convert the graph to an `igraph` object. igraph is a highly efficient graph library written in C, and though its Python front-end takes some getting used to (and is regrettably not too well-documented), it is very, very fast. We begin the conversion by relabeling our existing graph's nodes as integers, and setting these as node attributes.

```
osm_ids = list(graph.nodes)
graph = nx.relabel.convert_node_labels_to_integers(graph)

osm_id_vals = {k: v for k, v in zip(graph.nodes, osm_ids)}
nx.set_node_attributes(graph, osm_id_vals, "osmid")
```

Next, we construct the graph using `ig.Graph`, and add the vertices and edges from the NetworkX graph. We assign the edge lengths as edge attributes (again from the NetworkX graph), which will save us having to look up distances in the GeoPandas `GeoDataFrame`:

```
i_graph = ig.Graph(directed=True)
i_graph.add_vertices(graph.nodes)
i_graph.add_edges(graph.edges())
i_graph.vs["osmid"] = osm_ids
i_graph.es["length"]
 = list(nx.get_edge_attributes(graph, "length").values())
```

We can now go on and construct our functions to determine distances between nodes (`determine_distance()`) and obtain the distance to the nearest hospital:

```
def get_distance_to_nearest_hospital(source_node) -> float:
    return min([sum(list(map(
                    lambda x: i_graph.es[x]["length"],
                    i_graph.get_shortest_paths(v=source_node,
                                               to=i,
                                               weights="length",
                                               output="epath")[0]
                ))) for i in hospitals.nearest_node])
```

The function above bears some dissection. The innermost function is a `map` that applies a lambda, which obtains the length attribute of each edge, to the output of the `get_shortest_paths()` method of the igraph object. Because `get_shortest_paths()` by default returns a list of shortest paths, even if it only returns one, we select the first element (i.e., `[0]`). The `output="epath"` argument tells `get_shortest_paths()` to return a list of edges rather than nodes. This is convenient, because we happen to have the lengths of edges attached to the edges themselves, so we can retrieve it with the lambda above. The sum of these edge lengths is, of course, the total path length. Since this is the point's distance to all hospitals, we take the minimum to represent the distance to the nearest hospital.

Then, obtaining the actual distances is as easy as applying the `get_distance_to_nearest_hospital()` function to the `nearest_node` column of the `residential GeoDataFrame` that we created in the step before last:

```
residential["distance_to_nearest_hospital"] = \
            residential.nearest_node.apply
            (get_distance_to_nearest_hospital)
```

The isodistance is not the be-all and end-all of the spatial accessibility of healthcare, for instance, it matters whether public transportation supplies the facility in question, or whether there are chokepoints (such as a bridge or a heavily congested stretch of road) that make access quite difficult. Equally, isodistance is rather theoretical above a certain distance if appropriate transportation is not available, or if there are other spatial barriers (road blocks, unsafe neighborhoods &c.) in the way. Nevertheless, an isodistance map is a good first approximation for identifying areas that are at heightened risk of being relegated to the margins of healthcare response in an epidemic crisis. Mapping service distance from potential service users can assist in placing emergency expansion capacity in an epidemic, or provide outreach by mobile care teams. (See Fig. 8.8.)

A notebook implementing the contents of this Computational Note is available on the book's companion Github repository in the folder `/ch08/access_ heterogeneities/`.

8.2.5 Site selection and facility location

The optimization problem for selecting ideal sites for vaccination, testing, or treatment can be summarized as follows: given a graph of locations $x_1, \ldots, x_n$, which we assume are reflective of a roughly equal number of individuals per location, where ought we place our finite number k of facilities so as to satisfy an objective function? Typically, this objective function is distance: if $\delta(x_i, s_j)$ is the distance between the i-th location of service users and the j-th facility, we seek to optimize the locations of $s_1, \ldots, s_m$ so as to

$$\underset{k \in \mathbb{Z}[1,k]}{\text{minimize}} \sum_{i=0}^{n} \min(\delta(x_i, s_k)). \tag{8.15}$$

This is the problem known in operations research as the k-median problem (sometimes also referred to as p-median) for facility location [304], and can be solved given a number of candidates by building a distance matrix between each demand

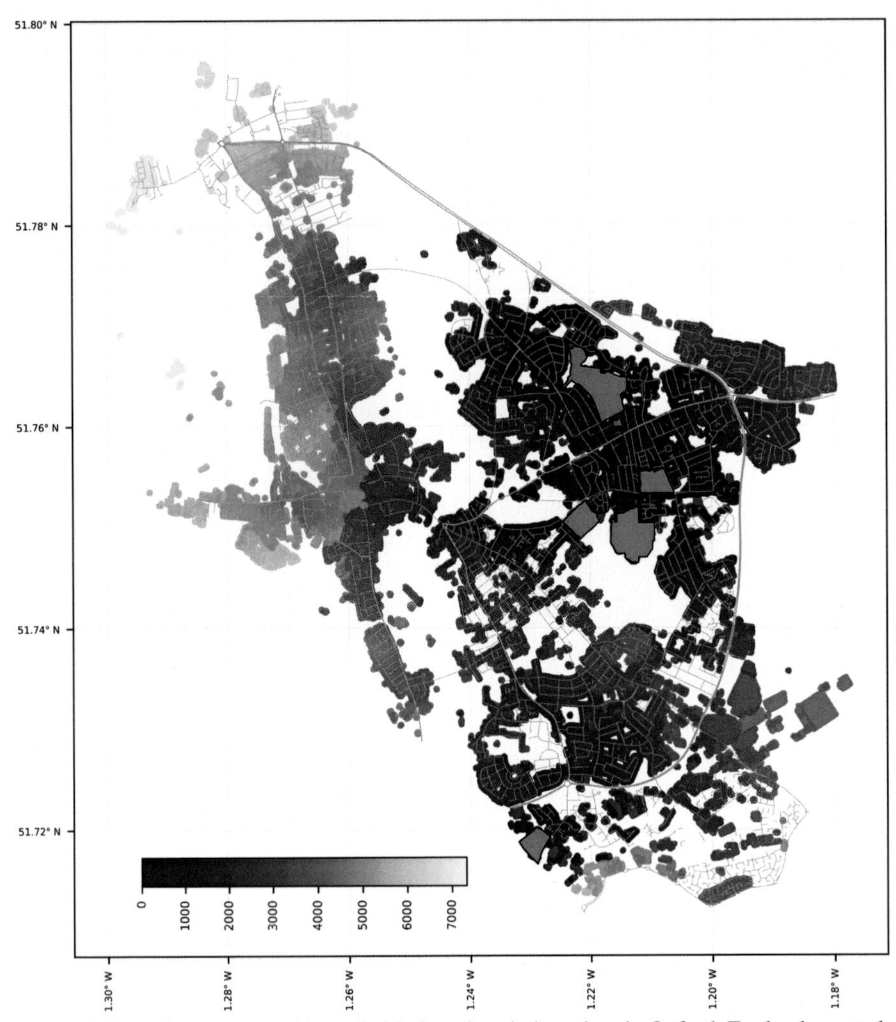

Figure 8.8 Isodistance map of households from hospital services in Oxford, England, created using the process in Computational Note 8.8. Residential buildings are colored according to their shortest-route paths to the nearest of the six hospital facilities (shaded in red). The town is separated into a Western and an Eastern part by the historic town centre, which is barred to automobile traffic. Only four traffic links connect the eastern part, where all hospitals are located, to the western neighborhoods. Distances are given in meters.

site (which in this case would be each household) and each potential supply site. The COVID-19 pandemic showed that in selecting candidate sites, thinking outside the box might be necessary to pick the right place: many testing (and later, vaccination) sites were established at parking lots for drive-through testing. Similarly, convention centres and sports venues were used to erect emergency hospitals in many crisis-hit areas.

Practice Note 8.4 The many and the few

The k-median solution delivers a compromise that generally privileges the majority. It may do so, however, at the expense of a small minority, and this expense may be quite significant. For instance, a k-median solution has no real incentive to take account of a small, sparsely populated village when there are cities full of people to cover. Inherently, k-median solutions marginalize those on the geographic, but also often social and economic, fringes of society.

k-center is a related solution, which is an inherently minimax approach: it seeks to minimize the maximum distance traveled by anyone to reach a facility. In addition, various optimization mechanisms exist that seek to locate facilities in a way to reduce an index of inequality, typically the Gini coefficient. Though the k-median problem is computationally vastly simpler to accomplish than other alternatives (in particular, minimax problems are difficult to solve computationally, because minimax objective functions are hard to optimize by the branch-and-bound method, which most solvers use under the hood), it is worth considering how a certain allocation could serve the interests not only of the many but also of the few on the margins of the picture.

It is worth noting that the k-median problem solves for the smallest sum of distances, not the smallest mean distance. The consequence is that in addition to distance optimization, it also provides a relatively balanced distribution of individuals assigned. A trivial nearest-facility assignment may allocate large numbers of individuals to a nearby facility, while allocating hardly anyone to a more outlying site. k-median solutions generally avoid this, except for large distance differences. As such, k-median solutions are likely to provide acceptable first approximations not just of distance-specific allocations but also a good balance of allocations to facilities assumed to be equicapacitant.

Computational Note 8.9 Placing COVID-19 testing sites in Manhattan

There are many considerations that flow into choosing the right location for mass medical interventions, such as an emergency quarantine facility, drive-through testing centres, or vaccination hubs. Distance, however, almost always proves to be a primary consideration, and more often than not, *the* primary consideration. Distance is a good proxy for public accessibility. In this Computational Note, we will be looking for suitable locations to set up COVID-19 testing sites in Manhattan.

As discussed in the foregoing subsection, we need, in general, three things to solve this type of problem:

- First, we need some candidate locations. In this example, suitable candidate locations will be hospitals only, as in the initial phases of an outbreak, the specialized HVAC facilities and separation facilities to manage large crowds are not available in a field-expedient setting. In practice, later on, community centres, parking facilities, event venues, as long as they cover a certain minimum size, could be selected. Any less, and it would be difficult to maintain social distancing, turning a public health intervention into a mass encounter event.
- Next, we will need to identify demand for the tests. In this case, we assume that all buildings have equal populations, which is often not very accurate. In practice, we would be using census data to help weight the spatial model. In the United States, a census block in urban areas roughly corresponds to a city block, or between 500 to 3000 people. This is often granular enough to sustain a detailed spatial model.
- Finally, we will need a distance metric, which in this case will be contingent on road travel distances. Thus we need to establish the weighted graph that governs the distance between points in our space.

We begin by obtaining the last of these, namely the graph, using the methods described in previous Computational Notes (see, e.g., Computational Note 8.7).

```
graph = osmnx.graph_from_place(location, network_type="drive",
                               simplify=False)
```

Next, we turn our attention to candidate sites, which we also obtain by osmnx [293,294]:

```
eligible_sites = osmnx.geometries.geometries_from_place(
        location,
        tags={"amenity": ["hospital"]})[["geometry"]]
eligible_sites["nearest_node"] = osmnx.distance.nearest_nodes(
        graph,
        eligible_sites.geometry.centroid.x,
        eligible_sites.geometry.centroid.y)
eligible_sites = eligible_sites[eligible_sites.geometry.apply(
    lambda x: x.type != "Point"
)]

eligible_sites = eligible_sites.reset_index()
```

We have accomplished four major tasks in this step: we obtained all geometries from OpenStreetMap that are tagged with the relevant amenity features, obtained centroids, and identified the nearest nodes. Finally, we have eliminated building footprints that are stored as points, as these are typically duplications of existing building footprints, where they coincide with a point of interest.

Finally, we need to retrieve the locations of demand, which in our instance will be all buildings. OpenStreetMap is good at providing building footprints, but it has significant trouble distinguishing between residential and nonresidential structures. In areas such as Fairfax, where most buildings can be safely assumed to be residential, we can simply measure for all buildings and take the resultant small error over ignoring entire neighborhoods that are not adequately mapped:

```
buildings = osmnx.geometries.geometries_from_place(
    location,
    tags={"building": True})[["geometry", "building"]]
buildings = buildings[buildings.building.isin(building_filter)]
buildings = buildings[buildings.geometry.apply(lambda x: x.type
                      != "Point")]
buildings = buildings.reset_index()
buildings["nearest_node"] = osmnx.distance.nearest_nodes(graph,
                          buildings.geometry.centroid.x,
                          buildings.geometry.centroid.y)
```

Now that we have the requisite data, we may begin the optimization process. We will use `spopt`, part of the PySAL spatial analytics library for Python, which serves as a front-end for various optimizers. Many of these optimizers require a license fee or are otherwise constrained by licensing requirements. In this case, we will use the built-in COIN-OR optimizer. This is an open-source and relatively good optimizer, although commercial optimizers often deliver better performance.

`spopt` provides a p-median solution out of the box, but it generally requires a cost matrix, which is essentially a matrix of distances from each demand location (i.e., each building) to each potential testing site. We obtain this by finding the shortest route between the building and each of the hospital facilities. Since this is a fairly computationally expensive step, it is useful to convert the graph into `igraph`, as we have done in Computational Note 8.8:

```
osm_ids = list(graph.nodes)
graph = nx.relabel.convert_node_labels_to_integers(graph)

osm_id_vals = {k: v for k, v in zip(graph.nodes, osm_ids)}
nx.set_node_attributes(graph, osm_id_vals, "osmid")

i_graph = ig.Graph(directed=True)
i_graph.add_vertices(graph.nodes)
i_graph.add_edges(graph.edges())
```

```
i_graph.vs["osmid"] = osm_ids
i_graph.es["length"]
 = list(nx.get_edge_attributes(graph, "length").values())
```

To construct the distance matrix, we first define the distance metric as a function:

```
def get_distance(source_node, destination_node) -> float:
 return sum(list(map(lambda x: i_graph.es[x]["length"],
                     i_graph.get_shortest_paths(v=source_node,
                                                to=destination_node,
                                                weights="length",
                                                output="epath")[0])))
```

We follow this up by building the matrix itself:

```
cost_matrix = np.zeros(shape=(len(buildings), len(eligible_sites)),
                       dtype=float)

for i in range(len(buildings)):
    for j in range(len(eligible_sites)):
        cost_matrix[i, j]
         = get_distance(buildings.loc[i, "nearest_node"],
                        eligible_sites.loc[j, "nearest_node"])
```

We initialize, then run, the *p*-median optimizer over this, weighting buildings equally (i.e., with a weights matrix of ones):

```
pmedian
 = PMedian.from_cost_matrix(cost_matrix,
                            weights=np.ones(cost_matrix.shape[0]),
                            p_facilities=p)

pmedian.solve(pulp.COIN_CMD())
```

This launches the solver, which may for a problem of this size take some time. In what is somewhat unusual for a Python package, spopt only populates the `cli2fac` property of the `PMedian` result object after the respective functions have been called:

```
pmedian.facility_client_array()
pmedian.client_facility_array()
```

Rather than returning any results, these functions merely serve to populate `cli2fac`, the property that associates each building with a testing facility, and also represents the p selected testing facilities:

```
selected_facilities = np.unique(pmedian.cli2fac).tolist()
```

All that is required from this stage onwards is to visualize the assignment of each building to each facility, and determine the distances. Fig. 8.9 shows the results for $p = 3$ in Manhattan. Allocation maps such as this allow us to analyze, and reason about, optimal site selection. However, it is worth noting that excessive reliance on mathematical optimization risks us falling into the trap of the McNamara fallacy, the erroneous belief that what cannot be calculated does not matter [305]. There are many factors that cannot be quantified, or are usually not quantified in such models: the impact on health equity (see Practice Note 8.4), the willingness of populations to take up a treatment or prophylaxis offered, and limitations on mobility conditioned by socioeconomic status, to name a few. Mathematical optimization is a good way to deliver a first approximation of the best places from which to deliver a service, but it is only as good as its weighting algorithm, which necessarily cannot comprehend the incredible variety of factors that flow into picking the best location for service provision.

A notebook implementing the contents of this Computational Note is available on the book's companion Github repository in the folder `/ch08/testing_site_location`.

8.2.6 Spatial graph interdiction

Closeness keeps us alive. As our simplified simulation by way of cellular automata has shown (see 8.1.1), connections keep us alive. Humans and animals need to be part of a larger whole as much as the dark (or "alive") cells on our imaginary infinite canvas die from loneliness if they are disconnected from neighbors altogether. Collaboration, foraging, protection, shelter and the emotional-psychological need for company are just a few ways in which we depend on connections for our survival.

At the same time, connections put us at risk. The same dynamics that keep us safe in a pack, herd, or society, and comfortable in our family, friends or neighbors also serves as a way for pathogenic transmissions. The warmth of a human dwelling or the immense complexity of a bee hive is also an opportunity for a pathogen to tap into a susceptible population. Network interdiction is a comprehensive name for algorithms intended to disrupt such connections.

Network interdiction refers to a category of problems in operations research: given a network, the attacker's problem is to sever the minimum amount of edges to reduce flow (or communication, interaction, infection,...) in the network, as defined by some particular metric. In spatial networks, the interdiction problem refers to the identifica-

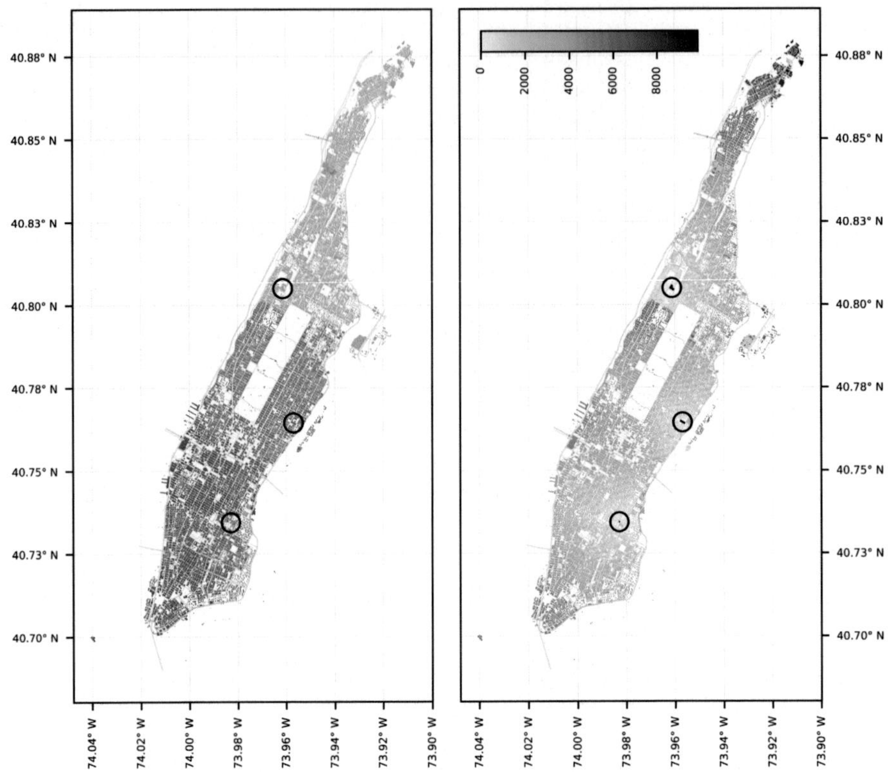

Figure 8.9 Optimization-selected sites for mass testing in Manhattan, New York City, New York. The three selected test sites are denoted by circles. The map on the left shows the allocation of each building to a testing site, whereas the map on the right indicates the distance to the allocated testing site. Gray-filled areas denote candidate testing facilities (hospitals). Distances are given in metres.

tion of vertices or edges that have a high impact on overall communication within the network.

The effect of interdiction is often best accommodated by comparing a certain variable that describes communication over the graph before and after the interaction, an approach that Latora and Marchiori [306] called a "delta model." To a delta model, the value of an edge or a node is the loss to the overall network that arises from losing a node or an edge. In Latora and Marchiori [306]'s original work, this was expressed in terms of network efficiency, which is the mean length of all shortest paths between all nodes (which is equivalent to the Wiener index divided by the number of nodes in the graph). This is a model, as the authors noted, of information flow, but it is equally appropriate for the modeling of the spread of a spatial phenomenon like infectious disease.

Unlike in spatial risk, which is inherently a metric of vulnerability (although the inverse of the way that term is used in graph theory, where being disconnected rather than being connected is considered as the loss), interdiction is more efficient if we know a location (often referred to as a "source") to assign to the infectious process. Thus we are not interested in all node-to-node shortest paths, but only the distance between the source node and possible susceptible nodes. Representing the infection as emanating from a single node s on the graph G, the infective efficiency over G from s is

$$I(G, s) = \begin{cases} \sum_{j \in G, j \neq s} \frac{1}{\delta_{s,j}} & \text{if } (s, j) \in E \\ 0 & \text{otherwise} \end{cases}, \tag{8.16}$$

where N is the total number of nodes, and $\delta_{s,j}$ is the distance between s and an arbitrary node j. Where s and j are not connected, $I(G, s)$ is defined to be zero.

We may now associate a value $\Delta_I(G, s, \epsilon)$ with edge ϵ in G's edge set, so that

$$\Delta_I(G, s, \epsilon) = \frac{I(G, s) - I(G \setminus \epsilon, s)}{I(G, s)}. \tag{8.17}$$

In reality, infections are rarely point processes, or at least infections for which an effort of interdiction is worth mounting. For a set of infected nodes S, that is, a subset of G, we may define $I(G, S)$ as

$$I(G, S) = \begin{cases} \sum_{j \in G, j \notin S} \max_{i \in S} \frac{1}{\delta_{i,j}} & \text{if } (i, j) \in E \\ 0 & \text{otherwise} \end{cases}. \tag{8.18}$$

It stands to reason that given S and a spatial connectivity graph G, with some distance metric δ (which may be distance or time, the latter being a very sensible choice indeed if there are various ways to traverse the graph at different speeds), we may associate with each edge the value Δ_I, which represents the infection's loss of relative ability to spread if that edge is severed. Higher values of Δ_I indicate edges that are more crucial to the infection, making good candidates for interdiction.

Though this model is premised on edge interdiction, it can be adapted to node interdiction equally well. Node interdiction is different in that it severs all edges connected to the node being interdicted. Just as blocking an entire airport has more severe effects than a storm disrupting a single route, node interdiction is inherently more effective but less selective. Edge interdiction models approaches, such as roadblocks and closure of air or sea routes, adequately, whereas node interdiction is a better proxy for scenarios that involve immunization or quarantine in graphs, where each node represents a community, conurbation, or household (indeed, the partitioning model of vaccine targeting proposed by Chen et al. [307] is an example of node interdiction).

It is worth noting that Δ_I changes if the network topology is altered, as is indeed the case if a road block is applied, for instance. In such cases, nested interdiction can be applied, selecting the highest-ordered edge, interdicting it, removing it from the graph and reevaluating Δ_I on the modified graph.

Computational Note 8.10 Nested interdiction: stopping SARS in Kowloon

Dr. Liu Jianlun, a physician from Guangdong Province, PRC, arrived in Hong Kong on 21 February, 2003 for his nephew's wedding. Unbeknownst to him, he harbored a dangerous fellow traveler, the causal agent of the epidemic of SARS raging in his home province. Dr. Liu stayed in Room 911 at the Metropole Hotel in Kowloon, which turned into the focus of Hong Kong's SARS epidemic. A day after his arrival, he became unwell and sought medical attention at the Kwong Wah Hospital. In that single day in the Metropole, he touched off chains of transmission that spanned the globe, from Canada to Taiwan, from Vietnam to Singapore.

In this Computational Note, we simulate an outbreak centering on what used to be the Metropole, and examine potential best locations for a limited number of roadblocks using iterative interdiction based on Δ_I. Densely populated urban environments, such as Kowloon, often facilitate rapid transmission, and methods of physical isolation, such as roadblocks or closure of major thoroughfares, are important tools in the public health arsenal.

We analyze each edge's suitability for interdiction by their effect on the network at large, using a modified efficiency metric. Though the efficiency metric described by Latora and Marchiori [306] describes the damage to the communicability of the overall network if a certain edge is lost, we are not particularly interested in this; rather, our concern is very specifically with the degree to which the loss of an edge makes infectious nodes less accessible. Thus for the set of infected nodes S, that is, a strict subset of G's node set, we determine loss of infectious network efficiency as the sum of the inverse of the distance to the nearest node.

Let us begin by importing the Kowloon street grid, converting it into `igraph` using the process described in Computational Note 8.8 and define the infectious nodes as the nodes that are directly connected to the node closest to the location of the Metropole (we will be using coordinates for this, as the site has since then been renamed, and thus does not georesolve):

```
sourcex, sourcey = 836151.13, 820010.00
source_node = osmnx.nearest_nodes(graph, X=sourcex, Y=sourcey)

infectious_radius = 1
infectious_nodes = i_graph.neighborhood(source_node,
                                        order=infectious_radius,
                                        mode="out")
```

If the coordinates look somewhat unusual, this is because we have converted the geographical coordinates to EPSG:2326, the Hong Kong coordinate grid. At a radius of 1 (which includes the source node and all other nodes within one hop), we get an initial infectious set of five nodes.

Next, we need to define a distance metric. We are using `igraph`'s `get_shortest_paths` method and selecting the first, weighed by length:

```
def get_distance(G, source_node, destination_node) -> float:
    return sum(list(map(lambda x: G.es[x]["length"],
                        G.get_shortest_paths(v=source_node,
                                             to=destination_node,
                                             weights="length",
                                             output="epath")[0])))
```

This function obtains the list of edges that constitute the shortest path between the `source_node` and the `destination_node`, and sums up their length. Now, given a list of infectious nodes, we can simply iterate over them, determine the distance for each point, and thus calculate $\mathrm{DNIN}(G, n, \mathcal{S})$, the distance to the nearest infectious node in $\mathcal{S}$ from node n on the graph G:

```
@np.vectorize
def dnin(G, node) -> float:
    """Distance to the nearest infectious node."""
    distances = []

    for infectious_node in infectious_nodes:
        dist = get_distance(G, source_node=node,
                            destination_node=infectious_node)
        if dist > 0:
            distances.append(dist)
        else:
            distances.append(np.nan)

    return np.min(distances)
```

In many use cases, a space-partitioning data structure, such as a ball tree or some other tree-based search algorithm, proves useful to decrease the time vis-a-vis taking the minimum of a brute force search. However, if the number of infectious nodes is sufficiently small (less than half a dozen in this example), it is often not worth the effort to construct a more complex data structure. Of course, if the number of infected nodes is larger, the case for a space-partitioning structure becomes much stronger.

Next, we need to define our metric for the prioritization of edges to be cut. This is the difference in network infectious efficiency, which in turn requires us to quantify the latter. Given the infectious subgraph, the nodes that are considered infectious; the network infectious efficiency is given by the function

```
def network_infectious_efficiency(G, infectious_subgraph) -> float:
    agg = 0

    for i in range(G.vcount()):
        if i not in infectious_subgraph:
            dist_nin = dnin(G, i)
            if not np.isnan(dist_nin) and dist_nin > 0:
                agg += 1/dist_nin

    return agg
```

In this case, since we do not envisage completely disconnecting any nodes, there is no point in implementing a penalty for disconnected nodes. The distance to a disconnected node cannot be ordinarily ascertained (unhelpfully, `igraph` returns a zero distance if it cannot calculate a shortest path). The typical way of penalizing for an entirely disconnected node is to consider its distance to be infinite, and hence the inverse of the distance converges to zero. In practice, this is implemented by dividing the sum of shortest paths by the number of uninfected nodes:

```
def network_infectious_efficiency(G, infectious_subgraph) -> float:
    ...
    uninfected_nodes = G.vcount() - len(infectious_subgraph)

    return agg/uninfected_nodes
```

Now, we can calculate the efficiency loss corresponding to the loss of each edge:

```
def delta_efficiency(G):
    base_efficiency = network_infectious_efficiency(G,
                G.neighborhood(source_node, order=infectious_radius,
                                     mode="out"))

    deltas = []

    for edge in G.es:
        modified_graph = G.copy()
        modified_graph.delete_edges(edge)
        gpeff = network_infectious_efficiency(modified_graph,
                G.neighborhood(source_node, order=infectious_radius,
```

```
                mode="out"))
        deltas.append((base_efficiency - gpeff)/base_efficiency)

    return deltas
```

It may at first glance be somewhat counterintuitive why we do not refer to the infectious nodes as a set of known nodes, but rather obtain them by taking the neighborhood of the source node over and over again. `igraph` is strictly indexed, that is, if an edge or a node is deleted, the rest may be renumbered to keep up a strict integer index. For this reason, the numbers that refer to the infectious nodes might change if the network is modified.

Finally, we can put this into a function that iteratively performs this process for each graph configuration:

```
def cascading_interdiction(G, k=3):
    g = G.copy()
    base_efficiency = network_infectious_efficiency(G,
                G.neighborhood(source_node, order=infectious_radius,
                                       mode="out"))

    interdictions = []

    for i in range(k):
        e, Eff = None, 0
        for edge in g.es:
            modified_graph = g.copy()
            modified_graph.delete_edges(edge)
            gpeff = network_infectious_efficiency(modified_graph,
                G.neighborhood(source_node, order=infectious_radius,
                                       mode="out"))

            if (base_efficiency - gpeff)/base_efficiency >
                Eff and gpeff > 0:
                e = edge
                Eff = (base_efficiency - gpeff)/base_efficiency

        interdictions.append(e.attributes()["uvk"])
        g.delete_edges(e)

    return interdictions, g
```

This returns a tuple object, consisting of a list of the interdicted edges' indices and the final graph. The latter is useful to obtain the configuration of the final graph after k optimized interdictions. Based on this, we can obtain the distances of points, e.g., buildings, from the infected nodes subject to our interdictions. Fig. 8.10 shows such a model for the area of Kowloon surrounding the former Metropole hotel. The impact of interdiction is quite apparent: the line of interdictions in the west, blocking Argyle street, Shantung street, and Waterloo road (from North to South) are effective at protecting the densely populated Mong Kok area (at the western edge of the figure), whereas the absence of interdictions to the east makes Kadoorie Hill (to the north of the sources of infection) and Ma Tau Wai (to the east) much more vulnerable, even though they are spatially further away.

A notebook implementing the contents of this Computational Note is available on the book's companion Github repository in the folder `/ch08/graph_interdiction`.

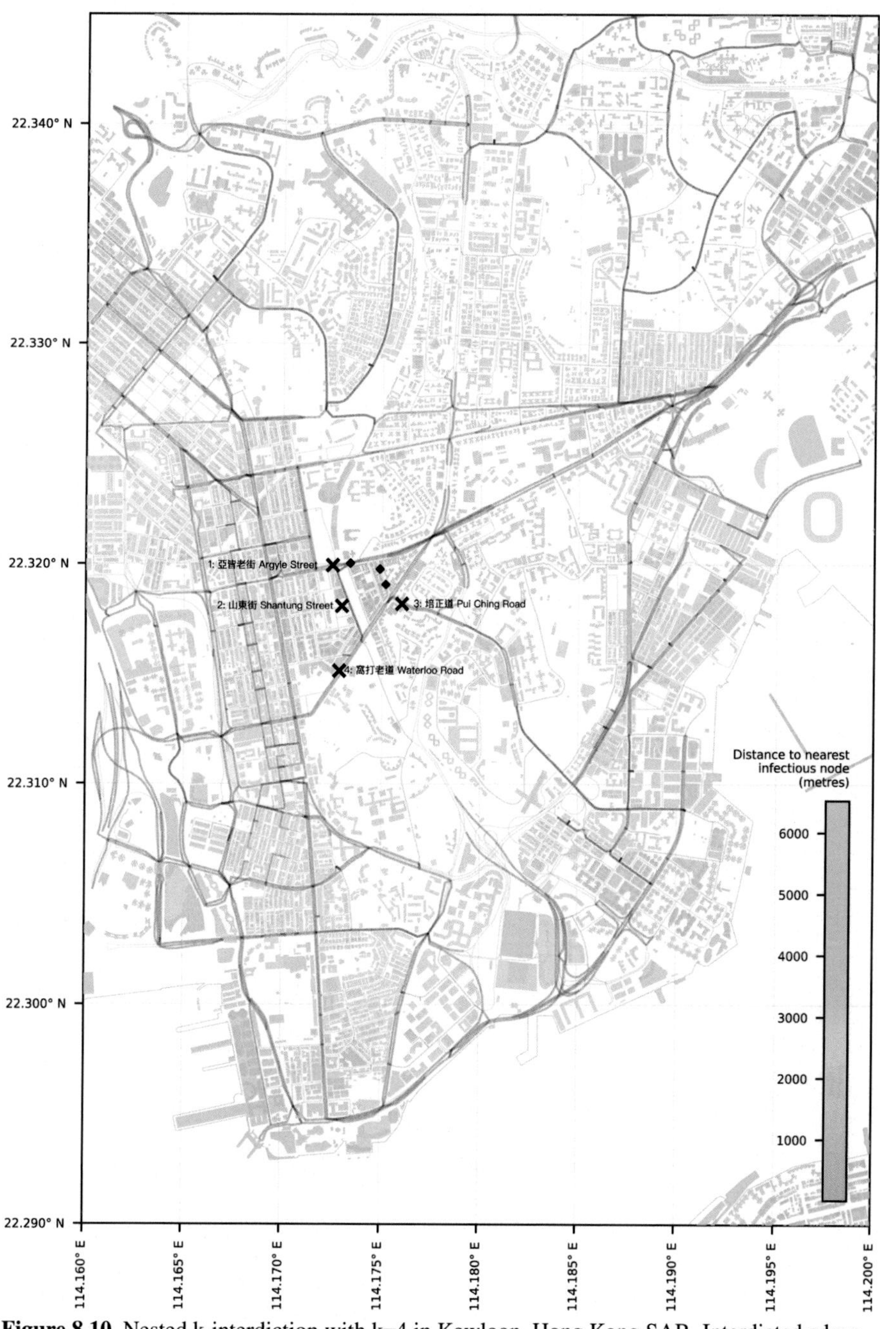

Figure 8.10 Nested k-interdiction with k=4 in Kowloon, Hong Kong SAR. Interdicted edges are marked by an X. The nodes closest to the origin of the infection, constituting the infectious subgraph, are noted by black diamonds.

Agent-based modeling
Simulating populations at scale

Imagine how hard physics would be if particles could think!

Murray Gell-Mann, quoted by Crooks et al. [308]

9.1 The fundamentals of agent-based modeling

One of the most complex computer games ever devised is called Dwarf Fortress. It is not much to look at: its graphics are the terminal-based structures that were in vogue in the 1980s. What makes Dwarf Fortress an extraordinary game is the depth of agent-based logic: every character, every enemy unit, even pets are endowed with a hugely complex agent-based behavioral model [309]. As an example, cats in Dwarf Fortress can stray into puddles of spilled beer, lick their paws later, and succumb to alcohol poisoning.

Yet agent-based modeling is about much more than belligerent dwarves and drunk cats. Agent-based models are powerful computational tools to simulate large populations of boundedly rational actors who act according to preset preferences, although often enough in a stochastic manner. They can simulate the complex human behaviors of quasi-rational decision-making, represent large populations and, through iterative simulation, highlight likely behavioral outcomes of crowds.

Most of the foregoing chapters described a kind of statistical mean-field dynamics of epidemiology—we might have known what a population does, but not much about any one individual in a population. At best, we could deduce the state or behavior of an individual in terms of likelihoods, e.g. if 30% of a population is infectious, there is approximately a 30% chance a randomly selected individual from the population will be infectious. Like statistical mechanics, it offers us the ability to reason about dynamics at the population scale without having to model each individual.

This chapter explores an alternative approach. Agent-based models are primarily inductive—we obtain information about the population by large-scale, repeated simulation of individual agents. Such models allow a different glimpse into the operation of an epidemic process. Many phenomena that would be challenging to model on their own, such as heterogeneous populations with multiple heterogeneities, some continuous and others categorical, become almost trivially easy to analyze using agent-based models. On the other hand, agents can adopt complicated behaviors and even very complex behavioral profiles are relatively easy to describe in the agent-based paradigm, because we only need to describe an individual rather than an entire population. This chapter discusses how we can leverage agent-based models for understanding infectious disease dynamics.

Computational Modeling of Infectious Disease. https://doi.org/10.1016/B978-0-32-395389-4.00018-9

Practice Note 9.1 Communicating agent-based results

While agent-based models are older than both the author of this book and most of its readers, their application to public health is of relatively recent vintage. As Maziarz and Zach [310] point out, ABMs have not traditionally been included on the 'evidence pyramid' of evidence-based medicine (EBM), leaving those unfamiliar with the field rather confounded as to what the relative strength of a model is. This is no careless omission by the EBM community: rather, it is a recognition that agent-based modeling is a metamodel, not a model in and of itself. It is an approach and a methodology to construct models rather than a class of models that can be narrowly circumscribed the way e.g. generalized linear models are.

An added problem is that agent-based models are often poorly understood by epidemiologists who have had limited or no exposure to computational methods. Thus, ABMs are often seen as less 'scientific', or as a kind of 'brute-forcing of reality'. These perceptions then migrate into the public consciousness, especially where ABMs are referred to as 'simulations', which carries the connotation of being distinct from 'real evidence'. Evaluating an agent-based model is complex—it requires the reader to understand the model's entire panoply of assumptions, including the assumptions it fails to make and the factors it does not consider. Good, thorough yet parsimonious documentations of a complex ABM can easily run into the hundreds of pages. ABM evidence is thus challenging for experts and often impenetrable for laypeople and policy-makers.

Paradoxically, ABMs are often the most valuable in contexts of limited information [311]. As Marshall and Galea [312] note, ABMs can contribute to areas such as causal inference, where there are underlying complexities that make less complex models unsuitable.

It is thus important that users of agent-based models communicate the model and its assumptions—including processes and factors that are known but intentionally ignored—in detail. All computational epidemiological models must be extensively documented to allow replication—as Janssen, Pritchard, and Lee [313] argue, computational models still often fall short of reproducible research due to missing data, and the proliferation of badly documented—or at times entirely undocumented—computational models during the COVID-19 pandemic has raised tough questions about the extent to which ABMs (and other complex models) ask policy-makers to make decisions based on opaque evidence from a black box that may be inscrutable to experts, too. Clear code, well-written documentation and a willingness to elucidate agent-based modeling and its differences vis-a-vis traditional approaches can go a long way in communicating and situating ABMs to less familiar audiences.

9.1.1 Case study: Schelling's surprising result

As humans, we are attracted to people like us: most of the social interactions we enter into by choice is with people who share attributes that we consider important, be they mutable or immutable. In the social sciences, this phenomenon is known as homophily. In 1971, Schelling explored what level of homophily is required for a system to trend towards segregation—and his result was quite surprising [19].

In Schelling [19]'s simplified model, agents were modeled on a rectangular grids and split into two groups—"blue" and "red". Each turn in discrete time, agents check their Moore neighborhood (the eight cells that have at least one point in common with the cell occupied by the agent). If, ignoring empty spaces, the fraction of the neighborhood occupied by agents of the same "group" fails to meet or exceed a homophily threshold, the dissatisfied agent will attempt to move to the first vacant spot in which the homophily threshold can be met or exceeded.

Schelling's perplexing finding was that given two groups of equal size, the degree of homophily required to achieve segregation is quite low (around 0.33). In other words, very small preferences for being surrounded by people like oneself can result in the development of a segregated environment. His study used agent-based modeling to answer a question that is rather less amenable to traditional analytical solutions.

While such models are obviously quite simplistic (for instance, Schelling's model assumes that all agents have the same threshold of homophily and the same resources to relocate), they can help us explore vistas of complexity that would otherwise be hardly accessible. In an assessment of agent-based models in population health, Silverman et al. [314] write that the impact of ABMs is most significant where their use seeks to answer questions to which the standard tools of epidemiology would not be able to respond. Agent-based models thus augment the more traditional methods of epidemiology with powerful new computational tools.

9.1.2 Simulating disease processes with ABMs

All agent-based modeling follows a certain internal logic. Understanding this logic is the key to implementing complex ABMs. It is a particular perspective on processes in the real world, and might take some getting used to, but rewards the user with a powerful tool to model complexity that eludes other instruments.

There are, in general, three questions to answer before defining an ABM.

1. **What** do we model? In other words, what are the states and properties our agents can have? What we used to think of as compartments are now properties that individual agents can assume (albeit subject to the same exclusionary rules that account for the partition property). A powerful feature of this is that the sets of states need not be mutually exclusive. We might, for instance, keep track separately of whether one is susceptible, infectious or recovered on one hand and whether one is or is not in quarantine on the other.
2. **Whom** do we model? That is, what are our agents representing: individuals, metapopulations, perhaps even individuals from different planes of infection? What are their behaviors and policies? There is virtually no limit to what an agent

can do, but we will have to specify each of those things. In short, when we define agents, we define a set of behaviors with them, too.

3. **Where** do we model? Agent-based models do not need to have a spatial dimension (although they do need some topology over which the concepts of distance and neighborhood can be defined), but they might also exist in networks or even in two- or three-dimensional space. The space may be an abstraction, such as an infinite sheet or a torus, or it may be a representation of real-world geographic features.

Computational Note 9.1 Using Mesa

In our applications of agent-based modeling, we will use a Python package called `mesa`. There is a not inconsiderable amount of computational housekeeping involved in running an ABM, and `mesa` handles most of this. It comes with 'batteries included', i.e. most of the methods used for managing ABMs are included and functional out of the box.

`mesa` provides five kinds of resources we will make extensive use of:

1. **The `Agent` class** is a base class that we subclass and expand to create our blueprint for an agent by defining its properties and behaviors.
2. **The `Model` class** is a base class that defines a model, including the space and the schedule of running the model.
3. **Spaces** (sometimes called **grids**, even if they are not, actually, a grid) represent the physical spatial positions of agents. This may be a grid, a network, continuous space or discretized space.
4. **Schedulers** describe the order in which each agent is activated to perform their step actions.
5. **Data collectors** capture information about the state of the model at each time, and keep track of the model's values.

The practical workflow of building a `mesa` model and running it successfully consists of three steps: defining states, defining agents and defining the model with its activation and its spatial aspects (its grid). When the model is executed, the process happens in reverse: first, `mesa` initializes the model, then populates it with instances of the class we used to define our agents and activates each of the agents. Activation results in calling the agent objects' `step()` method, so it will be crucial for us in designing agents to ensure that `step()` includes all we want the model to do at each step in discrete time.

The Computational Notes in this chapter describe, step by step, the construction of an agent-based SIRD model. It will serve as our point of departure for later, more in-depth models. Agent-based models can get almost arbitrarily complex, but good models are built from a combination of two things: a well-defined initial plan and iterative expansion towards complexity.

Working with Mesa is somewhat heavier on object-oriented programming (OOP) than our work has hitherto been. Section A.9 provides a primer to OOP in Python that ought to suffice for applications in this chapter.

9.1.2.1 Defining the model

In the abstract, an agent-based model essentially consists of four major parts:

1. The **population** is the set of agents that constitute the model. While we generally use this term for individuals, a population in an ABM may be anything: households, groups, even entire localities.
2. The **environment**, also called the **grid** or **space**, is the topological domain in which agents interact. It describes not only what space looks like, but also its rules with respect to movement and interaction. For instance, defining the environment as a graph means an agent on one node of the graph has access only to the agents who are on nodes of the graph directly connected to it.
3. The **processes** are the actions agents may take as a function of the system's state (or a subset thereof). These include **intrinsic processes**—sometimes called **update processes**—, which relate to an agent applying a function that maps an agent's state at t to its state at $t + 1$, and **extrinsic processes**, sometimes called **interactions**. The latter class involves processes in which the state, actions or properties of an agent affect other agents.
4. The **parameters** of the model, which govern the way processes are taking place. If we view processes as functions that take the model's previous state as inputs, the parameters are the additional arguments that parametrize the function.

By way of illustration, let us consider an agent-based version of the SIRD model described in Subsection 2.2.2. We consider a single state variable for each agent: they may be, at any given time, in no more and no fewer than one of five compartments (susceptible, infectious, recovered and deceased). With that, we have largely answered our first question as to the population.

Next, we will have to turn to whom we model. In this case, we are considering individuals, who start as susceptibles (except a small population of seed cases). At each time step in discrete time, our agents will undergo two processes: the contact event and the status update. The former represents the opportunities for contact during which our agent may sustain infection. We assume that each person gets in contact with each living member of their neighborhood, and passes the infection on with the likelihood of β.

Finally, we need to consider the space in which we model. In the majority of cases, networks (graphs) are better approximations of real social contexts than lattice grids. We interact in space, but more than that, we interact with individual people and things—our workplace, our family members and the places we frequent. Our typical interactions are rarely with all in a particular space, but with a small number of individuals and/or spaces. In the animal world, flocking and crowding behaviors mean that in a geographical area, most individuals' contacts are from within their pack, flock or other agglomeration. A benefit of graph representations is that if the collective does not substantially change, movement can often be disregarded, since translation in space generally does not change either the node set or the adjacency matrix of a graph (or, simply put: the birds in a flock are 'around' the same birds most of the time, regardless of where the flock moves).

Computational Note 9.2 Initializing the model

In Mesa, the model is a top-level container for all the objects that, together, make up the ABM. In practice, defining the model usually comes towards the later stage of building the model, after defining states and agents. However, cognitively, the model comes first, and for this reason, we will peek behind the curtain towards model definition before we tackle the individual elements that will flow into the model.

Models are classes that subclass (inherit from) the `mesa.Model` base class. A model contains

- the model's population: indirectly, through the scheduler and through the environment definition,
- the model's environment: as a property of the model, usually `self.grid` (even if it is not, actually, a grid),
- the model's processes: again somewhat indirectly, as processes belong to agents who perform them, which are stored in the scheduler, and
- the parameters of the model, which are typically taken as arguments of the `__init__()` constructor function and stored as properties of the model.

The two methods at the heart of a model are `__init__()`, which is responsible for taking the model's variable parameters, and `step()`, which governs what the model should do for each time step.

Below is the model specification for a relatively simple SIRD model. The Computational Notes in the rest of this subsection will elaborate on each of its elements, reconstructing the process of model design through this sample model.

```
class NetworkInfectiousDiseaseModel(Model):

    def __init__(self,
                 nodes=5_000,
                 mean_degree=12,
                 recovery_period=14,
                 beta=0.04,
                 CFR=0.05,
                 base_mortality_rate=1.25e-4,
                 I0=0.005):
```

```
        self.N_agents = nodes
        self.recovery_period = recovery_period
        self.beta = beta
        self.CFR = CFR
        self.base_mortality_rate = base_mortality_rate

        self.graph
         = nx.erdos_renyi_graph(n=self.N_agents,
                                  p=mean_degree/self.N_agents)
        self.grid = NetworkGrid(self.graph)

        self.schedule = RandomActivation(self)
        self.running = True

        for idx, node in enumerate(self.graph.nodes()):
            agent = Person(uid=idx + 1, model=self)
            self.schedule.add(agent)
            self.grid.place_agent(agent, node)

            if np.random.rand() < I0:
                agent.state = State.INFECTED

        self.datacollector
          = DataCollector(agent_reporters={"State": "state"})

    def step(self):
        self.datacollector.collect(self)
        self.schedule.step()
```

A notebook implementing the contents of this Computational Note is available on the book's companion Github repository in the folder `/ch09/sir_abm`.

9.1.2.2 Defining states

Compared to the abstractions of differential equations, which look at infectious processes as transfers of faceless quantities from one compartment to another, agents in ABMs have a much more complicated inner life—almost, at times, approaching features we would associate with personality [315,316]. Agents are capable of holding and keeping track of a number of variables that is only limited by the system's available memory. For instance, agents in an ABM may have an age variable that affects their response to vaccination and the likelihood of vaccination failure in the elderly.

An epidemic ABM model must, at the very least, have a variable that identifies those capable of passing on an infection and those to whom an infection can be passed. Without that, we cannot engineer the equivalent of mass action in an agent-based setting. Thus, just as every compartmental model must at least have two compartments, an agent-based model needs at least two states.

Computational Note 9.3 Using enumerations to define states

Simulating large populations means we will have to instantiate thousands, or sometimes even millions, of the agent objects. The consequence is that any memory-saving tricks are quite welcome. One such trick is to represent states that can be one of a relatively small number as an integer enumeration.

An integer enumeration essentially stores a set of constants that are referenced to integers. This means that inside the model, they are stored as a single integer and take up an integer's worth of memory space (which, in Python, is 4 bytes). However, they can be referenced in our code with a rather more legible name. Consider the following:

```
class State(enum.IntEnum):
    SUSCEPTIBLE = 0
    INFECTED = 1
    RECOVERED = 2
    DECEASED = 3
```

Once we have defined this enumeration, we can refer to `0` as `State.SUSCEPTIBLE`, for instance, or compare a variable to `State.RECOVERED` (which will yield `True` if the variable's value is `3`). This is convenient as we can refer to the symbols of these states by a more legible name. Consider using enumerations where you have variables that you want to store efficiently, and you know that the number of possible options is limited to a relatively small number of cases that you want to be able to invoke by name.

A notebook implementing the contents of this Computational Note is available on the book's companion Github repository in the folder `/ch09/sir_abm`.

9.1.2.3 Defining the agent class

Next, we need to define the agent's attributes and behavioral processes. The agent class is, essentially, a blueprint for what we want each agent to look and act like. These can range from something as simple as a single attribute or two—Schelling's model, discussed in 9.1.1, contained only the agent's 'color' (red or blue) and their position on the grid.

More complex models can have a wide range of attributes as well as behavioral profiles. The latter are often implemented using functions, to determine actions depending on various inputs. These are discussed in more detail in, among others, Section 9.2.

Computational Note 9.4 Creating the agent blueprint

In `mesa`, objects that define agents derive from the `Agent` class, which is created with two arguments: `uid`, a unique ID for each agent, and `model`, the state of the model at the time. The latter is quite useful to enable an agent to 'see' things other than their own state—the *sine qua non* of models of interaction between agents.

Thus, we begin by subclassing `Agent` to create the `Person` class. Using a `super().__init__()` call, we pass the `uid` and `model` arguments onto the constructor method of `Agent`. In addition, we set the state to susceptible and zero-initialize time at infection (since infection has not occurred yet).

```
class Person(Agent):
    def __init__(self, uid, model):
        super().__init__(uid, model)
        self.state = State.SUSCEPTIBLE
```

At every step, the model calls the `step()` method of every agent, in the order of activations. Thus, as we subclass `Agent`, we need to redefine the `step()` method to provide the agent's behavior and state changes.

A notebook implementing the contents of this Computational Note is available on the book's companion Github repository in the folder `/ch09/sir_abm`.

Agent-based models shine in being able to accommodate complex conditional behavior—including stochastic behavior. By behavior, we understand the agent's actions as performed iteratively throughout the simulation.

In a SIRD model, for instance, we want three main processes:

1. Any agent who has been exposed becomes infected and symptomatic.
2. Any agent that has been infected for longer than the sum of the latency period and the recovery period resolves their infection. Depending on the mortality rate, they may resolve by recovery or they may resolve by moving into the deceased compartment.
3. Any agent in any compartment may die of a natural death, with the probability of μ.

Computational Note 9.5 Probabilistic steps in ABMs

The common way of implementing probabilistic issues, such as whether an infectious agent who has reached the time of resolved infection will recover or die, is by using a (pseudo)random number generator (PRNG). For a fair PRNG returning a series of random numbers in the range [0, 1],

$$\lim_{n\to\infty} \frac{1}{n} \sum_{i=0}^{n} [x_i < \epsilon] = \frac{1}{\epsilon} \tag{9.1}$$

Therefore, over a sufficiently large number of executions, `np.random.rand() < epsilon` will yield `True` about $\frac{1}{\epsilon}$ of the time. Thus, for the base mortality rate μ as the `base_mortality_rate` property of the model, we may check for mortality at every step by

```
if np.random.rand() < model.base_mortality_rate:
    self.state = State.DECEASED
```

This is a common idiom in ABMs: it checks the model's `base_mortality_rate` against a random variable. The random variable will be smaller than `base_mortality_rate` around $\frac{1}{\mu}$ of the time, in which case the state of the agent is set to `State.DECEASED`.

For a SIRD model, we might want to cater for the additional steps, creating the `status_update` function:

```
def status_update(self):
    if self.state == State.INFECTED:
        if np.random.rand() < 1/self.model.recovery_period:
            if np.random.rand() < self.model.CFR:
                self.state = State.DECEASED
            else:
                self.state = State.RECOVERED

    if np.random.rand() < self.model.base_mortality_rate:
        self.state = State.DECEASED
```

This function updates the state of the individual agent according to its current state, and the model parameters. Note the use of properties of the `model` object: since the model is passed to each agent at creation, each agent can refer to any property of the model, as the agent's property `self.model`. This includes `self.model.schedule.time`, which is the current time step of the model's iteration.

A notebook implementing the contents of this Computational Note is available on the book's companion Github repository in the folder `/ch09/sir_ abm`.

The last remaining process is of course that of infection. ABMs shine by being able to consider neighborhoods, both spatial and graph theoretical. This is a powerful tool to reflect spatial or physical relationships. For instance, the total population of all the islands in an archipelago are not well approximated by the traditional mass action term if there is no homogeneous mixing. The traditional compartmental models assumed a homogeneity that often does not hold. Agent-based models break that homogeneity, as well as the isotropy of transmission that is implied by it.

Practice Note 9.2 Against the odds

In some cases, we know that a certain parameter is not randomly distributed. For instance, we might know reasonably informative priors of the mean and standard deviation of the recovery time, which in most diseases roughly follows a standard deviation.

The correct approach in such cases is to initialize a random distribution with the known parameters, then compare the PRNG's output with a random value sampled from the distribution. For instance, the correct way to manage a situation in which we expect the recovery rate γ to be distributed according to a normal distribution with a mean of $\frac{1}{7}$ and a standard deviation of 0.01 is as follows:

```
gamma = scipy.stats.norm(loc = 1/7, scale = 0.01)
...

if np.random.rand() < gamma.rvs():
    ...
```

In agent-based models, infection travels from agent to agent across a neighborhood. Unlike in spatial models (see Computational Note 8.4), the neighborhood in agent-based models may be defined in quite complex ways. It is particularly common to define the space of the model as a graph, in which case the neighborhood of a node consists of all reachable nodes within a particular distance from the node (which in turn may be defined with weighted or unweighted edges). It is possible to consider this a micro-mass action term: each susceptible individual (S) will be infected with the probability β by the single infectious individual (i.e. βSI for $I = 1$).

Computational Note 9.6 Creating the infectious process

In the infectious process, the infectious nodes traverse their neighborhood. The susceptible neighborhood is the subset of the neighborhood that consists of susceptible nodes. For each of the susceptible nodes, there is a β likelihood that the nodes will be infected. This moves them into the infectious state.

```
def contact_event(self):
    neighborhood = self.model.grid.get_neighbors(self.pos,
                                                 include_center=False)
```

```
    susceptible_neighborhood = [
        agent for agent in \
        self.model.grid.get_cell_list_contents(neighborhood) \
        if agent.state is State.SUSCEPTIBLE
                                ]
    for neighbor in susceptible_neighborhood:
        if np.random.rand() < model.beta:
            neighbor.state = State.INFECTED
```

The `.grid.get_neighbors()` method of the `model` object (which we have access to from every agent as the `self.model` property) takes the node's position (`self.pos`) as an argument, and returns a list of references to cell identifiers. The somewhat convoluted list comprehension in the row that follows is due to the way Mesa understands neighborhoods. To Mesa, a neighborhood is strictly a spatial thing (even for graphs)—it is a set of places to be, not the things that are in those places. The `get_cell_list_contents()` method takes a list of positions (in this case, generated by `.get_neighbors()`) and returns all agents with matching positions. We use the conditional in the list comprehension to filter out the susceptibles only.

It is worth noting that this method (which is part of the `Person` class we are building) does not say anything about which agents are doing the traversing. This is intentional: we do not want every node to call this function. Instead, we specify inside the `.step()` method that `.contact_event()` shall only be called if the node is in an infectious (i.e. infected or exposed) state.

```
def step(self):
    self.status_update()

    if self.state is State.INFECTED:
        self.contact_event()
```

In this case, we have used a traversal mechanism, in which every infectious agent visits all of their neighbors and stands a fixed chance of transmitting the disease. There are alternatives to this, explored in Subsection 9.1.3.

A notebook implementing the contents of this Computational Note is available on the book's companion Github repository in the folder `/ch09/sir_abm`.

9.1.2.4 Grids and spaces

Agent-based models can take account of space in remarkable ways. Space is to be understood quite expansively here: from graphs to multidimensional spaces, there are very few limitations as to where things can move and what they can affect. In our work as infectious disease modellers, there are two major factors in which spaces affect our agents.

First, infection (and interaction, in general), in an ABM, is a spatial phenomenon. It is a process in which an agent interacts with a spatially confined neighborhood, as we have seen in Computational Note 9.6. This is the model's topology of interaction.

Definition 9.1 (Topology of interaction). An ABM's **topology of interaction** describes the conditions that govern whether two agents may be able to interact. In general, topologies of interaction are defined as subspaces continuous space or discrete space. Topologies of interaction must form metric spaces.

In this context, discrete space means any space that can be exhaustively defined by a finite graph. This is so regardless of dimensions, although it is quite uncommon to see ABMs with topologies of interaction that are intended to reflect more than two spatial dimensions. Discrete topologies of interaction in two dimensions do include what we intuitively think of discretized space (a grid of equally-spaced nodes), but also discretizations that are non-rectangular regular tessellations of a plane (e.g. hexagonal tessellations, beloved of wargamers as it is the regular tessellation that best approximates a circle) and discretizations that are not regular tessellations of a space at all (graphs with nodes of inequal degree, i.e. the networks that we think of when we ordinarily discuss graph models).

This is to be distinguished from topologies of movement, which relate to an agent's ability to change its position Agents in ABMs may be able to move, and do so in ways that range from the relatively trivial to the fairly complex. In moving, they find and interact with new neighborhoods, carry infection to new places or become part of the susceptible pool in an infectious agent's neighborhood.

Definition 9.2 (Topology of movement). An ABM's **topology of movement** describes the conditions that govern whether an agent may move from one position to another, and may associate a cost with a pair of possible locations. Topologies of movement must also form a metric space.

These two phenomena do not exist in isolation: in fact, often, the topology of interaction is identical to the topology of movement.

The topologies of interaction and movement are fundamental to the dynamics of a disease in an ABM—and the evolution of the infectious process over time. An epidemic in continuous space may play out like a 'prairie fire' diffusion process, while the same epidemic over a network—such as the sexual contact networks examined in Chapter 3—will have to make its way through a complex structure. A disease spreads differently in the English countryside—which is largely passable—than it would from oasis to oasis in a desert with few routes. In ABMs, space is as much part of our operational reality as it was in dedicated spatial models discussed in Chapter 8.

In general, there are four major topologies used in ABMs:

- **Grids**, also called **discretized space**, are the archetypal space for ABMs. In discretized space, agents may occupy points on a typically rectangular, equally spaced grid. A question we encountered in Chapter 8 recurs here—what happens at the edges? Typical solutions are either infinite grids (which are appealing but computationally inconvenient as the size of the grid is not constant and hence it cannot be represented as a 3-dimensional tensor with two space dimensions and a time dimension) and toroidal grids, which join opposite edges (which can, and is therefore, much preferred).

- **Continuous space** reflects reality rather more accurately. At the macroscopic level, space is continuous, and we move through it with fluidity rather than hopping from one spot to another. However, continuous space is vastly more computationally expensive to model. At the right scale, a grid model will be just as accurate a reflection of reality as continuous space, and yet be computationally more efficient.
- **Unweighted graphs** are perhaps somewhat more abstract than grids. However, grids are very convenient spatial abstractions for the fact that not every point in the real world is connected. Some places, for instance, might be altogether impassable except by traversing other places. This yields a definition of neighborhood that perhaps reflects practical realities more.
- **Weighted graphs** take into account not only what point is connected to what, but also a weight metric of going over an edge. This may, trivially, be the distance or mean flight time between world cities, or the integral of energy expended to move across a route over the length of the route.

Practice Note 9.3 Distance, facts and fiction

An important argument to consider in favor of graph-based models is that geographical distance is somewhat of a fiction. The haversine distance between two points might be a more or less adequate metric of distance on the high seas, but in reality, the distance that governs epidemic processes has more to do with the time (and distance) to get from one place to another than with their respective positions. The towns of Sölden and Mittelberg—both in Tyrol, Austria—are separated by a little over five miles. To get from one to the other, however, takes almost tenfold that long: the two towns lie at the end of their respective alpine valleys, separated by a mountain ridge. Grid models and even discrete space might appear to be a neat approximation of reality, but while space might be continuous in mathematics, in geography it exhibits deep discontinuities, from impassable rivers to chains of tall peaks. Weighted graph models, where edge weights reflect the distance in terms of access and time, can assist in modeling these relationships.

Computational Note 9.7 Networks in Mesa

One of the strongest features of Mesa is that it is closely linked with `NetworkX`, the *de facto* graph and network library for Python. For this reason, any network created in `NetworkX` can be transformed into a `NetworkGrid` object in Mesa.

Thus, when subclassing `mesa.Model` for our own model class, we may write:

```
self.graph = nx.erdos_renyi_graph(n=self.N_agents, \
                                  p=mean_degree/self.N_agents)
self.grid = NetworkGrid(self.graph)
```

This first instantiates an Erdös-Rényi graph with the specified number (`n`) and average degree (`p`) of nodes, but this is only the beginning. `NetworkX` graphs can be created from various real-world spatial phenomena, such as physical distances, travel times or traffic flow capacities. Two towns might be the same distance apart from a major metropolitan centre, but the maximum number of individuals who can get to the town connected to the metropolis by a six-lane highway will be different from the number of those who can access the town connected by a poorly maintained second tier access road. Many tools of spatial analysis, such as osmnx [293,294], have algorithms that turn maps into `NetworkX` graphs (or at least objects like adjacency and weight matrices that can be used to create one). Even if, in the real world, space is continuous, networks—by definition discrete—might often prove to be better approximations of spatial relationships.

It is worth noting that defining a weighted graph according to one's own distance metric is much easier (and very, very much cheaper computationally) than creating a custom neighborhood function that uses a custom distance metric. Thus, if anything other than simple Euclidean distance is used, a weighted graph with the custom distances may be a significantly cheaper solution. This is so in particular for notions like Manhattan distance that are less amenable to simple trigonometric solutions.

A notebook implementing the contents of this Computational Note is available on the book's companion Github repository in the folder `/ch09/sir_abm`.

9.1.2.5 Activation and steps

Agent-based models always operate in discrete steps. In general, in any time step, every agent will be activated (its step function will be called) more or less once. However, the order of activations is absolutely crucial. Consider a graph-based ABM, in which a susceptible node A is connected to an infectious node B. The latter has reached the span of infection, and will therefore during this time step cease to be infectious and recover.

- If A is activated ahead of B, A is connected to an infectious agent, and therefore stands a chance of sustaining infection.
- If B is activated ahead of A, then by the time A's contact is assessed, B is no longer considered infectious. A therefore (assuming it is not contacted to any other infectious agents) cannot contract the infection from B.

At sufficiently large scales and over a sufficiently large number of iterations, these effects eventually disappear into noise. Nevertheless, the right activation is fundamental. In the vast majority of cases, random activation is a good first choice.

Computational Note 9.8 Activation models in Mesa

Mesa expects a model to have an activation as a subobject. The `mesa.time` module provides a range of pre-defined schedulers. The two most frequently used schedulers are `RandomActivation` and `SimultaneousActivation`.

- `RandomActivation` schedules the activation of each agent randomly, then shuffles them by each time step.
- `SimultaneousActivation` is somewhat more complex. It calculates the model's state for the next step, but does not apply it until the state for all agents has been calculated. The consequence is that from the perspective of each individual agent, it will appear as if they were the first agent to be activated.

Mesa requires the schedulers to be instantiated with the model itself as their argument, which may appear counterintuitive but is in line with Mesa's effort to make everything available to every object. In addition, it is often common to assign a `running` property to the model, which allows us to keep track of when the model stops:

```
class NetworkInfectiousDiseaseModel(Model):

    def __init__(self,
                 ...):
        ...
        self.schedule = RandomActivation(self)
        self.running = True
```

This initializes the model with a random activation, and prepares it for running.

A notebook implementing the contents of this Computational Note is available on the book's companion Github repository in the folder `/ch09/sir_ abm`.

9.1.2.6 *Parametrization, reporting and data collection*

Like our ODE-based models of infectious disease, agent-based models need parameters such as β, γ and so on. Uniquely, however, many things that are at least rather inconvenient to accomplish in ODEs becomes remarkably simple in an ABM. A simple example is transmission risk expressed as a continuous function of age. Since ABMs can keep track of a rather wide array of parameters such as age, and expressing parameters of infection as a function of an agent's value is fairly straightforward.

Computational Note 9.9 The model class and parameterizing the ABM

Mesa models inherit from the `Model` class. In general, a `Model` class must do three things:

1. Gather the parameters that govern the model, e.g. β, γ and so on.
2. Gather the space and schedule for the model.
3. Expose the `step()` method, which represents a time step in the model.

The `step()` method progresses time in the model according to the schedule. `step()` advances the time, then activates the agents in the schedule according to the schedule's activation logic. It is common to invoke the model's data collector before the step.

```
def step(self):
    self.datacollector.collect(self)
    self.schedule.step()
```

Note that you do not necessarily have to collect on each time step. A common pattern is to use the modulo operator to collect on, say, every 10th step. This is often used to reduce the volume of data being collected where lower temporal granularity is sufficient:

```
def step(self):
    if self.schedule.time % 10 == 0:
        self.datacollector.collect(self)

    self.schedule.step()
```

In general, Mesa requires that the parameters that agents and the model wish to use to be assigned as properties of the model class (the class that subclasses `Model`). The same goes for setting certain initial parameters, such as defining the parameters of the network for a network model. These parameters are available to agents as properties of `self.model`, e.g. if we assign `self.beta` during the `__init__()` function of the model class, the agents will be able to access it as `self.model.beta`.

```
class NetworkInfectiousDiseaseModel(Model):

    def __init__(self,
                 nodes=5_000,
                 mean_degree=12,
                 recovery_period=14,
                 beta=0.04,
                 CFR=0.05,
```

```
                base_mortality_rate=1.25e-4,
                I0=0.005):

    self.N_agents = nodes
    self.recovery_period = recovery_period
    self.beta = beta
    self.CFR = CFR
    self.base_mortality_rate = base_mortality_rate

    ...
```

This assignment passes the parameters into the object, where they become the object's properties.

A notebook implementing the contents of this Computational Note is available on the book's companion Github repository in the folder `/ch09/sir_abm`.

Practice Note 9.4 Stochasticity in ABMs

Often, we want to use ABMs to reflect a certain degree of stochasticity inherent in many processes. In general, this is accomplished by parameterizing a model with a probability distribution. It is important, however, to keep in mind when the probability distribution is sampled. A probability distribution that reflects something particular to an agent and does not change should be created in the `Model` object and sampled at the creation of each agent, and passed onto the agent's constructor. If the distribution describes something that varies over time, it should be created in the agent object and called in the agent's step function.

An antipattern that is sometimes seen in ABMs is to create, rather than merely sample, the probability distribution (such as a `scipy.stats` distribution) in the step function. Creating the probability distribution object each time for each agent and each step may very significantly slow down the simulation. The correct way is to create the distribution as a property of the agent class, then sample it (for SciPy distributions, using `.rvs()`) in the step function.

Where possible, distributions used in ABMs should be based on real-world data, and work using such models should make it explicit how the parameters were obtained [317] and where possible, critically reflect on the goodness of fit.

In most cases, with agent-based models, we are interested in what the model itself can tell us about disease dynamics. As such, we are less interested with what state each agent is in and rather more interested in the number of agents by state (corresponding to the size of compartments). Data collectors are routines within a model that take a snapshot at each time step and obtain this data.

Computational Note 9.10 Collection and export from ABMs

Mesa's primary data collector is the `mesa.datacollection.DataCollector` class. It is assigned as a property of a model, and takes a dictionary object called `agent_reporters`. The content of this dictionary tells Mesa what variables of each agent to keep track of—and what to call them. For instance,

```
self.datacollector
 = DataCollector(agent_reporters={"State": "state"})
```

creates the data collector as the property `datacollector`. Each time it is invoked, it stores the state of each agent, the time and each parameter specified in `agent_reporters`. The dictionary element `"State": "state"` tells the data collector to obtain the `.state` property of each agent, and store it as the variable `state`.

Once our model has run its course, the data collector's `get_agent_vars_dataframe()` method returns the data collector's output as a Pandas `DataFrame`. In epidemiology, we are typically more concerned with the number of individuals in a compartment than who those individuals are. The following pivot transform gives us compartment size snapshots at each time:

```
def to_df(self):
    state_profile = self.datacollector.get_agent_vars_dataframe()
    output_df = pd.pivot_table(state_profile.reset_index(),
                    index="Step",
                    columns="State",
                    aggfunc=np.size,
                    fill_value=0)
    return output_df
```

A mapping,

```
output_df.columns
 = [i.title() for i in State.__dict__.get("_member_names_")]
```

can translate the numeric states/compartments into verbose names, although this will fail if there is an empty compartment throughout. This is because the pivoting results in no result column for a state that does not occur at least once throughout the process.

A notebook implementing the contents of this Computational Note is available on the book's companion Github repository in the folder `/ch09/sir_ abm`.

9.1.2.7 Populating and seeding models

In Subsubsection 9.1.2.3, we created a blueprint for our `Agent` class—a definition of what our agent looks like, what properties it may have and what behaviors it may exhibit. However, this exists as an abstract template. We need to populate our model with individuals.

This is often an important step: the population that we select may have an important effect on the outcome of the model. A model that has age-differentiated contact rates, for instance, will behave differently depending on how age is distributed among the agents. In many cases, real world data can be helpful—fitting a distribution over real world data then feeding the best fit parameters to a random variable is a commonly used way to create distributions in agent based models that reflect a real-world distribution.

Computational Note 9.11 Creating seed populations

An agent-based model needs a population of agents to deliver results. In Mesa, agents have to be added both in time and space:

- in **time**, an agent has to be added to the model's `schedule`,
- in **space**, the agent has to be assigned a position.

The common way to accomplish this is to iterate over the number of agents to be created, create an agent, assign the resulting agent to a variable, then call the `.add()` method on the schedule and the `place_agent` method on the grid or graph. This is relatively straightforward for graphs, especially where each node represents an agent in a static network at the same time;

```
for idx, node in enumerate(self.graph.nodes()):
    agent = Person(uid=idx + 1, model=self)
    self.schedule.add(agent)
    self.grid.place_agent(agent, node)
```

For grids, the `.find_empty()` method returns a random empty cell. A common idiom at creating agents is to invoke this from `.place_agent()`:

```
self.grid.place_agent(agent, self.grid.find_empty())
```

It is often useful to initialize an infectious seed population in this step:

```
if np.random.rand() < I0:
    agent.state = State.INFECTED
```

This assigns nodes the `INFECTED` state for a proportion approximating $I(0)$ (`I0`), our initial parameter for the infectious population.

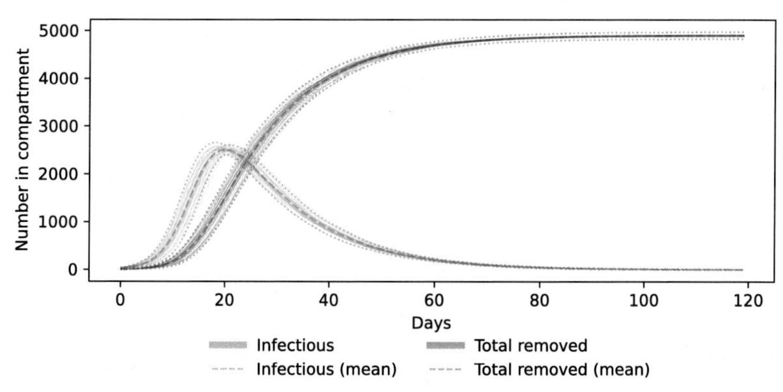

Figure 9.1 Results from 36 iterations of a SIRD ABM over a 5000-node Erdös-Rényi graph with mean degree of 12. The dotted lines represent the margins of the $\pm 1.96\sigma$ confidence band.

> *A notebook implementing the contents of this Computational Note is available on the book's companion Github repository in the folder* `/ch09/sir_abm`.

9.1.2.8 Batch running, iteration and space exploration

Agent-based models are enlightening, but the highly stochastic nature of the modeling approach means that good practice requires iterative execution and evaluation of models, as well as running with different parameters. In their sweeping review of agent-based models in obesity, Giabbanelli, Tison, and Keith [317] note the presence of a sensitivity analysis as one of the hallmarks of a good quality model. This is often best accomplished by exploring combinations of different initial parameters. In many cases, this may extend to not merely the quantitative initial parameters but also the model's design [317].

There is no magic number of iterations that will render an agent-based model 'sufficiently accurate'. Fig. 9.1 shows the results of iterative evaluation of a SIRD model over 36 iterations. As the figure shows, even a very deterministic ABM can give results that diverge rather considerably from run to run. Well-conceived models, however, should result in generally convergent model runs, and small differences in the parameter space resulting in large differences in the outcomes should be a warning that the system might be exhibiting complex non-linear behavior (see Subsection 7.3.5).

Computational Note 9.12 Iterative running of ABMs

Mesa provides the `batchrunner` submodule to manage iterative batch runs of a model, across a range of parameters:

```
br = batchrunner.FixedBatchRunner(NetworkInfectiousDiseaseModel,
                                  parameters_list=[{"beta": 0.2},
                                                   {"beta": 0.3},
                                                   {"beta": 0.4},
                                                   {"beta": 0.5},
                                                   {"beta": 0.6}],
                                  iterations=10,
                                  max_steps=60,
                                  display_progress=True)
```

The above will initialize a batch runner with the `NetworkInfectiousDiseaseModel` model (which must be a subclass of `mesa.Model`). On its own, the batch runner does not execute the model, but serves as an object storing all parts of the process. We run the model by calling the `.run_all()` method of the instantiated batch runner. This will, then, run the model ten times each for the five model parameters `beta`, for 60 steps each.

Parameters can be a list of values for a single parameter or combinations of parameters—but they must be provided as a list of dictionaries, where each dictionary is one parameter set. Thus, `{"beta": [0.2, 0.3, 0.4]}` would not be valid. Parameter sets do not, incidentally, have to have the same number of parameters—missing values are overridden by the class default.

A handy idiom for the frequent situation of when we have multiple parameters is using the Cartesian product function in `itertools`, a package of iterators that comes with Python by default:

```
betas = [0.2, 0.3, 0.4]
nodes = [200, 400, 500]

[{"beta": i, "nodes": j} for i,j in itertools.product(betas, nodes)]
```

This returns an array of dictionaries with the keys `beta` and `nodes`, for each permutation of the two lists.

`FixedBatchRunner` expects models to be completely parametrized. In our design of the model, we have specified defaults for all options, therefore we can simply specify the option that we want to change. An alternative to this is to provide a dictionary of parameter names and values as `fixed_parameters` to the model:

```
br = batchrunner.FixedBatchRunner(NetworkInfectiousDiseaseModel,
                             parameters_list=[{"beta": 0.2},
                                              ...
                                              {"beta": 0.6}],
                             fixed_parameters={"nodes": 250},
```

```
                     iterations=10,
                     max_steps=60,
                     display_progress=True)
```

The keys of this dictionary must be keyword arguments to the model's constructor. `display_progress=True` shows the time taken per iteration and the current iteration. Somewhat idiosyncratically, Mesa will refer to each run of each parameter as an iteration in this context, i.e. six different parameter sets run over ten iterations each will count up to sixty iterations.

In general, the `batchrunner` was devised to collect data only once—at the end of execution. Often, however, what we want is actually a time series of snapshots for each iterator. Achieving this is something of a challenge. As long as the model has an `agentreporter` that reports the variable we are interested in, we can supply the currently run model's `DataCollector` to `batchrunner`:

```
br = batchrunner.FixedBatchRunner(NetworkInfectiousDiseaseModel,
                     parameters_list=[{"threshold": 0.2},
                                      {"threshold": 0.3},
                                      {"threshold": 0.4},
                                      {"threshold": 0.5},
                                      {"threshold": 0.6}],
                     iterations=10,
                     max_steps=60,
                     model_reporters={"vals": lambda m:
                                               m.datacollector},
                     display_progress=True)
```

This will return a data frame in which the `vals` column will contain the entire data collector of the model, with all the time reports. Even though we are actually interested in agent variables, this data frame will be accessible only as `br.get_model_vars_dataframe()`.

A notebook implementing the contents of this Computational Note is available on the book's companion Github repository in the folder `/ch09/sir_ abm`.

9.1.3 Models of infection

The simple model built up throughout Subsection 9.1.2 used one of a number of possible schemes for modeling infection—traversal, in which each infectious individual visits their entire neighborhood. Alternative approaches—inspired to some extent by opinion dynamics—may reflect the peculiar characteristics of a pathogen or its host population more adequately:

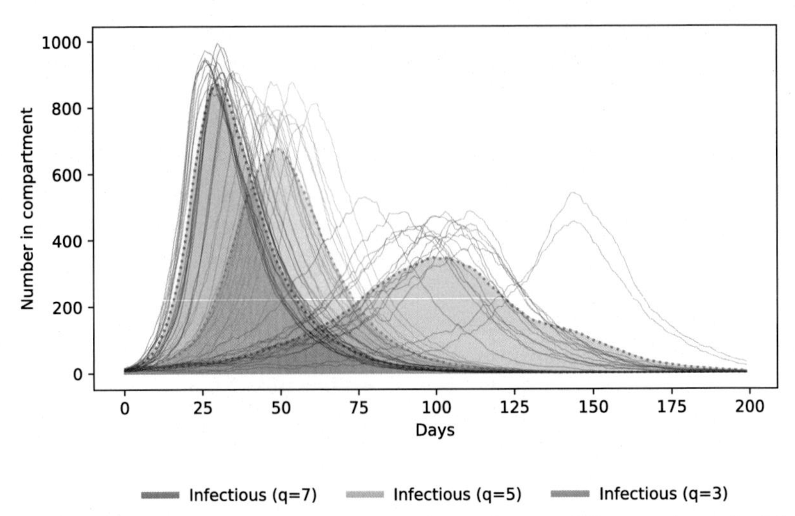

Figure 9.2 Infectious nodes over time in a q infector model over different values of q. The model was initialized over a 2000-node Erdös-Rényi graph with mean degree of 12. The dotted lines supported by the shaded areas denote the means over 15 iterations.

- In **q infection**, an infectious patient will identify q of its neighbors (or all of them, if there are q or fewer). It will then make contact only with the identified subset of neighbors. This reflects the fact of life that we are, regardless of the size of our social circle, limited by certain bounds, such as time. We have limited capacity to interact with others, especially compared to our 'neighborhood': our social network might involve hundreds of people, but we rarely see more than a handful of them per day. Fig. 9.2 shows the effect of different q parameters: lower values for q result in lower maximum numbers of infected and a prolonged but less intense epidemic. q infection is thus a good model to explain why specifically restricting the average number of interactions with different individuals is a successful public health measure in reducing the peak impact and overall burden of an epidemic alike.
- In **proportional infection**, infectious agents visit not a fixed number (like in q infection) but a proportion of neighbors. For grid models (where neighborhoods have a fixed size) and graph models with a relatively narrow distribution of node degree, this is largely equivalent to q infection.
- The **maximum-bounded model of infection** is a useful proxy to model disease-avoidant behavior. Infection will take place at the rate of β_0 until the proportion of infectious agents in the neighborhood are below a critical density ϱ. Thereafter, infection will take place at the rate of β_1. The modulation of transmission behavior based on the prevailing perception.
- **Disgust** as a disease-avoidant behavior has been described by Williams and Mikler [318] in the agent-based context. In an agent-based model of disgust, apparent symptoms modulate the likelihood of encounter, so that the likelihood of a susceptible individual seeking an encounter with a symptomatic agent is reduced by

a discount factor ψ. This represents the reduced propensity of the healthy to associate with those showing symptoms.

Computational Note 9.13 The q infector

Adapting an ABM to a different mechanic of infection is not particularly difficult. The adaptations required to implement the q infector can be easily done by adding the stochastic element of q infection partner choice to the function in our agent specification that we used to define interactions:

```
class Person(Agent):
    ...
    def contact_event(self):
        neighborhood
        = self.model.grid.get_neighbors(self.pos,
                                        include_center=False)
        susceptible_neighborhood = [
            agent for agent in
            self.model.grid.get_cell_list_contents(neighborhood)
                if agent.state is State.SUSCEPTIBLE
                                                  ]

        q_choice = random.sample(susceptible_neighborhood,
                                 min(self.model.q,
                                 len(susceptible_neighborhood)))

        for neighbor in q_choice:
            if np.random.rand() < self.model.beta:
                neighbor.state = State.INFECTED
```

`random.sample` (not to be confused with the similarly named NumPy function) draws a specified number of samples from a collection without replacement. Since the logic of the q infector is that it draws q of its neighbors or as many as possible (when the neighborhood size is smaller than q), we set the number of samples to q or the neighborhood size, whichever is smaller. Fig. 9.2 shows the results of this change over different values of q.

A notebook implementing the contents of this Computational Note is available on the book's companion Github repository in the folder `/ch09/q_ infector`.

9.1.4 ABMs of heterogeneous populations

ABMs can model a wide range of heterogeneities among agents, including both continuous and non-continuous heterogeneities. From the agent-based perspective, a heterogeneity is simply a population with different values of an agent-specific property. The effect of these heterogeneities can be wide-ranging, including complex behaviors. An agent might, for instance, use their age variable to determine whether to get vaccinated or not for a disease. Two major approaches exist for modeling such individual characteristics:

- Continuous modeling expresses a quantity as a function of the parameter of heterogeneity. This only works if the heterogeneity itself is expressed as a continuous variable, e.g. age. For instance, the case-fatality ratio of an infection may often assume a beta distribution over age (as e.g. influenza did in most cases, the influenza pandemic of 1918–1919 being a notable exception). We can fit a distribution over the values of a parameter and obtain a functional form that will allow us to calculate the effective value of that parameter for the individual agent.
- Discrete modeling uses a categorical variable to determine a value. By way of example: a study by Altuntas et al. [319] showed that recipients of haematopoietic stem cell transplants (HSCTs) in Turkey had a statistically significantly higher case-fatality ratio (15.6%) than both patients with haematological malignancies who have not undergone HSCT (11.8%) and patients without cancer (5.6%). In this case, the patients' status with respect to HSCT and malignancies is a categorical variable (although in practice, it would be represented by two Boolean variables in a computational setting, for reasons of memory efficiency), and case-fatality ratios would be drawn from a subpopulation-specific distribution.

9.1.5 Population generation and synthesis

Agent-based models are only as good as their assumptions. In their commentary on the quality of ABMs, Giabbanelli, Tison, and Keith [317] correctly pointed out that high-quality ABMs use parameters drawn from a distribution fit on the modeled population, wherever possible. The process of population synthesis is a modeller's effort to supply the agent-based model with a realistic reflection of the population it seeks to represent. There are three major steps to population synthesis:

- Determination of relevant variables: unlike in pure population synthesis, we are not interested in comprehensive simulation of all facets of the population. Rather, our focus is solely on the variables that are involved in our model's description of interactions and behaviors. If our model is invariant as to age (as may be the case e.g. for veterinary epidemiological models of short-lived animals), there is no reason to add it as an agent property.
- Determination of distribution: the right statistical distribution is a matter of both fit and type of distribution. There are few hard and fast rules as to what makes a distribution better than others for the synthesis of a particular parameter in a population. Comparative fitting over real-world data from the population being modeled is often a productive way of identifying the most suitable distribution and its parameters. This assumes, of course, that the data exists, or at least that we can make some educated guesses—in Bayesian terms: informative priors—about their distribution. The sources of such insights might be complex and elusive, and an

infectious disease modeller's key asset in this field is an open mind. Anthropology, field research, population surveys—of both humans and non-human animals—and, often enough, a modicum of shoe-leather, door-to-door epidemiology frequently prove valuable sources of data that is otherwise not easily accessible.

- Derivation of population: since we now know what variables matter and, moreover, also know how they are distributed, we can now model these variables if the right data sources exist.

Reality does not always align neatly with statistical models, no matter how much we would wish it to. For this reason, it is crucial to keep track of the goodness of fit of the representative distributions and factor its results into our sensitivity analysis and report it accurately.

Practice Note 9.5 Reflecting populations and topologies

All models are based on assumptions, and ABMs are no exception. The assumptions that have the most profound effect on an ABM are the population and the topology of interaction.

Many of the models used in this book use Erdös-Rényi graphs, which are random graphs that can be created in a computationally relatively inexpensive manner, and for most social interactions, they are suitable starting points. On the other end of the spectrum, real-world contact networks, such as those obtained by contact tracing, are much harder to obtain but undoubtedly more accurate (although note the caveat described in Practice Note 3.4 about the limits of traced networks).

In some theoretical works, e.g. Chen et al. [307], public health interventions were modeled on networks from other domains, e.g. networks of devices on the internet or collaboration networks in high-energy physics. The advantage of using these proxies is that massive graphs derived from real-world interactions are available, e.g. via the Stanford Large Network Dataset Collection [320]. This is a valuable approach if no real-world data could be found to represent the topology.

At other times, creating the topology of interaction might be feasible from spatial data, assuming that dissemination is isotropic in space or over a vector discretization of e.g. travel time. Isotropy, in this context, means that an infectious person has equal chances to infect two susceptible individuals living at a certain distance or a certain time-distance (isochrony). In some cases, the topology of interaction might be quite small, e.g. in the analysis of nosocomial outbreaks, where the topology of interaction typically involves a hospital ward [321], hospital [322] or treatment unit/emergency field hospital [323]. In such instances, the topology is often determined by the physical space to be modeled and ways that an infection might travel (e.g. the location and connection of air ducts, wards sharing the same corridor versus wards on different floors, and so on). For such micro-localized models, thorough knowledge of the model's real-world counterpart is indispensable.

9.1.6 Host-vector models

As we have seen throughout Chapter 4, differential equations of host-vector relationships can get unwieldy rather fast. ABMs have the advantage that adding multiple vectors, hosts, intermediate hosts and other classes from the spectrum of host-vector interactions is relatively easy to accomplish. Most ABM frameworks allow for different types (or classes) of agents to interact according to their own rules.

This becomes particularly pertinent when it comes to pathogens with complex life-cycles. Dicrocoeliasis is the disease caused by *Dicrocoelium dendriticum*, a liver fluke that lives a particularly complex life cycle (Fig. 9.3) [324]:

- Infected humans or hosts shed eggs in their excrement.
- The eggs are eaten by snails that serve as intermediate hosts. They mature to the stage of cercariae in the snail [325], and are then excreted.

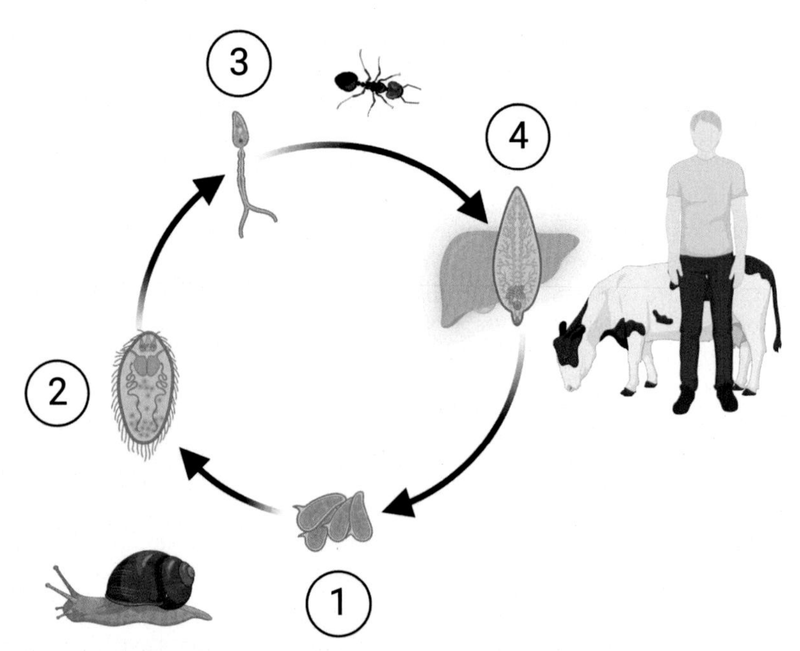

Figure 9.3 Life cycle of the lancet liver fluke *Dicrocoelium dendriticum*. Infected humans and cattle shed eggs (1), which are ingested by snails, where they mature into miracidia (2) and eventually, cercariae (3). Ants ingest the cercariae, where they encyst and become metacercariae. When humans and cattle ingest infected ants, the metacercariae become fully fledged adults that lodge themselves in the bile duct (4) and produce eggs that are excreted with faeces, beginning the cycle anew.

- Ants eat the cercariae, in whom they continue to mature to metacercariae [326].
- Humans and grazing animals ingest the infected ants [327], where the metacercariae grow into fully-fledged adults in the bile duct. Eggs are then transmitted into faecal matter.

The pure mathematical modeling of a life cycle this complex is a nightmare that might hold little appeal even to most mathematicians. On the other hand, for an agent-based model, it simply introduces two new agents—the ants and the snails—, with the respective probabilities of ingestion and transmission. One strength of such models is that they can be spatialized quite effectively, so that e.g. where a vector is associated with wetlands and swamps, agents who are present in the vicinity of such areas are more likely to come into contact with the vector than others.

Consider the following model for the transmission of cercariae:

1. A fraction of hosts—which includes humans, sheep, goats and other herbivorous mammals—are infected in every grid cell. They secrete eggs in their excrement at a rate proportional to their population every turn.
2. Each grid cell has a population of snails that ingest the eggs with a probability corresponding to the number of eggs in the grid cell.
3. There is no necessary reason to model eggs separately, but we can model them as properties of the snails: each snail has a property that stands for the number of eggs and the number of cercariae in them. At every turn, eggs are ingested, and go through the maturation process from miracidium to sporocyst to cercaria. Cercaria are then shed through the respiratory pores of the snail. It is often enough to model intake, maturation and release.
4. A grid cell also has a population of ants. These encounter cercariae at a rate corresponding to the number of cercariae in the grid cell, and in them, the cercariae mature into metacercariae.
5. The host rate of ingesting infected ants then governs the rate at which hosts are infected, who then begin the cycle of infection anew.

This may be worth modeling because there is a spatial aspect to the number of ants, snails and human and non-human ultimate hosts. By creating spatial separations, these specific and tightly interdependent processes can be isolated from each other. Tverdokhlebov [328], for instance, notes that the high density of terrestrial snails as well as ants on the steppes of South Kazakhstan resulted in high infection rates of cattle. Infections can be reduced by reducing populations using pesticides or molluscicides, but also by isolating potential hosts from areas with higher prevalence of the intermediate hosts (i.e. ants and snails). Notably, the chain of transmission requires all elements to be present—hosts, ants and snails. Agent-based models can help model such complex lifecycles and analyze the effect of various interventions in a spatialized manner. This is, of course, merely a single example of the way the agent-based paradigm does not share the limitations of simple compartmental models. In host-vector interactions, the complexity of compartmental models rises rapidly, while agent-based analyzes can model complex internal dynamics and complicated lifecycles with ease.

Practice Note 9.6 Proportional modeling for abundant populations

Agent-based models' resource demands are a function of the number of individuals modeled. This figure can rapidly go beyond our ability to sustain if we are to model every single ant or snail or ruminant in a grid square. A compromise sometimes adopted is to consider agents to reflect a larger population, e.g. one 'ant' agent in the *Dicrocoelium* model could represent a million ants. This is a useful simplification, and one we get to make because there are relatively few variabilities between individuals (the likelihood of an ant to pick up a cercaria, for instance, is not considered to be a property of the ant but merely the function of the cercarial load of the grid square where the ant resides). This agent-based mobile mean-field compromise is often a very useful method for creating complex models without needing to represent abundant agent classes in detail.

Computational Note 9.14 An ABM for pure vector-borne disease

Computational models of host-vector relationships involve two agent classes—a `Host` class and a `Vector` class. Where there are only two entities, it is common to model interactions in only one of them, so as to keep all code relevant to transmission in one class.

In this example, the `Host` class is going to be primarily passive: other than clearing the infection after a period of time and natural mortality, the host does not do much:

```
class Host(Agent):
    def __init__(self, uid, model):
        super().__init__(uid, model)
        self.state = State.SUSCEPTIBLE
        self.type = "host"

    def status_update(self):
        if self.state == State.INFECTED:
            if np.random.rand() < 1/self.model.recovery_period.rvs():
                if np.random.rand() < self.model.CFR:
                    self.state = State.DECEASED
                else:
                    self.state = State.RECOVERED

        if np.random.rand() < self.model.host_mortality_rate:
            self.state = State.DECEASED
```

```
    def step(self):
        self.status_update()
```

The vector, on the other hand, will carry out the contact process: it will randomly pick a host from its cell of residence and its Moore neighborhood, and interact with it (thus, we have a bite rate of unity per unit time per individual). Transmission occurs if either the vector or the bitten host are infectious (see Fig. 4.2).

```
class Vector(Agent):
    def __init__(self, uid, model):
        super().__init__(uid, model)
        self.state = State.SUSCEPTIBLE
        self.type = "vector"

    def contact_event(self):
        neighboring_hosts
         = [agent for agent in self.model.grid.get_neighbors(
            self.pos, include_center=True, moore=True)
            if isinstance(agent, Host)]

        if len(neighboring_hosts) > 0:
            target = random.choice(neighboring_hosts)

            if target.state is State.INFECTED and self.state is
            State.SUSCEPTIBLE:
                self.state = State.INFECTED
            elif target.state is State.SUSCEPTIBLE and self.state is
            State.INFECTED:
                target.state = State.INFECTED

    def step(self):
        self.contact_event()

        if np.random.rand() < self.model.vector_mortality_rate:
            self.state = State.DECEASED
```

A notebook implementing the contents of this Computational Note is available on the book's companion Github repository in the folder `/ch09/host_ vector_ abm`.

9.1.7 Multiple concurrent epidemics

Many studies have observed a marked drop of sexually transmitted infections (STIs) during the COVID-19 pandemic (see e.g. Pagaoa et al. [329] for nationally notifiable STIs in the United States in 2020, Bonato et al. [330] for syphilis in Italy and Johnson et al. [331] for chlamydia, gonorrhoea and syphilis in California), which may be partly accounted for by NPIs and movement restrictions, but also by disease-avoidant behavior and an altered risk perception.

Agent-based models can represent the complex interactions between multiple pathogens in staggering detail. The matter of concurrent epidemics has occupied our attention in Chapter 5, where we discussed compartmental approaches to the subject. However, agent-based simulations, too, can be helpful in analyzing multi-pathogen systems. From an agent-based perspective, there is no hard limit on the number of different infections we can track, and most importantly, these can have different dynamics and different interactions. An interesting model of coexistence emerges where disease-avoidant behavior is less feasible with respect to one pathogen than another, e.g. because one pathogen shows pronounced external symptoms while the other does not. Agent-based models permit for the modeling of disease-avoidant behavior with respect to the more obviously symptomatic infection, or for differential avoidance.

Practice Note 9.7 Perspectives on disease-avoidant behavior

In modeling, we are, from time to time, assuming a clinical tone that might from time to time gloss over certain unwelcome facets of human behaviors. Thus, when we speak about disease-avoidant behaviors, we must admit we are often talking about a kind of stigma directed at the sufferers of a particular illness. Disease-avoidant behavior can rapidly spill over into overt discrimination, hatred and even violence. The tragic story of HIV/AIDS stigma all over the world reflects this [332,333].

Disease-avoidant behavior is by definition marginalizing. In the Middle Ages, lepers were obliged to wear a bell around their necks, to alert townspeople to stay clear [334]. This clearly facilitated disease avoidance by the healthy, but undoubtedly served as a means of othering the sick, who were shunned despite having committed no crime against the community. They lost their integral human value—indeed, their very membership in society—to protect society. Epidemics and pandemics have, from age to age, recapitulated this tendency [335]. Often, this extends beyond infectious status, and spills over into xenophobia and racism, as the attacks on, and hostility towards, persons of Asian extraction in Western countries during the early days of COVID-19 has shown [336]. There may be societal benefit in such actions (as e.g. argue with respect [337] argue with respect to the ‘utility’ of HIV/AIDS stigma among children), but these benefits often come at the cost of serious injuries to individual rights and dignity that ultimately diminish society.

Disease-avoidant behavior does not exclude taking precautions with respect, respecting the inviolable human dignity of the individual regardless of their health status. The line separating stigma from precautionary conduct may be narrow, and it is undoubtedly a challenge to any society threatened in their very existences by an epidemic to avoid moving across that dividing line [338]. Yet stigmatization is not only detrimental to the individual's psychosocial health [339], it may be epidemiologically counterproductive: Kolte et al. [340]'s study of the Guarani-Kaiowá of Brazil, a group disproportionally affected by TB, showed that increased stigma associated with TB led to delays in seeking treatment and thus increasing societal disease burden. Rational disease avoidance can easily spill over into dehumanization, stigmatization, discrimination and, indeed, outright violence [341–343]. It is particularly harmful where groups of society with a relatively higher disease burden, e.g. the LGBTQ+ community in the early days of the HIV/AIDS pandemic, have to contend not only with a disease running rampant but also with being stigmatized and associated with the disease (e.g. the colloquialism of "gay plague" for HIV/AIDS in its early days, despite the fact that many other groups were at risk) [344]. Often, this stigmatization reflects an underlying belief of the illness being the result of the moral or ethical weakness of an ethnicity, nationality or a group, as e.g. the HIV/AIDS was seen as 'divine punishment' on the LGBTQ+ community [345]. Preventing such adverse manifestations of disease avoidance and fighting disease-related prejudice as well as disease-associated xenophobia, racism and other exclusionary beliefs are as much an integral duty of public health as preventing the spread of disease.

The agent-based modeling of multi-pathogen systems requires an understanding of the intrinsic characteristics of each pathogen, in particular, what model best describes it, and the conditions of infection. Computational Note 9.15 describes the contemporaneous operation of two pathogens in a population: a SI system (analogous, perhaps, to herpes simplex in human populations, which causes life-long infection) and a SIRD system (which would be more characteristic of an influenzavirus outbreak). In particular, it is important to understand how each of the two pathogens affect their host population, and how that affects the other pathogen's ability to infect susceptible individuals.

A simple multi-pathogen system assumes mostly no interaction. This is its crucial difference vis-a-vis epidemic competition (see Subsection 9.1.8), a distinct but closely related phenomenon. Epidemic competition involves both behavioral and non-behavioral, immunological factors, such as cross-immunity, while pathogenic coexistence may occur without any effects at all except for mortality (an individual succumbing to one pathogen will be permanently inaccessible to all pathogens).

Computational Note 9.15 A SI+SIRD two-pathogen system

In this Computational Note, we will create an ABM for a two-pathogen system, where pathogen A causes a lifelong non-fatal infection while the expiry of pathogen B's infectious period results in recovery or death. First, we define the enumerated states:

```
class StateA(enum.IntEnum):
    SUSCEPTIBLE = 0
    INFECTED = 1
    DECEASED = 2

class StateB(enum.IntEnum):
    SUSCEPTIBLE = 0
    INFECTED = 1
    RECOVERED = 2
    DECEASED = 3
```

Even though we are modeling a SI+SIRD system, succumbing to pathogen B will make the individual permanently unavailable to pathogen A as well, so it is appropriate that we model death in respect of both pathogens. Where the pathogens have the same states, the same enumeration can be used (in fact, strictly speaking, the `StateB` enumeration could be used for state with respect to A as well, but separate states generally make a model tidier). Next, we initialize our agent with an initial susceptible state in respect of each:

```
class Person(Agent):
    def __init__(self, uid, model):
        super().__init__(uid, model)

        self.state_A = StateA.SUSCEPTIBLE
        self.state_B = StateB.SUSCEPTIBLE
```

Recall that the contact event in the traversal model consists of every infected agent visiting every susceptible agent within their neighborhood, and passing the pathogen with a fixed chance β_A and β_B, respectively. In the single-pathogen version, we used the `.step()` method to execute the contact event only if the agent was in the infected state. In the present case, we will have to run the contact event if the agent is in the infected state with respect to either pathogen. Consequently, the contact event function will have to differentiate two neighborhoods (susceptible with respect to A and susceptible with respect to B), and perform the appropriate traversal(s):

```
def contact_event(self):
    neighborhood = self.model.grid.get_cell_list_contents(
        self.model.grid.get_neighbors(self.pos,
                                      include_center=False)
                                                          )

    susceptible_to_A = [agent for agent in neighborhood if
                        agent.state_A is StateA.SUSCEPTIBLE]
    susceptible_to_B = [agent for agent in neighborhood if
                        agent.state_B is StateB.SUSCEPTIBLE]

    if self.state_A is StateA.INFECTED:
        for neighbor in susceptible_to_A:
            if np.random.rand() < self.model.beta_A:
                neighbor.state_A = StateA.INFECTED


    if self.state_B is StateB.INFECTED:
        for neighbor in susceptible_to_B:
            if np.random.rand() < self.model.beta_B:
                neighbor.state_B = StateB.INFECTED
```

Since there is no recovery from infection with pathogen A, and also no infection-related fatality, we only need to account for B in our `status_update()` method:

```
def status_update(self):
    if self.state_B is StateB.INFECTED:
        if np.random.rand() < 1/self.model.recovery_period:
            if np.random.rand() < self.model.CFR:
                self.state_A, self.state_B
                  = StateA.DECEASED, StateB.DECEASED
            else:
                self.state_B = StateB.RECOVERED

    if np.random.rand() < self.model.base_mortality_rate:
        self.state_A, self.state_B = StateA.DECEASED,
                                     StateB.DECEASED
```

Where multiple pathogens are involved, and where each have different parameters, it may be sensible to store the parameters—other than state, which we want to be able to access as a simple property of the agent object—as a collection. Thus, `model.mu` would be a list $[\mu_1, \mu_2, \ldots, \mu_n]$ for n pathogens in

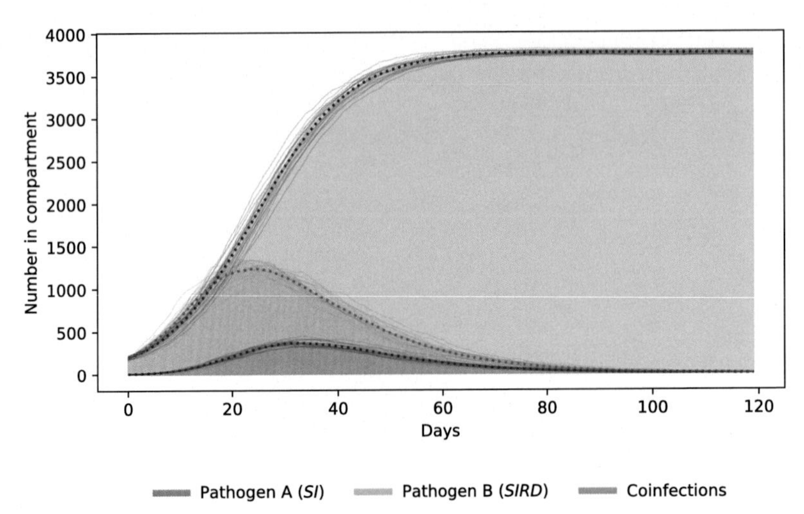

Figure 9.4 The infectious compartment sizes in an ABM of two pathogens in the same population. Pathogen A (blue) follows a SI model with $\beta = 0.004$. Pathogen B (red) follows a $SIRD$ model with $\beta = 0.008$, $\gamma = \frac{1}{14}$ and $\mu_I = 0.05$. The model was initialized over an Erdös-Rényi graph of 2000 nodes with mean degree of 30 with an infectious seed population of 5% each.

the model. This makes models much more easier to interpret and manage when multiple pathogens are involved.

The result of this model, together with the number of coinfections, is displayed as Fig. 9.4. The power of such models is that they give a good estimate of the likely numbers of not only infections in one or the other compartment, but also the dynamics of coinfections. We will expand on this in Subsection 9.1.8, where the pathogens do not merely play operate in the same population but also affect transmission likelihoods.

A notebook implementing the contents of this Computational Note is available on the book's companion Github repository in the folder `/ch09/epidemic_competition`.

9.1.8 Epidemic competition

As we have discussed at length in Chapter 4, pathogens compete for susceptible hosts. For a simple two-pathogen system, infection with one pathogen might render the individual temporarily unavailable (through reduced contact rates as a result of illness, e.g. staying home from a workplace or school where the second pathogen circulates) or permanently inaccessible (through a fatal outcome of the infection by the first pathogen) to the other. On the other hand, pathogens may also facilitate each other, the way e.g. the effects of symptomatic HIV infection on the immune system facilitates coinfections by tuberculosis [346] or Hepatitis C virus [347]. There are multiple dynamics involved in epidemic competition:

- As mentioned before, active disease with a pathogen that produces overt symptoms reduces interactions [318]. Obvious signs of sickness (skin lesions in smallpox, haemorrhage in viral haemorrhagic fevers, the characteristic rash in measles) creates an aversion response that is strongly connected to the emotion of disgust [348].
- As Gul et al. [349] showed, disease avoidance during COVID-19 has been associated with a reduction in mate-seeking behaviors. This is in line with the findings of reduced STI incidence during the pandemic. The perception of risks from interaction may outweigh the normal human reproductive drive.
- Pathogenic processes, at least during their active stages, exact a physical toll from the individual that even in the case of relatively mild disease might be considerable enough to inhibit social activities.
- Prosocial behaviors in a disease-aware agent (e.g. voluntary social distancing upon exposure or symptoms) [350,351] may generally reduce an individual's interactions of the same modality as the primary illness's mosde of transmission. Thus, a welcome side-effect of social distancing, increased emphasis on hygiene and the widespread use of face masks during the COVID-19 pandemic was a remarkable decline of other respiratory infections [352], most notably influenza [353–355].
- Cross-immunity prevents epitopically similar pathogens from infecting the same individual effectively. Immune activation by different pathogens may also prevent infection by a different pathogen, a phenomenon commonly referred to as superinfection exclusion and widely used in the biological management of insect-borne illness [356,357].

Formally, let us consider the population of all agents $x_1, \ldots, x_n$ to exist under two pathogens, A and B. Those susceptible to pathogen A comprise all living individuals who are not recovered from, or infectious with, A. These may, with respect to B, be susceptible, recovered or infectious. Epidemic competition is then modeled when the rate of transmission from $x_i \in I_A$ to $x_j | x_j \in I_B \cap S_A$ is different from the rate of transmission to $x_j | x_j \in (S_B \cup R_B) \cap S_A$. The difference between the two rates of transmission represents the degree to which active infection with B forecloses infection with A, compared to non-infectious but living states with respect to B. We can account for this by way of a discount factor ψ_{β_B}, so that

$$\beta_A(x_i, x_j) | x_i \in I_A = \begin{cases} \beta_A & \text{if } x_j \in (S_B \cup R_B) \cap S_A \\ \psi_{\beta_B} \beta_A & \text{if } x_j \in I_B \cap S_A \\ 0 & \text{otherwise} \end{cases} \tag{9.2}$$

Strictly speaking, ψ_{β_B} is an intrinsic factor with respect to β, i.e. it only modulates the rate of transmission once interaction occurs. In practice, however, we know that there are two factors at play here—reduced encounter rates as well as reduced likelihood of transmission. It is tempting to combine these into a single term, and often enough, it will deliver similar or comparable results. However, a much clearer approach is to introduce a separate variable that directly modulates the infectious process, or even adopt a differential infectious process approach (see Subsection 9.1.3). The latter is quite suitable to model certain kinds of settings in infectious disease hospitals: a susceptible or recovered individual may interact with their entire network (traversal model of infection) but shift to a q-infector model during active infection, where q represents the number of visitors allowed to an infectious patient.

A comparison of Fig. 9.4 with Fig. 9.5, which is identical to the former except for the introduction of a discount factor of 0.1 and a q susceptibility with $q = 2$ for those infectious with B, elucidates the scale of this effect. Note that modal shifts (changes in the model of infection) do not have to be symmetrical: where two pathogens coexist that have different forms of transmission and different symptoms, infection with one might reduce transmission by the other, but not transmission by the first pathogen.

Computational Note 9.16 Competing pathogens with a modal shift

We are expanding on Computational Note 9.15 by introducing both a discount factor and a shift from traversal to a q susceptibility model of infection for those who are in the infectious phase of pathogen B. In q susceptibility, an individual infectious for one pathogen only makes contacts capable of also infecting them with the other pathogen with q members of their neighborhood. In this model, interactions that can transmit B occur regardless of status with respect to A, but A-transmitting encounters are q-limited.

First, we need to store each B-infected individual's q neighborhood, which we randomly determine at every time step. This is accomplished by providing the `Person` class with a class property, `q_neighborhood`. Next, we create a method for the same class that determines the q neighborhood:

```
def set_q_neighborhood(self):
        neighborhood = [agent for agent in
            self.model.grid.get_cell_list_contents(
            self.model.grid.get_neighbors(self.pos,
                                          include_center=False)
                ) if agent.state_A is not StateA.DECEASED]

        self.q_neighborhood
         = random.sample(neighborhood,
                         min(self.model.q, len(neighborhood)))
```

Next, we need to modulate the infectious dynamics from traversal to q infection with a discount factor—which we supply, along with the q value, to the `Model` class as a property:

```
def contact_event(self):
```

```
...
if self.state_A is StateA.INFECTED:
    for neighbor in susceptible_to_A:
        if neighbor.state_B is StateB.INFECTED and self in
            neighbor.q_neighborhood:
            if np.random.rand() < self.model.beta_A *
                self.model.discount_factor:
                neighbor.state_A = StateA.INFECTED
        elif neighbor.state_B is not StateB.INFECTED and
            np.random.rand() < self.model.beta_A:
            neighbor.state_A = StateA.INFECTED
...
```

This accomplishes two things: it checks if a potential new infectee susceptible to A is infected with B. If so, it determines if the infector is in the infectee's q neighborhood. If it is, the infection will take place, but with a penalty (the `discount_factor`) that represents the reduced infectiousness of the encounter. Otherwise, the infection proceeds at the normal parameters. The `.status_update()` method is updated to select the new q neighborhood of every B-infected node at every turn.

As this Computational Note shows, it is relatively trivial to implement even complex behavioral changes, such as changing the entire model of infection of an agent altogether, in dependence to various state variables. It is noteworthy that modal changes can by applied to much more than co-existing epidemics. In a single pathogen epidemic, many factors—such as lockdowns, quarantines (which are equivalent to a q=0 q infector) and other movement restrictions—can be adequately modeled by a transition from normal, high-encounter behavior in a low-risk regime to a more restrictive mode in times of a higher risk regime.

A notebook implementing the contents of this Computational Note is available on the book's companion Github repository in the folder `/ch09/epidemic_competition`.

9.1.9 Opinion dynamics

How rational agents around us behave influences how we behave. Because opinions influence how agents—and, more importantly, people—respond to particular situations, they influence epidemic processes where they affect human behavior vis-a-vis an infectious disease (e.g. the decision to vaccinate or the decision to quarantine). A surprising—perhaps unsettling—amount of human decision-making in a space of considerable uncertainty, such as the emergence of a novel pathogen, is driven by the social diffusion of ideas, attitudes, hopes, fears and information [358–360], which in turn affects the epidemic process. The COVID-19 pandemic has focused acute at-

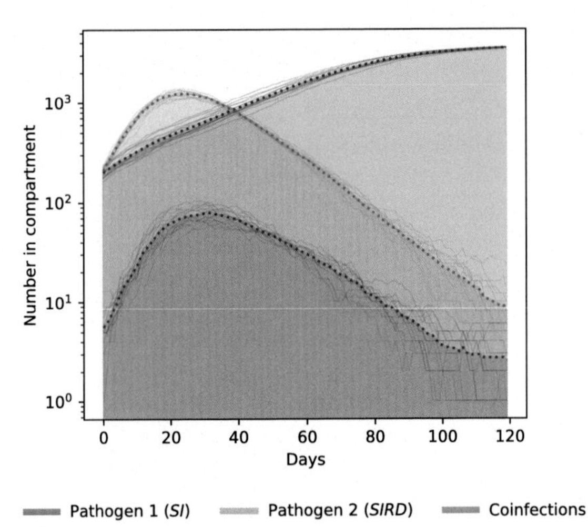

Figure 9.5 Modulation of the two-pathogen ABM depicted in Fig. 9.4 by a shift to q susceptibility. Agents infected with B move to a q susceptibility model with $q = 2$ and $\psi = 0.1$.

tention on the way an individual's response—or, indeed, that of whole societies—is affected by such processes (see e.g. Ali, Rubin, and Sarkar [361], Ye et al. [362]).

Most approaches to quantify opinion dynamics fall into one of two categories: discrete and continuous. In a discrete model of neighborhood effects, an agent will hold an absolute position (e.g. "vaccinate" versus "do not vaccinate"), and act always according to its current position. In continuous models, agents hold a sentiment and their actions at a time are determined either by a threshold (e.g. "vaccinate if sentiment reaches 0.4") or stochastically changes from step to step, so that the sentiment refers to the likelihood an agent will act a certain way. Table 9.1 lays out some of the most important models of opinion dynamics that are used in ABMs.

Frequently, opinion dynamics operates almost like a second infectious process (see Subsection 9.1.7). This was particularly evident during the COVID-19 pandemic, when vaccine misinformation in particular has created immense costs to public health [372]. In this sense, information behaves quite similarly to pathogens, a phenomenon now widely recognized. The epidemiological approach to analyzing transmission phenomena have been applied to a range of real-world concerns, from stock markets [373] through cultural traits [374] to internet memes [375]. It is equally apt to represent opinions with analogous formalisms [376].

Hongmei et al. [377] argue that an epidemic of misinformation would largely take the shape of a SICR model, incorporating a critical stage. Most models use a SIR [378,379] or SIS [380] configuration, while an interesting contribution by Mathur and Gupta [381] uses a SEIZ structure, with the last compartment representing skeptics, who are akin to an immunized compartment. A SI model is also often appropriate: the confirmation bias inherent in human thought, reinforced by the plan continuation bias, means that in the short to medium term, sea-changes to a person's belief in misinformation are quite unlikely.

Table 9.1 Major models of opinion dynamics.

Discrete	Simple voting (Clifford-Sudbury) [363]	Agents adopt the majority position of their neighborhood.
	q-voting [364]	A generalized version of simple voting. Agents poll q neighbors. If the neighbors are in agreement, the agent will adopt their position. Otherwise, it may randomly change its position with a set probability.
	Cluster majority	A number of neighbors are chosen at random. All neighbors adopt the majority position. This differs from simple voting in that the voting clusters are reshuffled at every turn.
	Sznajd [365,366]	Pairs of neighbors are chosen at random. If they agree, all their neighbors adopt the position they agree on.
Continuous	Threshold [367]	Agents determine their actions based on a threshold value, compared against their neighbors.
	Peer pressure	Agents determine their actions based on a changing sentiment value. Individual sentiment is updated in the direction of the position held by the majority of neighbors.
	Hegselmann-Krause [368,369]	Agents select their neighbors whose position is within a tolerance value ϵ of their own, and change their opinion by the mean difference between their opinion and their neighbors'.
	Algorithmic bias [370,371]	Agents select a neighbor with the probability decreasing with difference. If the difference is below a threshold value, both adopt the mean of the positions.

Agent-based models can track an arbitrary number of interacting infectious and infection-like processes in agents, and for this reason, they are particularly useful to interpret the effect of opinion dynamics on the underlying epidemic process. Human ideas spread faster than disease, and the experience of the COVID-19 pandemic highlights the importance of taking the socio-psychological dimensions of (mis)information into account.

9.2 Agent-based models of disease control

The control of disease in a population is a complex process that includes nuances ordinary modeling often cannot replicate adequately. Populations consist of individuals who—at least in the case of human populations—make their own complex assessments about risk and reward based on the information available to them, and who often behave in complex manners in the face of uncertainty and risk.

9.2.1 Case study: the anti-vaccine epidemic

Public and individual sentiment skeptical of vaccinations emerged almost contemporaneously with vaccines themselves. Like every great advancement in science, it required a cognitive shift to accept that infection with one pathogen (the vaccinia virus) can be protective against another (smallpox). Early objections often originated from the perception of vaccines as 'unnatural' and thus offensive to the divine order, of which presumably diseases were but part and parcel [382]. Fear concerning the safety and efficacy of vaccines has also always been a present phenomenon. However, the emergence of social media has turned a slow-moving endemic of vaccine hesitancy into a devastating wildfire.

The analogy of vaccine hesitancy (which generally refers to a legitimate concern, whether well-founded or not, as to the risk-benefit ratio of vaccines) and vaccine refusal (which is a more entrenched, often ideologically motivated position) to an infectious disease is not accidental. Information often exhibits the same dynamics as pathogens, transmitted by those who are aware of it (the equivalent of the infectious compartment) to those who are not (susceptibles). These may or may not adopt the position or accept the information (the transmission coefficient). The emergence of social media has vastly increased the range of susceptibles an infected individual can reach.

During the COVID-19 pandemic, anti-vaccination groups on social media were an early source of (mis)information while public health authorities were still struggling to get their bearings [383]. Mckinley and Lauby [384] pointed out that social media was more heavily weighted as a source of information by individuals who exhibited vaccine-specific views. The emergence of 'influencers'—individuals with large online followings—means that even a very small fraction of influencers who are opposed to vaccinations can spread this harmful message to millions of followers [385]. Thus, in a sense, COVID-19 was 'an epidemic over an epidemic', one a traditional physical epidemic operating in the spatio-temporal plane, and another taking place in discussion groups, Telegram channels, group chats and social media platforms. The dynamic interaction between the two epidemics was evidenced by lower vaccination rates in countries where anti-vaccine sentiment was more prevalent on social media.

Public health in human populations takes place in multifaceted social contexts governed by various dynamics of fear, trust and perceptions of risk. In the information age, the layer of the 'infodemic', the epidemic process that governs the transmission of information, is often just as much at the centre of the human story of health and disease as the physical disease.

9.2.2 ABMs of quarantine

Quarantine, as we have seen in Section 6.3, can take many forms. The power of ABMs of quarantine is that they allow for a wide range of quarantine behaviors to be modeled. This includes, among others,

- different ways of triggering quarantines (e.g. circuit breakers, alternating-day quarantines, post-exposure isolation),

- different ways of leaving quarantine (e.g. biomarkers of pathogen clearance, minimum asymptomatic period, set period regardless of clinical indicators), and
- different effects of quarantine (e.g. reduced transmission risk, personal protective equipment use, movement restrictions).

In addition, agent-based models are capable of reflecting different attitudes towards quarantines: agents may be more inclined to shoulder the economic and emotional burdens of quarantine if they perceive the risk from the disease to be particularly high, e.g. if there has been a recent spike in cases or deaths in their immediate environs. On the reverse, agents may refuse to quarantine for a range of reasons: imperfect quarantine enforcement, high economic costs vis-a-vis the agent's loss-bearing ability or a convergently stable Nash equilibrium, in which the marginal return of quarantining converges on zero while the marginal cost of quarantining remains constant.

Computational Note 9.17 Quarantine modeling

Modeling quarantines depends on the specific form of quarantine considered. A frequently used model, which gives a rather good approximation of reality in most cases, is the SEIRD+Q model, in which individuals become asymptomatically infectious (exposed, E) for a short period of time (the latency period), after which they become symptomatic (I). A key feature of quarantine is that in general, it requires some level of symptomatic markers to identify whom to quarantine. Whether these are biomarkers (e.g. the presence of antibodies) or overt symptoms of illness, conventional quarantine—unlike post-exposure medical isolation—affects only the symptomatically ill.

We begin by creating a separate compartment (state) for the quarantined:

```
class State(enum.IntEnum):
    SUSCEPTIBLE = 0
    EXPOSED = 1
    SYMPTOMATIC = 2
    RECOVERED = 3
    DECEASED = 4
    QUARANTINED = 5
```

Next, we initialize agents with a property to keep track of how many days remain from their quarantine. We set this to `None` initially, which we are going to be able to distinguish from zero values.

```
class Person(Agent):
    def __init__(self, uid, model):
        ...
        self.days_remaining_of_quarantine = None
```

After amending the contact_event() method of the Agent class to include both exposed and symptomatic individuals as possible infectors, we must adapt the status_update() method to accomplish two things: quarantine individuals at a certain rate (a given fraction of all infectious individuals at any given time), and release them into the recovered state once their quarantine elapses:

```
def status_update(self):
    if self.state == State.SYMPTOMATIC:
        if np.random.rand() < 1/self.model.recovery_period:
            if np.random.rand() < self.model.CFR:
                self.state = State.DECEASED
            else:
                self.state = State.RECOVERED
        elif np.random.rand()
         < self.model.quarantine_capture_fraction:
            self.state = State.QUARANTINED
            self.days_remaining_of_quarantine
             = self.model.quarantine_length + 1

    if self.state == State.EXPOSED:
        if np.random.rand() < 1/self.model.latency_period:
            self.state = State.SYMPTOMATIC

    if self.state == State.QUARANTINED:
        self.days_remaining_of_quarantine -= 1
        if self.days_remaining_of_quarantine < 1:
            self.state = State.RECOVERED
            self.days_remaining_of_quarantine = None

    if np.random.rand() < self.model.base_mortality_rate:
        self.state = State.DECEASED
```

The days_remaining_of_quarantine property is used as an exact agent-based countdown, which is rather more accurate than the generalizing method used in differential equations, where outflows from quarantine are a constantly applied $\frac{1}{\tau_q}$. In this instance, every agent will have an exactly τ_q-day quarantine. This is a very good approximation for situations where the quarantine duration is set regardless of symptomatic evolution or subsequent testing, although it may be more appropriate to specify this value as a probability distribution where the quarantine exit criterion is microbiological clearance as evidenced by a given number of negative tests. Since time to clearance varies between individuals, a distribution over the mean, with a heavy right tail (representing the small number of individuals who only achieve microbiological clearance after a relatively long time), is often more apposite for situations with a biomarker-based exit from quarantine.

Finally, we expand the model to accept the variables on the quarantine capture fraction and the length of quarantine:

```
class NetworkInfectiousDiseaseModel(Model):

    def __init__(self,
                 ...
                 quarantine_capture_fraction=0.1,
                 quarantine_length=7):
        ...
        self.quarantine_capture_fraction
         = quarantine_capture_fraction
        self.quarantine_length = quarantine_length
```

Fig. 9.6 shows the result of iterative evaluation of a SEIRDQ network model. Note that the lag between peak symptomatic cases and peak quarantines recapitulates the lag between exposure and symptomatic presentations. This illuminates a fundamental feature of quarantines of the symptomatic—they operate at a lag both with respect to infectious cases (since they do not ordinarily capture every new infectious case right away) and at a greater lag with respect to exposures.

A notebook implementing the contents of this Computational Note is available on the book's companion Github repository in the folder `/ch09/quarantine`.

Where quarantine is specifically applied to the symptomatic, it may make sense to differentiate the likelihood of capture by the level of detectable symptoms. For instance, poliovirus infection causes abortive poliomyelitis, a clinically unremarkable minor illness if it is symptomatic at all, in about 99.5% of infections [386]. Only in about 0.5% of cases does the disease manifest as the well-known presentation of flaccid paralysis. Similarly, infection with ebolaviruses does not always result in the characteristic and easily detectable haemorrhagic fever presentation, but may manifest as mild, non-specific illness [387,388] or be even asymptomatic altogether [389,390]. The rates of detection, and thus the likelihoods of capture, may vary depending on presentation, which in turn may also be conditioned by time since infection: many diseases follow very clear temporal sequences of symptoms that typically become more obvious over time. In ABMs, this can often be accommodated by charting the clinical course of an individual over time, and adjust capture likelihood based on the level of symptoms they express.

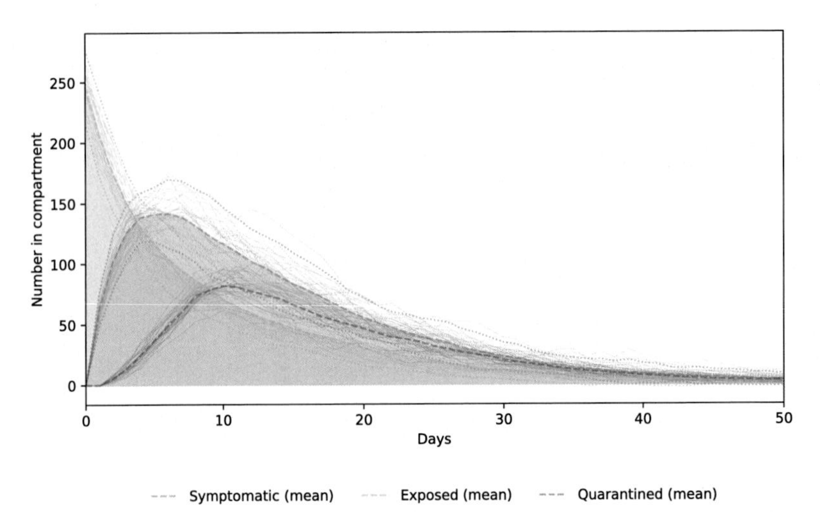

Figure 9.6 Simulations of an agent-based SEIRDQ model over an Erdös-Rényi graph of 5000 nodes with mean degree of 12 and an infectious seed population of 5%. The quarantine capture fraction was set as 0.1 and the quarantine length as 7 days, equal to the recovery period. Dashed lines denote the mean of 35 iterations. Dotted lines denote upper and lower edges of the 95% confidence interval.

9.2.3 ABMs of vaccination

Agent-based models can shed light on much more than vaccination alone. In a human social context, vaccination takes place amidst complex dynamics of opinions, fears and varying assessments of competing risks—from the vaccine and from the disease. We have already traversed this terrain once from the perspective of game theory (see Subsection 6.1.4). While game theory deals with strategies and payoffs as a mean-field model of a population, agent-based models can accommodate more granular perspectives.

Agents may derive their decision as to vaccination from a range of sources:

- Spontaneous opinion: agents will vaccinate or not vaccinate based on a personal preference value, notwithstanding any other factor.
- Risk responsive vaccination: agents will associate the vaccine with a fixed risk, ρ_ν. Their decision to vaccinate or to forego vaccination will depend on whether, within their horizon of awareness, the epidemic process behaves in a way that indicates a higher risk from not vaccinating. In short, there is an inflection point at $\rho_\nu = \rho_I p_I$, where ρ_I is the risk from infection (i.e. the cost of sustaining an infection) and p_I is the probability of sustaining an infection. p_I is assessed with reference to the agent's horizon of awareness, which comprises the neighborhood which the agent has the requisite information to reason about. This may range from the immediate vicinity to a relatively large scale, to reflect the way information about epidemic prevalence has become widely available in a timely manner and at a large scale, to the public.

- Opinion dynamics: in this case, agents hold a discrete (yes/no) or continuous attitude towards vaccination. The surrounding agents' attitudes affect the agent's own attitude. Since vaccination is a binary variable—one may decide to vaccinate or not—but attitudes may be continuous, there is often a need to discretize these continuous attitudes by way of a threshold.
- Non-Nash prosocial vaccination behavior: we have seen in Subsection 6.1.4 that a convergently stable Nash equilibrium, at which point no pair of agents could change their position without reducing total utility, may not be sufficient to achieve collective immunity. In prosocial vaccination, unvaccinated agents accept an excess risk of ϵ_ρ as long as their vaccination has marginal utility, which is the case as until the total vaccination fraction has reached the collective immunity threshold. In an adaptation of this model, agents are willing to take a risk commensurate to the marginal utility of vaccinating: as the returns that arise from vaccinating converge to zero as the vaccination fraction converges to the collective immunity threshold, the maximum excess individual risk the agent is willing to accept also converges to zero.

The beauty—and power—of agent-based models is that these approaches are not mutually exclusive, i.e. they can be combined to reflect the fact that decisions about individual health and welfare are often multifactorial. They include individual but also social considerations as well. Often, such decisions are a function of the dynamics that prevail in the current society, of individual preferences (which may vary significantly from agent to agent—it is not unrealistic to model an agent's altruism, risk-taking or risk-aversion as a quite broad distribution over the population), of group biases, of true risk and of risk as it is perceived by the individual. It is therefore often reasonable to incorporate a combination of multiple influences. A popular approach is to consider an opinion dynamic process, the diffusion of social opinions or attitudes, over an epidemic dynamic. The two exist in interaction: how people feel about the risk posed by the epidemic affects how they relate to it (and to vaccination as an alternative thereto), while the attitudes that people hold about vaccination in turn affects the vaccination rate, and by that, the overall dynamics of the infectious process.

Computational Note 9.18 Vaccination and peer influence

In this Computational Note, we will model vaccination and infection in a network with Hegselmann-Krause opinion dynamics and a SIRVD pathogen model. Each agent x_i is associated with an individual critical threshold value $\hat{\nu}_i$ and a time-varying attitude towards vaccination, $a_i(t)$, as well as a tolerance level ϵ_i. The critical threshold value represents the agent's perception of the relative risk/benefit ratio of the vaccine versus the infection. If x_i is alive, then if $a_i(t) > \hat{\nu}_i$, x_i gets vaccinated at t. We will, for the purposes of this model, assume that the vaccine is sterilizing, i.e. vaccinating an infectious individual immediately puts an end to the infection.

The opinion dynamics are governed by the agent's immediate neighborhood, $\mathcal{N}_{x_i}$. Specifically, following Hegselmann-Krause filtering [368], we are only concerned with the qualified subset of the neighborhood, which we will refer to as $\mathcal{N}'_{x_i}$:

$$\mathcal{N}'_{x_i} = \left\{ n \in \mathcal{N}_{x_i} \;\middle|\; |a_n(t) - a_i(t)| < \epsilon \right\} \tag{9.3}$$

The tolerance level here is an indicator of the homophily of influence, or rather, the inverse of the extent to which an individual agent disregards opinions coming from those who already strongly disagree. This is a reflection of the human tendency for confirmation bias that manifests in seeking out opinions that match one's own [391]. Our model specification is therefore

```
class Person(Agent):
    def __init__(self, uid, model, initial_attitude, tolerance,
    threshold):

        ...

        self.state = State.SUSCEPTIBLE
        self.attitude = initial_attitude
        self.tolerance = tolerance
        self.threshold = threshold
```

x_i will traverse his qualified neighborhood, and update $a_i(t+1)$ so that

$$a_i(t+1) = \frac{1}{|\mathcal{N}'_{x_i}|} \sum_{j \in \mathcal{N}'_{x_i}} a_j(t) \tag{9.4}$$

We implement this as the method `.adjust_attitude()` of the agent class, which we then call later in the step function. Since dead neighbors cannot really exert much of an influence, we will limit this process additionally to the non-deceased. Finally, living agents whose attitude exceeds their threshold will vaccinate:

```
class Person(Agent):

    ...

    def adjust_attitude(self):
        neighborhood = self.model.grid.get_neighbors
                       (self.pos, include_center=False)
        qualified_neighborhood = [agent
            for agent in self.model.grid.get_cell_list_contents
```

```
                (neighborhood)
            if (np.abs(agent.attitude - self.attitude)
                < self.tolerance and
                    agent.state is not State.DECEASED)
                                ]

        if len(qualified_neighborhood) > 0:
            self.attitude = np.mean([
                neighbor.attitude for neighbor in
                qualified_neighborhood
                ])


    def status_update(self):
        if self.attitude > self.threshold:
            self.state = State.VACCINATED

        ...

    def step(self):

        ...

        if self.state is not State.DECEASED:
            self.adjust_attitude()
```

This model confirms an important insight: population attitudes towards vaccinations modulate the infection's evolution in the population by affecting the size of the susceptible subpopulation. Fig. 9.7 exemplifies this phenomenon for various mean threshold values, under the assumption that the number of vaccines available is effectively unlimited when compared to the population to be vaccinated.

A notebook implementing the contents of this Computational Note is available on the book's companion Github repository in the folder `/ch09/blocking`.

9.2.4 Vaccine and prophylaxis allocation

During the COVID-19 pandemic, the initial rollout of vaccines faced a problem of significant demand but rather limited supply. Where the prophylactic measures are relatively limited in relation to the susceptible population, the question of allocation (or prioritization) arises. In the case of the COVID-19 vaccines, vaccination was prioritized by vulnerability (beginning with the over-65 age group at the highest risk for

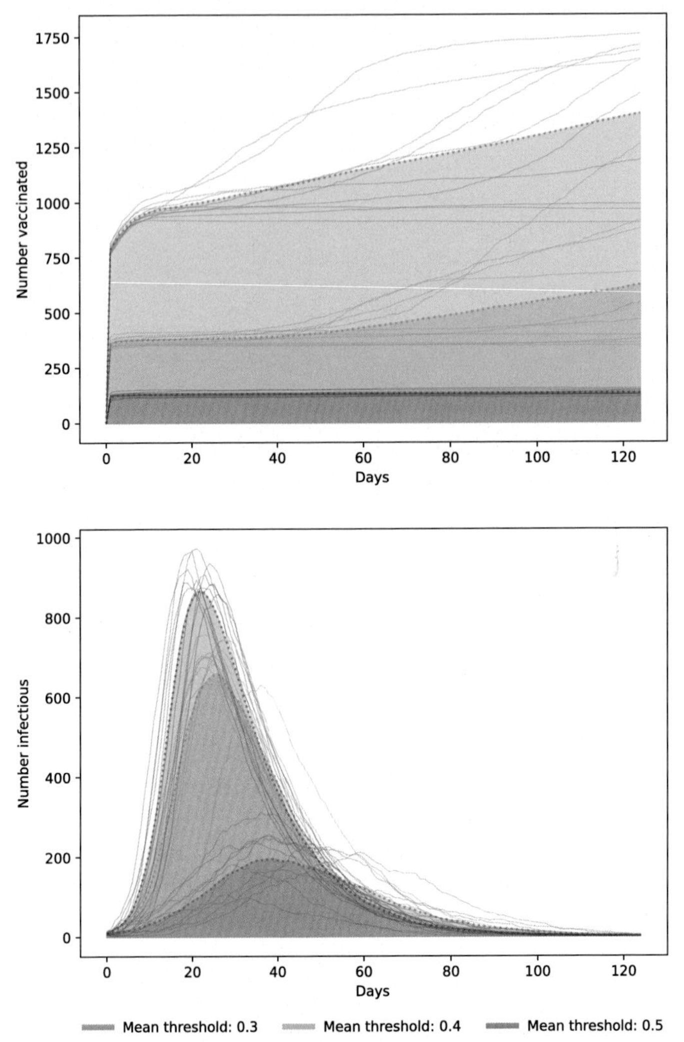

Figure 9.7 The number of vaccinated and infectious individuals in a threshold peer influence model over different mean threshold values over an Erdös-Rényi graph of 2000 nodes with mean degree of 30 and an infectious seed population of 5% each. The threshold values were distributed according to a beta distribution with the specified mean and a standard deviation of 0.05. Initial attitudes were $\sim$ Beta(3, 8), and opinion dynamics were driven by a Hegselmann-Krause model with $\epsilon \sim$ Beta(2, 20).

adverse clinical outcomes and inhabitants of long-term care facilities) and healthcare workers whose were subject to unavoidable exposure (e.g. physicians, allied health-care workers and first responders). Eventually, eligibility gradually expanded in line with supply. This was an example of what could be modeled as an agent-inherent vaccination policy, where the key determinative factor of who would be vaccinated was a factor internal to the agent—their age, their profession or their location in a long-term care facility.

Definition 9.3 (Agent-inherent vaccination policies). An **agent-inherent** vaccination policy determines vaccination allocation by reference to a property of an agent, such as their age or their comorbidities.

In many cases, however, this information may not be available or determinative. In such instances, the key purpose of allocating prophylactic measures is to get the pathogenic evolution under control. These approaches use not only the agent's own inherent features, but rather, the overall topology and the position the agent occupies in that topology.

Definition 9.4 (Topology-inherent vaccination policies). A vaccine allocation policy is considered **topology-inherent** if a governing feature of vaccine allocation lies in the topological features of an agent's neighborhood rather than an agent-inherent feature (e.g. age).

Of course, these approaches are not mutually exclusive—a policy may be both topology- and agent-inherent. Agent-based modeling can assist in improving our understanding of the impact of various policies in allocating prophylactic measures where such medications or vaccines are limited in supply or ability to effectively allocate at scale. The key to such models is to understand what the objective—to use the terminology of mathematical optimization, the goal function—is. This may be a pure objective (a single objective) or an objective hierarchy (where, once one objective is attained, the next highest ranking objective is prioritized).

- Often, a priority is to protect healthcare capacity. As we have noted earlier (REF!), overload of the healthcare system increases mortality. Healthcare-shielding prophylaxis policies have two main targets: healthcare workers and those who are more likely to suffer serious enough illness to necessitate the use of healthcare services. Modeling such a policy would have to take account of age, pre-existing conditions that predispose to adverse clinical outcomes and various other risk factors.
- Similarly to this, the prevention of mortality and, potentially, serious illness, may be a priority. In this case, subsets of the population who are relatively less likely to suffer fatal outcomes are ranked behind those who are more likely to develop a fatal clinical outcome.
- Protection of special groups whose treatment would be difficult includes providing prophylactic resources in areas that could not easily access healthcare resources in case of serious illness. This is the case in particular in places where access to healthcare resources is limited by distance.
- Exposure risk based prophylaxis, which prioritizes professions and groups who are at higher risk of exposure (e.g. physicians, EMTs, allied healthcare workers) or whose continued work in an epidemic situation is necessary for the functioning of society (e.g. law enforcement, military, firefighters, essential retail) is often useful not only from a risk perspective, but also to reduce what would be entirely

understandable absenteeism in these fields in response to their higher risk without adequate safeguards.

- Firebreak vaccination serves to isolate a geographically spreading disease process by creating a barrier band (the firebreak) through which the spatial diffusion of the disease would be greatly impeded or even altogether halted. This strategy is widely used in rabies prevention in foxes by creating band-like zones ahead of the diffusion front of the epidemic. This area is then densely seeded with airdropped oral vaccine bait pellets. The benefit of this strategy is that seeding a small and contingent area with the oral vaccine bait pellets is much more economical than seeding a wider area where the infection is already endemic. Its main risk is that if the firebreak is not wide enough, it may altogether fail to stop or slow down the diffusion. Murray [392] describes the calculations for estimating the width of firebreak zones in the rabies vaccination context in detail.
- The strategy described by Chen et al. [307] is altogether similar to firebreaks, except in this case, the separation is not spatial but based on contact networks. Its main drawback, in opposition to the spatial firebreak idea, is that it requires such networks to be adequately mapped, which is a non-trivial challenge.
- Finally, the equity of access to prophylactic treatments is crucial. Computational Note 9.19 explores an equitable policy, where nodes that have the fewest vaccinated neighbors are vaccinated first. Such policies might be difficult to implement in practice, but indicate that focusing vaccination on areas that are surrounded by communities with relatively lower vaccination rates is not just morally right, but also makes practical sense, creating islands of immunity that eventually create a mesh of impedance against the spatial diffusion of infection.

Agent-based models serve a unique purpose beyond identifying the most suitable policy for a population: often, a mixture of these policy objectives delivers better results than any one policy. "Stacking" policies (using different allocation strategies for fractions of the available prophylactic/vaccination capacity) can approximate various policy mixes. The simulation of policy options gives infectious disease modelers and the decision-makers who are often the consumers of their work output valuable insights into the effectiveness of policy choices. In a space of considerable uncertainty, as is the case in the early days of an outbreak, simulation can fill the gap between missing information and inadequate guesswork, placing policy on a scientific computational basis.

Practice Note 9.8 Equity and ethics in the allocation of prophylaxis

The issue of prioritizing limited stocks of vaccines and prophylactic chemotherapeutics (e.g. the antibiotic ciprofloxacin in the context of an anticipated dissemination of *B. anthracis*) is an issue with complex ethical, economic, social and political manifestations [393]. An important pitfall to avoid is the illusion of apparent objectivity. During the COVID-19 pandemic, age was one of the leading factors of vaccine prioritization, since older individuals tended to suffer worse clinical outcomes. Yet relying on this indicator alone, while giving the illusion of objectivity and rationality, in fact ignored what Smith [394] called the plurality of objectives in a vaccination campaign: vaccinating by age may be useful to satisfy one objective (the reduction of severe, hospitalized and fatal cases), but there are often other interests that are worth considering – such as reciprocal social duties towards those who incur an unavoidable risk in service of the community (HCWs, first responders, essential workers).

In addition, such an allocation may often be inequitable where subgroups of the population have a significantly worse overall health status: not only are they less likely to reach an age-specific vaccination threshold due to lower life expectancies, they are also likely to have the same risks at a younger age than individuals from groups with a better overall health status.

Finally, vaccinating by age—or even risk or susceptibility—might not even always be the best method of reducing overall mortality. As Chen et al. [307] showed, rational vaccination informed by the topology of interaction is often more efficient when the capacity to vaccinate is limited. A practical corroboration of Chen et al. [307]'s theoretical work lies in the success of ring vaccination for the control of smallpox [395,396]. Ring vaccination, the targeted vaccination of the contacts of known cases, is a feasible strategy where vaccine supply is limited or where general vaccination is undesirable (given the adverse effect profile of e.g. the smallpox vaccine, sometimes both of these are the case) and detection of cases is rapid and straightforward. In a sense, it is a simplified version of the more sophisticated graph partitioning approach of Chen et al. [307], and its practical success is proof that topology-responsive modeling may often deliver superior effects when compared to simple prioritization. Given the ethical and practical complexities inherent in vaccine prioritization, it is often a good idea to think beyond simple indicators and explore in agent-based models whether more complex allocation mechanisms. (See Fig. 9.8.)

Computational Note 9.19 Modeling different vaccine allocation policies

There are, as the foregoing subsection has shown, a nearly infinite variety of vaccine allocation schemes and policies. In the case of COVID-19, by the time the vaccines were available, the significant age-dependent heterogeneity in terms of outcomes, hospitalization rates and short-term post-discharge mortality were relatively well known. This is not always the case, and may not be the case at all where an existing vaccine or chemotherapeutic agent shows a prophylactic effect. In those cases, topology-inherent factors are often helpful.

In the present example, we will implement four strategies to allocate a supply of n vaccines per time step:

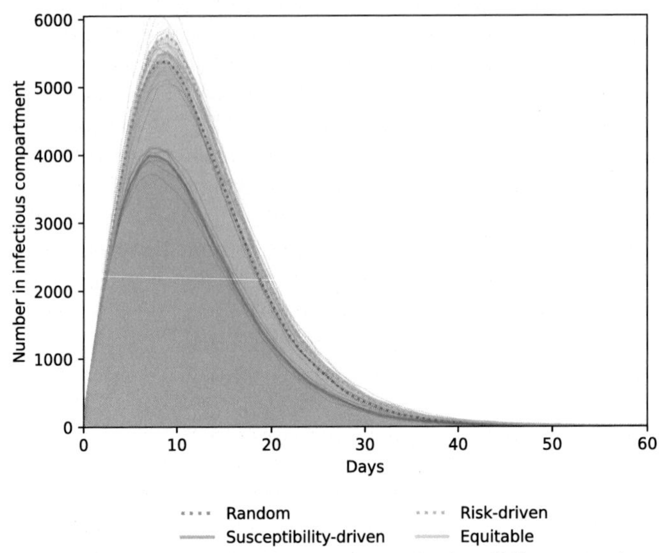

Figure 9.8 Evolution of the infectious compartment under four different vaccine allocation policies.

- Random allocation (`random`): at each time step n nodes are randomly selected for vaccination.
- Susceptibility-driven allocation (`susceptibility`): ranks unvaccinated susceptible nodes by the absolute number of their susceptible neighbors and vaccinates the n nodes with the largest number of susceptible neighbors. This reflects the fact that vaccinating a node connected to a large number of susceptibles delivers a risk reduction to a larger group.
- Risk-driven allocation (`risk`): ranks unvaccinated susceptible nodes by the number of infectious neighbors. An agent with a large number of infectious neighbors is at a high risk of infection.
- Equitable allocation (`equitable`): agents with the lowest number of vaccinated neighbors are prioritized.

To facilitate this, we are going to rely on agents to know their neighborhoods, but on the model to facilitate the vaccination policy execution. At each `.status_update()` method call, we determine the agent's neighborhood and count susceptible, infectious, vaccinated and total livig neighbors:

```
def status_update(self):
    ...
    neighborhood = self.model.grid.get_cell_list_contents(
                self.model.grid.get_neighbors(self.pos,
                include_center=False))
```

```
    self.susceptible_neighborhood_size = sum(
        1 for agent in neighborhood if agent.state is
        State.SUSCEPTIBLE)
    self.infectious_neighborhood_size = sum(
        1 for agent in neighborhood if agent.state is
        State.INFECTED)
    self.vaccinated_neighborhood_size = sum(
        1 for agent in neighborhood if agent.state is
        State.VACCINATED)
    self.living_neighborhood_size = sum(
        1 for agent in neighborhood if agent.state is
        not State.DECEASED)
```

Next, we establish a method inside the model class that obtains a list of susceptible agents. This helps us avoid having to repeat boilerplate code.

```
class NetworkInfectiousDiseaseModel(Model):
    ...
    def get_susceptibles(self) -> List[Agent]:
        return [agent for agent in self.schedule.agents if
                            agent.state is State.SUSCEPTIBLE]
```

Next, we implement the four policies as functions that each take an integer argument `n`, which defines the number of candidates the function is intended to identify and vaccinate:

```
def vaccinate_randomly(self, n: int):
    for i in random.sample(self.get_susceptibles(),
                            min(n, len(self.get_susceptibles()))):
        i.state = State.VACCINATED


def vaccinate_by_number_of_susceptible_connections(self, n: int):
    candidates = self.get_susceptibles().sort(
                    key=lambda x: x.susceptible_neighborhood_size,
                        reverse=True)
    if candidates:
        for i in candidates[:n]:
            i.state = State.VACCINATED


def vaccinate_at_highest_risk(self, n: int):
    candidates = self.get_susceptibles().sort(
        key=lambda x: x.infectious_neighborhood_size/  \
            (x.living_neighborhood_size + 1), reverse=True)
```

```
    if candidates:
        for i in candidates[:n]:
            i.state = State.VACCINATED

def vaccinate_by_equity(self, n: int):
    candidates = self.get_susceptibles().sort(
        key=lambda x: x.vaccinated_neighborhood_size/
            (x.living_neighborhood_size+1))
    if candidates:
        for i in candidates[:n]:
            i.state = State.VACCINATED
```

Since we want to be able to call different policies from the batch runner, we implement a resolver object. This is a dictionary that takes the keyword we associate with a policy, and associates it with the name of the function that implements that policy. We assign this to the model class during initialization:

```
class NetworkInfectiousDiseaseModel(Model):

    def __init__(self,
                 ...
                 policy="random",
                 vaccines_per_step_limit=25):
        ...
        self.policy = policy
        self.vaccines_per_step_limit = vaccines_per_step_limit

        self.policy_names = {
            "random": self.vaccinate_randomly,
            "susceptibility":
              self.vaccinate_by_number_of_susceptible_connections,
            "risk": self.vaccinate_at_highest_risk,
            "equity": self.vaccinate_by_equity
        }
```

Finally, we amend the model-level `step` function to call the specified policy function by taking the keyword provided to the constructor, resolving it to a function by keying the `self.policy_names` dictionary with it, and supplying it with the argument for the number of available vaccines per step:

```
class NetworkInfectiousDiseaseModel(Model):
    ...
    def step(self):
```

```
        self.datacollector.collect(self)
        self.schedule.step()
        self.policy_names[self.policy](self.vaccines_per_step_limit)
```

As this example shows, not only agents can have policies and behavioral profiles—the action of a wider system upon the individual can be modeled from the model object.

A notebook implementing the contents of this Computational Note is available on the book's companion Github repository in the folder `/ch09/targeted_prophylaxis`.

9.2.5 ABMs of treatment effects

As we discussed in Subsection 3.1.5, we generally model treatment for two reasons:

- Some treatments are fully or partially sterilizing, meaning that treated cases are less infectious than untreated cases. Often, this is almost trivially true—symptomatic treatment often reduces effects that the pathogen evolved to produce because they are beneficial for spread. Thus, for instance, antitussives reduce droplet emission, which is essential for respiratory pathogens to spread [397]. However, the effects tend to be quite modest for most treatments that do not target pathogens directly. Sterilizing treatments significantly interfere with a pathogen's lifecycle or its ability to spread. Thus, for instance, initiation of antibiotic treatment for many common bacterial pathogens results in cessation of infectiousness within 24–48 hours, even if symptoms take longer to resolve.
- Treatments may or may not affect infectiousness, but may confer a reduced likelihood of succumbing to the illness and/or a shortened infectious period. Reducing the likelihood of severe illness may be relevant if modeling limited healthcare resources, such as intensive care capacity.

Consequently, the best way to model treatment effects depends on our particular interest in such effects to begin with. In general, treatment effects do not manifest until a patient is symptomatic. This is important for models with an asymptomatic exposed/latent phase, as well as for models of diseases that may not be symptomatic in a part of the population at all. Symptomatic disease must then be recognized and treated, which in turn leaves some cases undiagnosed. Fig. 9.9 describes this continuous decrease of the volume of patients who are effectively covered by treatment. While this treatment funnel could no doubt be replicated by a well-designed (albeit tedious) model of stagewise compartments, the strength of ABMs is that it can accommodate complex conditionalities as to who will fall into which group:

- Seroconversion upon exposure depends on a wide array of factors, such as immune competence, age [398,399], socioeconomic status [400], general health status and, for some infectious diseases, sex [401].

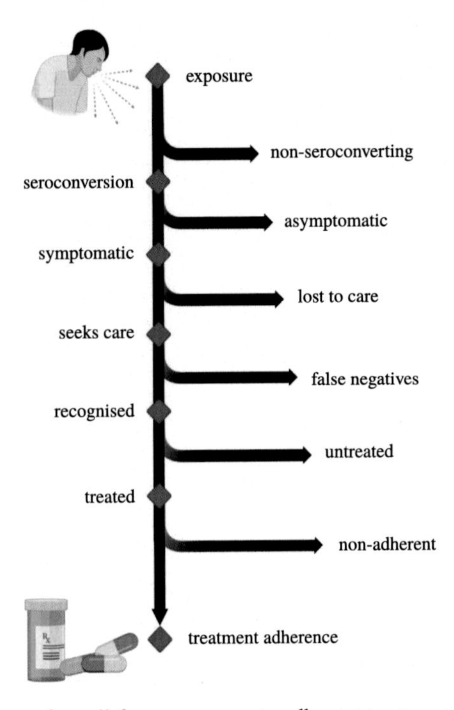

Figure 9.9 The 'treatment funnel' from exposure to adherent treatment.

- Care-seeking is a function of the availability of care to the individual. There is a strong association between higher socio-economic status (of the index patient or of the parent in the case of young children and infants) and an increased likelihood to seek care [402]. In addition, spatial factors may play a role: the cost of seeking care often includes a non-trivial cost of accessing such care. Thus, agents who are particularly far from the nearest site of medical care are less likely to seek care, especially in resource-constrained settings [403,404]. In addition, factors like gender [405,406], race/ethnicity [407] and the stigma associated with the disease [408] affect the decision whether to seek care.
- Recognition, especially if the symptoms of the disease are fairly non-specific, depends on the quality of care, as well as on a range of factors. Certain underserved populations, for instance, suffer from high rates of underdiagnosis. These individuals might not receive the adequate diagnosis that could transition them to the right treatment.
- Treatment may not be universally available. It may be locally unavailable, as is the case for limited stocks of medication in a public health emergency, or it may not be affordable to certain patients.
- Finally, adherence is one of the most complex phenomena in quantitative public health. Adherence is governed by many factors, including adverse effects (which may have to be considered if multiple cohorts receive different treatments with different adverse effect profiles) [409], health literacy [410], socio-economic sta-

tus, co-morbidities that affect adherence (e.g. cognitive dysfunctions [411]), age, ethnicity and stigma surrounding the condition [408]. In the infectious disease context, non-adherence is a particular concern, since incomplete antibiotic treatment contributes significantly to the development of antibiotic resistance [412].

Agent-based models can represent these disparities better than most alternative modeling approaches. In particular, it is quite feasible to create simulated populations based on real-world data, similar to the way we created simulated populations for the estimation of contact matrices in Computational Note 3.2. The chain of phenomena from infection to recognition to adherent treatment is, as noted above, multifactorial. Agent-based models can reflect the time-, space- or characteristic-dependence of these factors. Thus, for instance, an agent-based model may account for the fact that agents residing in an urban environment are likely to face fewer constraints on their ability to obtain a treatment in limited supply than agents in a rural area. Similarly, a treatment may be localized to a particular area (see the notion of vaccination 'firebreaks' in Subsection 9.2.4).

Computational Note 9.20 Treatment effects

Treatment effect models are similar to quarantine insofar as in both cases, candidates are recruited from the symptomatic fraction of the population. There are two crucial differences:

- In quarantine models, it is generally assumed that the quarantined have no interactions with any non-quarantined individuals. Treated individuals can not only have interactions with non-treated members of the population, but can also pass the disease on to them. Typically, this is at a reduced likelihood: the discount factor ψ_β represents the reduction of infectiousness that arises from treatment. In certain cases, ψ_β is highly time-dependent—typically, it takes a minimum time until antimicrobial treatment is absorbed and achieves its effect in individuals. In this example, we will ignore this time delay, but where therapy is relatively long and only manifests a depression of the coefficient of transmission after a while, it is worth taking into account in the model itself.
- Quarantine generally assumes no difference in terms of outcomes, especially mortality. Treatment models generally assume that a treated sick individual will experience a benefit in terms of mortality risk from the disease.

We implement this by creating a `TREATED` compartment:

```
class State(enum.IntEnum):
    SUSCEPTIBLE = 0
    EXPOSED = 1
    SYMPTOMATIC = 2
```

```
    RECOVERED = 3
    DECEASED = 4
    TREATED = 5
```

Most of the model's mechanics for treatment take place in the `status_update` method of the agent class. First, symptomatic agents can be moved into treatment:

```
class Person(Agent):
    ...
    def status_update(self):
        if self.state == State.SYMPTOMATIC:
            if np.random.rand() < 1/self.model.recovery_period:
                if np.random.rand() < self.model.CFR:
                    self.state = State.DECEASED
                else:
                    self.state = State.RECOVERED
            elif np.random.rand() < self.model.treatment_fraction:
                self.state = State.TREATED
                self.days_remaining_in_treatment
                 = self.model.treatment_length + 1
```

The exclusionary branching is used to ensure that an individual who has cleared the infection is not also placed into the treatment compartment. In the next step, we keep track of treatment time and ensure the differentiation of case-fatality ratios between treated and untreated populations:

```
class Person(Agent):
    ...
    def status_update(self):
        ...
        if self.state == State.TREATED:
            self.days_remaining_in_treatment -= 1

            if np.random.rand() < 1/self.model.treatment_length:
                if np.random.rand() < self.model.treatment_CFR:
                    self.state = State.DECEASED

            if self.days_remaining_in_treatment < 1:
                self.state = State.RECOVERED
                self.days_remaining_in_treatment = None
```

This is, as mentioned before, quite analogous to the quarantine model, with the principal differentiation being the differentiated case-fatality ratio. Finally, we need to account for ψ_β in the transmission mechanism, the `contact_event()` method:

```
def contact_event(self):
    neighborhood = self.model.grid.get_neighbors(self.pos,
                                                  include_center=False)
    susceptible_neighborhood = [
        agent for agent in self.model.grid.get_cell_list_contents
          (neighborhood) \
        if agent.state is State.SUSCEPTIBLE
                                ]

    for neighbor in susceptible_neighborhood:
        if (self.state in [State.SYMPTOMATIC, State.EXPOSED] and \
            np.random.rand() < self.model.beta) or \
            (self.state is State.TREATED and \
            np.random.rand < self.model.treatment_beta):

            neighbor.state = State.EXPOSED
```

This is a branched condition that tests transmission from symptomatic or exposed individuals against `model.beta` and treated individuals against `model.treatment_beta`, the discount factor adjusted value of β (i.e. $\beta\psi_\beta$).

As Fig. 9.10 shows, increasing the capture rate for treatment can significantly reduce the size of the infectious subsystem over time, as well as the prevalence time of the epidemic process itself. Thus, where treatment has the effect of reducing transmission, early and aggressive intervention can forestall the rapid growth of the infectious subsystem, even when applied randomly. Of course, the observations made in the prophylactic context in Subsection 9.2.4 apply to treatment as well where treatment has an effect in reducing transmission. Where treatment is limited, it is often worthwhile to explore which subpopulations are transmission 'chokepoints' for the pathogen that will result in lower overall spread over a network of transmission through presenting a barrier to the spread process.

A notebook implementing the contents of this Computational Note is available on the book's companion Github repository in the folder `/ch09/treatment_ abm`.

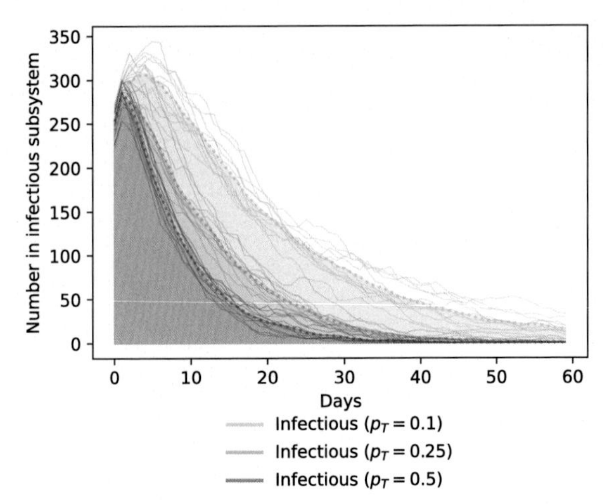

Figure 9.10 The evolution of the infectious subsystem (exposed and symptomatic) of a SEIRD+T model under different treatment capture fractions. The model was initialized over a 5000-node Erdös-Rényi graph with mean degree of 12. The dotted lines supported by the shaded areas denote the means over 12 iterations. The length of treatment is 7 days. Treatment confers a discount factor to the coefficient of transmission of $\psi_\beta = 0.2$ and a discount factor to the case-fatality ratio of $\psi_{\text{CFR}} = 0.2$.

9.3 Agent-based models of mobility

Among the many unique features of agent-based models is the nearly infinite array of mobility paradigms that we can assign to our agents. This includes mobility over a network, mobility that follows a pattern (e.g. commuting between a home location and a place of employment, encountering places in-between that are of special interest, such as grocery stores, but not generally diffusing into areas that are outside one's own spectrum of spatial encounters) and mobility where the distance follows certain distributions (as will be discussed in Subsection 9.3.1). Patterns of mobility are often quite complex, and often not amenable to simpler models. Agent-based mobility models thus bridge a crucial gap in our understanding of how humans and non-human animals interact in space and over time according to an often complicated array of interests, needs and objectives.

9.3.1 *Case study: MEV-1*

MEV-1 is the nominal (and, thankfully, fictional) protagonist of the 2011 bio-disaster flick *Contagion* by Steven Soderbergh. In one of the rare instances where $\mathfrak{R}_0$ gets the popular acclaim it deserves, the pathogen is described as a highly virulent encephalitic-respiratory complex (albeit not in those terms) with an $\mathfrak{R}_0$ of approx. 4. It is clearly inspired by henipaviruses, specifically Nipah encephalitis, as well as SARS. The index patient Beth, played by Gwyneth Paltrow, contracts MEV-1 during a business trip to Hong Kong, then returns to her home in the United States, where she becomes the index case of the ensuing pandemic. I shall not spoil the rest of the movie except to highlight that Beth's behavior reflects something we experience in practice: human motion is not disorganized, but obeys a specific distribution of distance traveled over time. We are drawn to our settlements and gravitate about a few central locations. Agent-based models that can reflect this behavior will perform better than diffusion models that conceptualize humans as milling about in a form of Brownian motion.

The business trip to Hong Kong is a great example of two features:

1. Homesickness: the propensity of agents to return to certain locations, typically their place of abode/shelter. There is an evolutionary imperative in both humans and non-human animals to establish a relatively constant place of safety. Note that this does not have to be a single place, but may be a sequence of safe locations, as is the case with e.g. nomadic tribes moving between relatively fixed places over time, or, in the world of non-human animals, the annual migration of species.
2. Lévy-like distribution of distances: the heavy-tailed distribution of trip distances. We may imagine most of the protagonist's movement through space would be over small distances: shopping for groceries, picking the kids up from school, and so on. These constitute the bulk of one's mobility interactions. In a long but heavy tail of movements, however, a person may travel thousands of miles, as Paltrow's character indeed did for her business meeting in Hong Kong.

One might, of course, see this reflected in one's own life. The number of instances where one returned to one's 'home base' are vastly more frequent than the few instances, such as moves or permanent changes of station, when one does not. Equally, if one were to obsessively plot every single trip one takes, one would find most would be to one's immediate neighborhood (such as commuting to work), but a fair number would be at greater distances, and after a point, increasing the distance would not decrease the likelihood significantly.

Truth is often stranger than fiction, but in the case of the fictitious MEV-1 pandemic, it gives a fairly close approximation of a real-world phenomenon of mobility.

9.3.2 Spatial graph agent-based models with mobility

Mobility-based models are among the prize jewels of epidemiology that have, until rather recently, largely been foreclosed to epidemiologists. Partly, this is because gathering location data is difficult. Movement data, which is at least an order of magnitude easier to obtain, has generally been regarded as equally difficult to obtain reliably. Highway use, for instance, may be relatively easy to measure using inductive loops or pneumatic counters, but this gives at best a rough estimate of population flows: the inductivity of a car with a single occupant is the same as that of a fully loaded car pool. For this reason, the introduction of mobile phone records as a proxy for human whereabouts in societies where both mobile phones and mobile phone base stations (masts) are ubiquitous, has been a groundbreaking development for epidemiology. For instance, the work by Hadachi, Pourmoradnasseri, and Khoshkhah [413] to discern commuting patterns can equally well be adapted to monitor movement between areas of different epidemic incidence. Already before the COVID-19 pandemic, several authors (see e.g. Mamei et al. [414]) have considered real-time mobility data as part of a wider 'smart city' strategy to combat epidemics. Nevertheless, the COVID-19 pandemic has perhaps been the first large-scale outbreak where mobility data reached the requisite maturity to be part of the public health response, being used to assess both adherence to public health measures [415] and the effect of spatial mobility on infectious dynamics [416].

Practice Note 9.9 CDRs and other proxies of movement

CDRs are highly convenient and available proxies for human movement. CDRs record the originating phone number—which can be associated with an account billing location, i.e. a 'home base'—, as well as the location of the cellphone mast to which the device is connected when receiving or initiating a call. Somewhat misleadingly named, CDRs include information not only on calls but also on text messages and, in recent generations of mobile telephony (3G and beyond), CDRs (which in this wider context are expanded to Charging Data Records) also include information on mobile internet use, which for many users is more or less permanent. However, CDRs have to be treated with care.

There is significant evidence of a "mobile gender gap", in which females—especially of older generations—have a significantly reduced access to mobile telephony [417]. A 2016 study by Gapsiso and Jibrin [418] showed a meaningful gender inequality in mobile telephony in Nigeria, and a study by Fatehkia, Kashyap, and Weber [419] based on Facebook's ad impact—arguably one of the widest used global services—has confirmed the existence of a mobile gender gap.

The consequence is that the degree to which mobile phone CDR data reflects real mobility is, in many countries, subject to limitations due to local attitudes towards gender, social and ethnic groups. This operates as a separate layer atop of the issue of access to mobile devices, which in many countries continues to be luxury goods rather than every-day implements.

Spatial graph ABMs recapitulate certain crucial features of reality that are otherwise quite difficult to model:

- Spatial graph ABMs can take account of the carrying capacity of paths between locations. The level of transmission between two villages connected by an ill-maintained dirt path is going to be rather significantly different from the level of transmission between two towns adjoining the same six-lane interstate highway.
- *Tempora mutantur*, and to paraphrase Ovid, so do human processes (as well as processes amongst non-human animals). Mobility follows a complex superimposition of diurnal dynamics ('rush hour'), weekly dynamics (weekdays *versus* weekends) and annual dynamics (holidays). The effect of unpredictable factors, from snow storms to an unseasonably warm February weekend, compounds this effect.
- As the COVID-19 pandemic has evidenced [415], mobility is a factor of risk perception. An individual's understanding of risk, coming from both personal per-

ception and public perception of both the risk of contracting an infection and of suffering non-trivial harm from it, mediates the decision whether to engage in mobility behaviors or not.

- Public health responses alter mobility. Quarantines restrict movement directly, while lockdowns and shutdowns of certain facilities eliminate viable targets of mobility.

Computational Note 9.21 A spatial graph with movement

Graph representations are eminently suited to reflect traffic networks like roads and their intersections, with roads being the edges and intersections the nodes of the graph. If the lengths of edges are generally similar, a movement from any node to any of its neighbors can be considered to constitute a step in unit time. This is very often the case for urban road networks.

In this Computational Note, we simulate an infection starting at a randomly selected node of the road network of Budapest. osmnx [293,294] is a powerful tool for the creation of road networks based on data in OpenStreetMap. To generate the graph, we call the `graph_from_place()` function in osmnx with a filter specification to limit us to motorways, trunk roads, primary, secondary and tertiary roads:

```
road_filter = "['highway'~'motorway|motorway_link|trunk|primary|
                secondary|tertiary']"

graph = osmnx.graph_from_place("Budapest, Hungary",
                               network_type="drive",
                               custom_filter=road_filter)
```

osmnx automatically geocodes the location and obtains the street network. The result is a NetworkX-like graph object, which means we can supply it to our model just as we supplied a random graph in previous examples. In order to implement mobility, we assume all agents to have a given likelihood p_m of moving to a neighboring node of the graph. We realize this as a method of the agent class that obtains the neighborhood and if there are connected nodes, randomly selects a destination:

```
class Person(Agent):
    ...
    def movement_event(self):
        neighborhood = self.model.grid.get_neighbors
                           (self.pos, include_center=False)
```

```
        if len(neighborhood) > 0:
            self.pos = random.choice(neighborhood)
```

Note that in a network grid in Mesa, a 'position' is not a set of coordinates but a node index. Thus, randomly picking from the list returned by the `.get_neighbors()` method of the network grid returns a destination's 'position', in terms of a node identifier. We call this function during the step function depending on a random value check:

```
def step(self):
    ...
    if np.random.rand() > self.model.movement_probability:
        self.movement_event()
```

We construct the model similarly to previous examples, except by supplying it with a `movement_probability` parameter that describes the likelihood of an agent moving on to a neighboring node. The graph we generated from OpenStreetMap using osmnx can now be assigned to be the graph of the `NetworkGrid` object:

```
class NetworkInfectiousDiseaseModel(Model):

    def __init__(self,
                 ...
                 movement_probability=0.85,
                 I0=100):
        ...
        self.movement_probability = movement_probability

        self.graph = graph
        self.grid = NetworkGrid(self.graph)
        ...
        self.epicenter = random.choice(list(graph.nodes))
```

In addition to assigning the road graph as the `NetworkGrid`'s underlying graph, we have also randomly chosen an `epicenter`, an initial node where the entire initially infectious population will be located. This is provided in absolute numbers rather than a population fraction. We will have to slightly amend our way of initially locating agents: first, we disperse all but `I0` agents randomly across the network. These are our initial susceptible population ($S(0)$). Then, we assign the remainder an infectious state and move them onto the epicenter:

```
for i in range(number_of_agents - I0):
    agent = Person(uid=i + 1, model=self)
```

```
        self.schedule.add(agent)
        destination = random.choice(list(self.graph.nodes.keys()))
        self.grid.place_agent(agent, destination)

    for j in range(I0):
        agent = Person(uid=number_of_agents - I0 + j + 1, model=self)
        self.schedule.add(agent)
        destination = self.epicenter
        agent.state = State.INFECTED
        self.grid.place_agent(agent, destination)
```

Fig. 9.11 shows a process at 1, 7, 14 and 21 days. The power of this technique is that the resulting network will be highly reflective of the spatial topologies of the chosen location. A pathogen's spread through the geometric grid of Washington, D.C. will be different from the same pathogen tackling the undulating streets and multi-way junctions of historic Prague. While the locations of our agents at intersections of roads (our nodes) is somewhat notional, they serve as close enough approximations to model spatial relationships if applied at suitable density.

A notebook implementing the contents of this Computational Note is available on the book's companion Github repository in the folder `/ch09/network_movement`.

9.3.3 Homesick agents

Homesickness was first described by Fujihara and Miwa [420] in the context of Lévy-distributed walks. It recapitulates the almost ubiquitously known human phenomenon that most of our human existences gravitate around certain locations that act as our 'home base': our homes, our places of shelter and our spots of safety. The same applies to non-human animals, who have their own requirements for shelter, their own roosts, caves or burrows. In fact, as Sugihara and Hayashibara [421] showed, such a movement strategy often strikes a crucial balance between safety (shelter) and forage (movement to areas least likely to be exploited by competitors).

Homesick walks are thus pseudo-random: agents may set random destinations, but they are constrained by the deterministic limitation of the maximum distance they can travel before having to chart a course for home. The importance of this, from an epidemiological perspective, is twofold:

- The movement away from the home base serves to spread the conditions of the home base to a different point in space—thus, if the home base's conditions are characterized by high prevalence of an infection, the movement may spread those conditions to the target location, as well as to any location the agent traverses on the way.

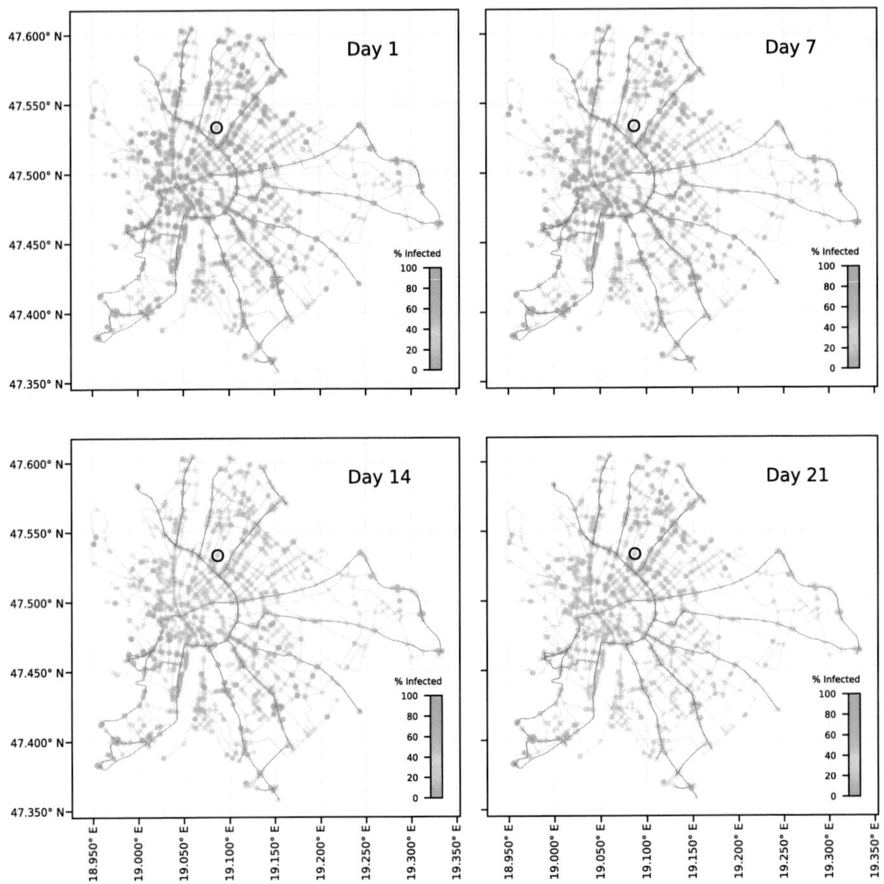

Figure 9.11 The evolution of an agent-based network over a real-world graph topology—in this case, the road network of Budapest, Hungary, restricted to primary, secondary and tertiary roads. The origin node of the infection is marked by a solid black circle.

- The return to the home base serves to bring the conditions of the target location, and of locations encountered on the path thereto, to the home base. If an agent ventures out from a location of relative epidemic safety and is unfortunate enough to set their course to an area that is affected by the epidemic, it may carry the epidemic home. In practice, this is often witnessed with healthcare workers and school teachers, who encounter relatively high infectious exposures during their workdays and bring it home to their families, who might otherwise not have been connected to the original infectious individuals.

A corollary of this interchange is that one agent's home base may be another agent's destination, a path on their journey towards their destination or, indeed, their own home base (e.g. multiple members of the same household). In this case, the inopportune coincidence of an infectious agent returning home and a susceptible agent passing by or leaving that location may create a transmission event that will begin to create cases at the formerly susceptible agent's destination. This, of course, is often how chains of transmission emerge.

Computational Note 9.22 Homesick random-destination walks

In the homesick random-destination walk, agents select a destination and move towards it on the shortest path. When they have traveled a certain distance, they begin to return to their home base. Once they have reached their home base, they move on to the next destination, until they move outside their homesickness range again.

In a graph model, it is often easiest to implement this by creating a route for each agent to follow. A route is, essentially, just a sequence of nodes on a graph that are sequential (i.e. there's a path from each node on a route to the next). We provide for this by equipping the agent class with a `route` property (initialized at creation as `None`) and a route plotting method:

```
class Person(Agent):
    def __init__(self, uid, model, homebase, homesickness_threshold):
        super().__init__(uid, model)

        self.distance_traveled = 0
        self.homebase = homebase
        self.homesickness_threshold = homesickness_threshold
        self.destination_node = None
        self.route = None

    ...
    def plot_route(self, destination):
        route = self.model.determine_shortest_path(self.pos,
                                                    destination)
        self.route = route
        route.pop(0)
```

This calls the model's method `determine_shortest_path`:

```
class HomesickWalkModel(Model):
    ...
    def determine_distance(self, node1, node2, *args) -> float:
        return self.edges.loc[(node1, node2), "length"][0]

    def determine_shortest_path(self, node1, node2) -> list:
```

```
        return nx.shortest_path(self.graph,
                                node1,
                                node2,
                                weight=self.determine_distance,
                                method="dijkstra")
```

The advantage of route plotting is that our agent can now have a relatively simple `move` method: retrieving the next destination and keeping track of the distance traveled.

```
def move(self):
    next_node = self.route.pop(0)
    self.distance_traveled += self.model.determine_distance
                              (self.pos, next_node)
    self.pos = next_node
```

In the `step` function, we determine between four actions:

- The agent is at its home base and has no new route, or its current route terminates in the home base: in this case, we reset the distance traveled, pick a new destination and plot a route to it.
- The agent has arrived at a destination: in this case, we plot a new route to a new destination.
- The agent is at any node other than its home base, and has a set destination that is not its current location: in this case, the agent will continue to advance towards the destination node.
- The agent is at any node other than its home base, and the agent's cumulative travel distance has reached the homesickness threshold: in this case, the agent will plot the shortest path to its homebase.

This is articulated in the `.step()` method of the agent class:

```
def step(self):
    if self.pos == self.homebase and (self.route == None or \
                                      self.destination_node ==
                                      self.homebase):
        self.distance_traveled = 0
        self.destination_node
          = random.choice(list(self.model.graph.nodes.keys()))
        self.plot_route(self.destination_node)
    elif self.pos == self.destination_node:
        self.destination_node
          = random.choice(list(self.model.graph.nodes.keys()))
        self.plot_route(self.destination_node)
    elif self.distance_traveled > self.homesickness_threshold and \
```

```
                                self.destination_node is not
                                self.homebase:
            self.plot_route(self.homebase)
            self.destination_node = self.homebase
            self.model.homesick_points.append(self.pos)
        else:
            self.move()
```

Fig. 9.12 shows an example of homesick motion over a real-life street network (of Budapest, Hungary).

A notebook implementing the contents of this Computational Note is available on the book's companion Github repository in the folder `/ch09/homesick_agent`.

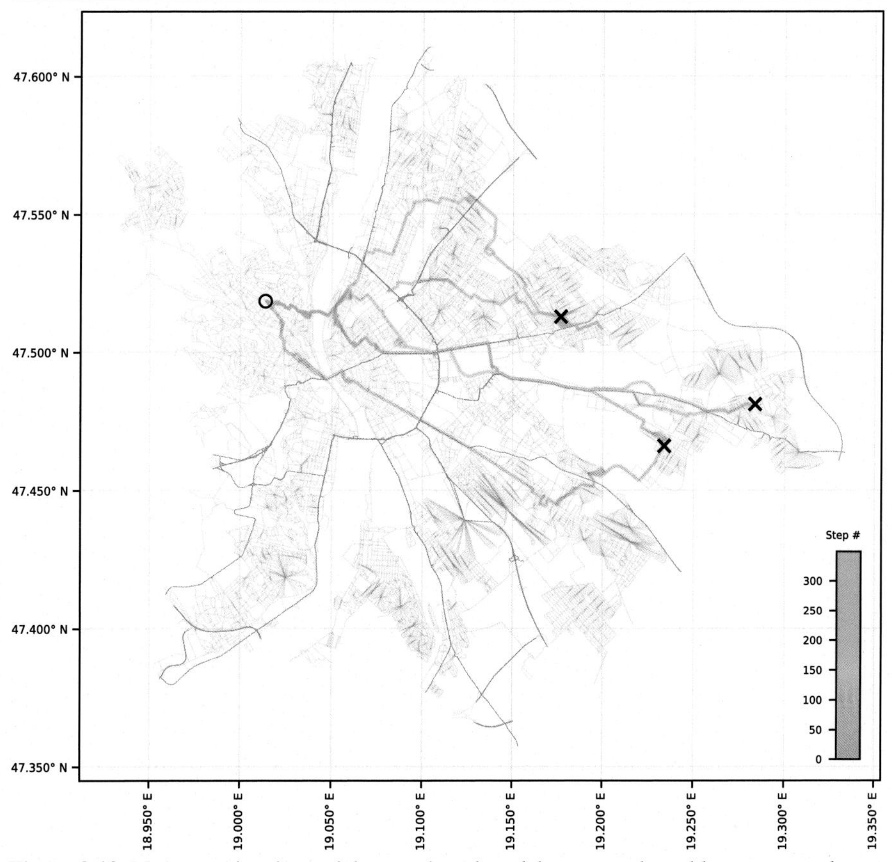

Figure 9.12 Movement in a homesick agent-based model over a real-world route network based on the street grid of Budapest, Hungary. The circle denotes the agent's home base. Crosses note points where target-directed motion is supervened by homesickness and thus return to the home base. Arrows are added every fifty steps to denote the direction of travel.

Fundamentals of Python syntax

Python is both an immensely powerful and highly versatile computing language. This book uses Python for computational examples, for two principal reasons.

First, Python has become the most widely used language of data science. A wide and flourishing ecosystem has come into fruition over the last few decades, with an abundance of tools for reading, manipulating, aggregating, and transforming data.

Second, Python is very legible, to the point that some have referred to it as "executable pseudocode." The consequence is that a competent programmer can look at Python code and transpose it into their language of choice with relative ease and speed.

This Appendix lays out the fundamental syntax of Python, as well as some fundamentals of the most widely used packages for representing data structures we have been using throughout this book.

A.1 Executing Python

Python is an interpreted language. This means that code is not compiled, and it is possible to have ongoing interaction with code, which is quite useful for the scientific process that relies so heavily on experimentation and gradual development.

Python files have the extension `.py`, and virtually all programmer's text editors (such as Notepad++ for Windows or TextMate and Sublime Text on OS X) offer Python syntax highlighting out of the box. So do the more old-school Linux-based text editors `vim` and `emacs`. If you are looking for a deeper IDE experience with more features, PyCharm community edition is a free IDE designed specifically for Python. Visual Studio Code also supports Python by default.

There are, in general, three ways of executing Python code.

1. Python code can be called from the terminal, using `python my_script.py`. This executes the script in the argument, in this case, `my_script.py`.
2. You can also interact with the Python interpreter through the read-evaluate-print loop (REPL). When calling `python` from your terminal without specifying a script, Python instantiates a REPL prompt for you. By convention, a REPL loop is indicated by "triple arrows" (>>>). Within the REPL, press Shift+Enter to break lines. Once you press Enter, all code past the last triple arrow prompt >>> will be executed.
3. Notebook environments, in particular Jupyter Notebooks, are popular with data scientists. They allow for extensive experimentation and line by line development, usually via a web frontend.

A.1.1 REPL

A REPL environment comes prepackaged with Python, and launching Python from the command line will invoke this REPL environment. Additionally, you can install `ipython`, which offers an enhanced REPL experience.

It is important to note that REPL, by default, only prints the return of the last operation. For this reason, it may be useful to explicitly print values of interest, using the `print()` function.

A.1.2 Notebook environments

Notebook environments offer a rather more ergonomic experience than using the REPL. In addition to being able to edit, rerun, and change your code, you can also add large noncode fragments. This makes notebook environments a good way to do "literate programming."

There are various hosted notebook environments, such as Google Colaboratory or JetBrains Datalore, both of which offer free options. Alternatively, you can run a Jupyter server on your own computer. Simply enter `pip install jupyter[notebook]` in your terminal to install the Jupyter notebook environment. You can then launch Jupyter by entering `jupyter` in your command line. Jupyter runs as a web service from your computer, and you would typically access it from your browser. If you have installed Python using Anaconda, this will be installed for you by default.

A.2 Basic syntactic rules

A.2.1 Spaces and indents

Python is remarkably sensitive to indentation. Unlike languages, such as C or JavaScript, which use braces to delineate code blocks, Python uses only indents for this purpose.

It does not, actually, matter how many spaces you use for your indentation, as long as you are consistent within the code block (by convention, one level of indentation is one tab or four spaces, but this is entirely arbitrary). Anything belonging to the same code block must have the same number of indents. As such,

```
if a > 0:
    print("A is positive.")
    print("The square root of A is:")
    print(math.sqrt(a))
```

is correct, whereas

```
if a > 0:
    print("A is positive.")
```

```
    print("The square root of A is:")
     print(math.sqrt(a))
```

would result in an `IndentationError`.

A.2.2 Multi-line code

Python uses the backslash (\) as a line break. A line broken by a backslash is interpreted to continue.

A.2.3 Comments

The hash (#) symbol introduces a single-line comment. In addition, multiline comments can be delineated by triple apostrophes:

```
# This is a comment.

'''
This is a
Multi-line
Comment.
'''
```

A.3 Identifiers and variables

Python instantiates variables by assignment. Therefore

```
a = 3
```

instantiates the variable `a` and assigns it the value `3`.

Python variables are not type-stable. The upside of this is that there is no need to specify a type at the time of assigning the variable, but Python also makes no promises that an object will be of any specific type at any specific time. In the same vein, a value of a different type can be assigned to a variable identifier:

```
a = 3
a = "baked potatoes with tuna salad"
```

is perfectly valid Python.

Python supports chained assignments, but these are somewhat tricky when they refer to a mutable object. Consider the chained assignment `a = b = []`. In Python's interpretation, two variables, `a` and `b`, are created, which both point at the same object. Altering either will affect this single object.

A.3.1 Valid names

There are very few rules as to naming variables (and functions) in Python.

- A variable name must begin with a letter or an underscore (_).
- A variable name cannot contain a space or a hyphen (-).
- Python has two dozen or so keywords, which cannot be the whole of a variable, but may be part of it, e.g., `if` is not a valid variable name, but `_if` is.

Note that `lambda` is a protected Python keyword. For this reason, it is common to use `lambdda` or, as it is in this text, `blamda`, for the variable λ in a model.

Python can theoretically use any Unicode character as, or in, a variable name that is part of the alphabet of any written language. In practice, that means that "θ" and "π" are valid variable names, but emoji are not. This book does not use non-Latin Unicode characters as variable names.

By convention, variables are written in "snake case," i.e., all lower case with underscores to replace spaces (e.g., `my_variable`). Constants are typically written in all upper case (e.g., `MY_CONSTANT`). Python does not enforce immutability of constants, however, so this is entirely syntactic.

A notable exception to this rule are single-letter variables, e.g., `X`. In particular, it is common in scientific Python to use `X` in upper case and `y` in lower case.

The variable name _ (and multiple iterations of it) is used as a throwaway. For instance, if the function `ternary_func(a, b, c)` returns a tuple of three values, but we are only interested in the second of these, the following assignment is common:

```
_, val, __ = ternary_func(a, b, c)
```

Other conventions include

- `i` as the iterator in a loop (and if another is needed, `j`, followed by `k`, `m`, and `n`),
- `k, v` as the key and value, respectively, in an iteration over a dictionary's keys, and
- `idx, val` for iterations over enumerations.

A.3.2 Assignment

The single equality sign `=` is the assignment operator. As of version 3.8, Python also supports the walrus operator, `:=`. The two are mostly functionally equivalent. An exception is the use of the walrus operator for "instant assignment." Instant assignment takes precedence over other operations, whereas regular assignment does not. Thus

```
a = [y = f(x), 2*y, 3**y]
```

would fail, since `y` is not known at the time. However,

```
a = [y := f(x), 2*y, 3**y]
```

is valid. The assignment `y := f(x)` is carried out first, then the whole line is interpreted in that context.

A.3.3 Typing

Python generally eschews the need for typing of functions, arguments, and variables alike. However, in recent years, optional typing has become increasingly prevalent. Following PEP 484, this has been implemented following Python 3.5.

Python uses the colon (`:`) as a type specification operator. To specify a variable's type, you may simply use

```
my_variable: int = 3
```

The same is used within a function's arguments. The arrow (`->`) symbol is used to specify the return type. A function that takes an integer and a floating point argument and returns a dictionary (`dict`) would thus be described as

```
def my_function(arg_1: int, arg_2: float) -> dict:
    (...)
```

It might be worth importing the `typing` package, which is part of the standard distribution of Python and includes a number of type specifications, such as `Iterable` (any iterable object), support for abstract type variables, and the ability to create unions of types (e.g., if a variable can be a string or an integer, `typing`'s `Union` class allows the on-the-fly creation of a type `Union[int, str]` that comprises both).

Note that Python itself does not, at this time, enforce typing. Thus referring to it as "type hinting" or "type annotation" is perhaps more appropriate than "typing." IDEs can use type annotations to check code against type violations, and external tools, such as `mypy`, can check code automatically. This book uses explicit typing only sporadically so as to make code more concise and legible, but in your own work, you may want to use typing extensively, both to facilitate a proactive planning process for your code and to allow future developers to better understand your work.

A.3.4 Numbers

Python provides a wide array of numeric types, the most important of these being `int` (integers) and `float` (floating-point values). There is also `complex`, a type for complex numbers, such as $3i + 4$. As far as Python is concerned, it will select the "right" size in terms of memory to fit the input, so there is rarely a need to specify a type of numeric variable, such as `int8`.

Floating point values support scientific notation, e.g., `3e2` is equivalent to 3×10^2. A value created in scientific notation will automatically be created as a `float`, even if it represents an integer.

Since Python 3.6, numeric literals can include underscores to aid visibility (PEP 515). These do not affect the value nor its equality properties, so `10_000` is equal to `1_00_00`. However, this is limited to single underscores: `10_000` is a valid numeric literal, `10__000` is not.

A.3.5 Strings

In Python, a string may be delineated using '', ' or ''', as long as it closes with the same delimiter. If the string was introduced using ''', the string may comprise multiple lines. Within strings, the backslash (`\`) acts as an escape character.

In Python, the basic string format is Unicode, and this does not need to be expressly specified, but can be by appending `u` at the beginning, e.g., `u"This is my song."`. Byte strings are identified by appending `b` to the beginning. Finally, regular expressions are identified by `r` in the same vein.

A.3.6 Booleans

The `bool` data type creates a boolean variable, which may take the values `True` or `False`. These are case sensitive.

Most objects in Python have a truth value, which can be obtained by calling `bool()` over the object itself. An object that evaluates to `True` is called truthy, whereas an object that evaluates to `False` is called falsy. Since there are fewer falsy objects, it is easier to list those. Among those you might encounter with any frequency are:

- `None`
- `False`
- zeroes: the integer zero `0`, the floating point zero `0.0`, and the complex zero `0j`
- empty collections: empty lists (`[]`), empty tuples (`()`), empty dictionaries (`{}`) and empty sets (`set()`)
- empty strings (`""`)
- empty ranges (`range(0)`, for instance)

Everything else can be assumed to be truthy, unless a class specifically provides for the object to be falsy in some or all circumstances.

A.4 Functions

A.4.1 Function syntax

In Python, a function is a variable of the `Function` type assigned to its name. A function is instantiated using the `def` keyword:

```
def my_function(a, b):
    return (a + b) ** 2
```

The function's return value is specified by the `return` keyword. Note that this is a keyword, not a function, and consequently it does not need parentheses.

A function may return multiple values. If this is the case, they are returned as a tuple. For instance,

```
def my_function(a, b):
```

```
        return a + b, a * b, a ** b
```

returns a tuple of three elements.

There is a relatively common idiom to create a multiline string right after the function definition to use for documentation:

```
def my_function(a, b):
    '''This function returns a + b, a * b and a ** b.'''
    ...
```

This becomes the function's `__doc__()` attribute, commonly known as its "docstring." Documenting functions is always useful, and there are a few conventions for it. PEP 257 is the venerable but somewhat outdated convention for Python documentation. In the scientific field, one is more likely to encounter "NumPy style" docstrings:

```
def my_function(a, b):
    '''This is an example function.

    Parameters
    ----------
    a: int, float
        The first parameter.
    b: int, float
        The second parameter.

    Returns
    -------
    tuple
        A tuple of a + b, a * b and a ** b.
    '''
    ...
```

Type annotation has generally made annotating parameter and return types somewhat obsolete, but the documentation of their meaning remains essential.

A.4.2 Positional arguments and keyword arguments

In Python, a function's signature may include keyword arguments, which become optional. For instance,

```
    def run_disease_model(s, i, r = 0.0):
        ...
```

can be called by specifying only `s` and `i`. In this context, the equality sign acts to assign a default value, unless otherwise specified at call. In Python, all arguments that do not have a default value must be listed before those with default values in the function definition.

In functions, order is prescriptive. Consider the following example:

```
def calculate_area(width, height):
    return width * height
```

Both of the following are valid calls to this function:

```
calculate_area(width = 2, height = 3)
calculate_area(height = 2, width = 4)
```

However, the call

```
calculate_area(1, 5)
```

will necessarily assign `1` to `width` and `5`, respectively, to `height`.

A.4.3 Unpacking parameters

In addition to the above, Python takes two special arguments, typically called `*args` and `**kwargs`, but any name may be used, as long as the number of asterisks is maintained:

```
def run_model(iterations = 10, *args, **kwargs):
    ...
```

These arguments must always be at the end of the function's signature, and in the specified order (i.e., the `*` argument before the `**` argument).

The `*args` parameter holds all other arguments as a tuple, whereas the `**kwargs` argument holds all named arguments as a dictionary. The tuple and the dictionary would be named `args` and `kwargs`, respectively. The naming is, of course, entirely up to you.

For the function `run_model`, described above, the call

```
run_model(1, 3, 5, 9, alpha = 4.0, beta = 9.2)
```

would result in `args` containing `(1, 3, 5, 9)` and `kwargs` containing `{"alpha": 4.0, "beta": 9.2}`. This is called variable unpacking; it is quite useful when there are many parameters, or the parameters could not be anticipated.

A.4.4 Destructuring assignment

A destructuring assingment assigns each of the elements of a function that returns a tuple of multiple results to different variables. For instance, for the function above, a destructuring assignment

```
added, multiplied, powered = my_function(a, b)
```

will assign `a + b` (the first of the result tuple) to `added`, `a * b` (the second of the result tuple) to `multiplied`, and so on.

A.4.5 Side effects

In Python, functions may create side effects. As a general rule, a function has access to all variables in the global scope, and can change them permanently.

A.4.6 Lambdas

A lambda is the Python equivalent of anonymous functions. It may also be assigned to a name, e.g.,

```
add_two = lambda x: x + 2
```

is equivalent to

```
def add_two(x):
    return x + 2
```

The most widespread use of lambdas is inside other functions, e.g., mapping an array to a function without previously defining it, using the `.map()` method of iterables that takes a function object or a lambda and returns the result of applying the function or lambda element-wise to the iterable.

A lambda may have multiple arguments, which are separated by commas:

```
add_and_add_two = lambda x, y: x + y + 2
```

A lambda differs from functions in three main respects.

- Lambdas cannot include type annotations.
- Lambdas must be a single expression.
- Lambdas are immediately invokable.

A lambda, put within parentheses, can be simply argumented in the same line:

```
>>> (lambda x: 2*(x ** 2) + 5)(2)
13
```

A.4.7 Global variables and variable scope

In general, a function's scope is local. The `global` keyword allows the declaration of a global reference within a function scope. Consider the following function:

```
def my_function():
    what = "scary"
    print("Infectious diseases are " + what + ".")
```

What would the outputs be if we defined `what` outside the function scope?

```
what = "manageable"

my_function()

print("Infectious diseases are " + what + ".")
```

The outputs of this example are

```
Infectious diseases are scary.
Infectious diseases are manageable.
```

In other words, the assignment we make within the function `my_function()` does not affect the same variable outside the scope. The `global` keyword allows a variable to change this if we define our function as

```
def my_function():
    global what
    what = "scary"
    print("Infectious diseases are " + what + ".")
```

A.5 Control flow and operations

A.5.1 Conditional statements

In Python, the conditional syntax uses the keywords `if`, `elif`, and `else`:

```
if a in "aeiou":
    print(a + " is a vowel.")
else:
    print(a + " is not a wovel.")
```

The flow of control is branching execution, consequently the above is equivalent to

```
if a in "aeiou":
    print(a + " is a vowel.")
print(a + " is not a wovel.")
```

Until recently, Python did not have an equivalent to the `case` statement that might be familiar to users of C/C++ and similar languages. As of Python 3.10.0, Python supports a similar syntax called `match`/`case`:

```
match a:
    case a %% 2 == 0:
        print("A is even.")
    case a %% 2 == 1:
        print("A is odd.")
    case _:
        print("A is neither even nor odd.")
```

In the `case-match` syntax, an always-true statement (by convention, using simply an underscore, `case _`) at the very end catches all remaining statements.

The `case-match` syntax is still new, and will not be recognized by older versions of the Python interpreter. For this reason, use it with caution, making sure that users of your code use a recent enough version of Python.

A.5.2 Iterative (looping) statements

Python primarily uses the `for` construct to iterate over an iterable object. In Python, that comprises lists and dictionaries, but also strings (where the unit of iteration is each character) and a wide range of other objects. Integers are not iterable, however, the `range()` function returns an iterable from 0 to the specified integer. Unlike the C/C++ syntax of a `for` loop, the Python syntax is significantly simpler, and does not require explicit indexing:

```
>>> cities = ["Budapest", "London", "New York"]

>>> for city in cities:
>>>     print(city + " is a city.")

Budapest is a city.
London is a city.
New York is a city.
```

Where indexing is desired, the `enumerate()` function wraps around an iterable to yield pairs of ordinal indices and values from the iterable:

```
>>> cities = ["Budapest", "London", "New York"]

>>> for index, value in enumerate(cities):
>>>     print(value + " is number " + str(index) + " on the list.")

Budapest is number 0 on the list.
London is number 1 on the list.
New York is number 2 on the list.
```

Note that `enumerate` indexes from zero.

An alternative is the `while` construct, which continues iteration as long as the condition is met. This is useful if the iteration is not intended to be over the elements of an iterable object. Note that unlike in the `for` construct, a `while` construct can go on indefinitely.

A.5.3 Transfer of control flow

Python has three transfer-of-control keywords: `break`, `continue`, and `pass`. These differ in subtle ways.

1. `break` terminates an entire loop. For instance, if applied within a `while` loop, `break` is equivalent to the loop condition becoming false.
2. `continue` shifts to the next item. In iterations, it is largely equivalent to `pass`.
3. `pass` is similar to continue in that it carries on the iteration. However, the role of `pass` is to generally instruct the interpreter to do nothing. As a matter of syntactic propriety, `pass` is used to fill a space where you need to enter something for syntactic reasons, but do not want anything to actually happen. `continue`, on the other hand, is the syntactically correct choice for advancing an iteration.

A.5.4 Exception handling

Where things go wrong, Python raises an exception. Fortunately, this means that you can provide for what to do in a certain circumstance. Say you had a list of friends, `friends`, and wanted to look up their phone numbers in a phone book, `phonebook`, which is a dictionary that pairs the name of a person with their phone number. However, some of your friends might not be in the phone book. If you try

```
for i in friends:
    print("The phone number of " + i + " is " + phonebook[i])
```

you may well encounter a `KeyError` if Python is trying to retrieve a key that is not within the phone book. Exception handling allows you to manage behavior in that circumstance:

```
for i in friends:
    try:
        print("The phone number of " + i + " is " + phonebook[i])
    except KeyError:
        print("Sorry, " + i + " is not in the phone book.")
```

All exceptions inherit from `Exception`, and you can catch any error using `except Exception:`, which is a catch-all. This is generally considered bad practice, as it catches all exceptions, which may thus be overly broad, but when used to debug code, it may be quite helpful.

If you are catching all errors or a superclass of errors, it may be worthwhile to see what that error is. `except Exception as e:` catches any exception that inherits from `Exception`, and assigns them to a variable, `e`. You can then use this object in the following code block to investigate what the exception is, by, e.g., printing it.

A.6 Collections and iterables

A.6.1 Iteration and indexation

A collection is a data structure in Python that contains multiple values, such as a list, a tuple, or a dictionary. Some of these are iterable, meaning that you can set up an

Table A.1 Collections in Python: an overview. Note that while dictionaries are considered unordered, there exists a special class, `OrderedDict`, for ordered dictionaries, and since Python 3.7, dictionaries preserve order.

	Iterable	**Indexable**	**Sliceable**	**Ordered**	**Mutable**
Tuples	Yes	Yes	Yes	Yes	No
Lists	Yes	Yes	Yes	Yes	Yes
Dictionaries	Yes	Yes	No	No	Yes
Sets	Yes	No	No	No	Yes
Generators	Yes	No	No	Yes	No

iterative loop to go through the constituent elements of the collection. A subset of iterable collections are indexable, meaning that individual values can be retrieved using an index or key. Every indexable collection is iterable, but not every iterable collection is indexable (sets are an example of a nonindexable iterable). In addition, not every indexable iterable can be sliced (dictionaries, which have indices that do not have to obey any order that permits for a comparison of magnitude, can be iterated by their keys but cannot be sliced). Table A.1 shows the most frequently encountered collections in Python and their key properties.

A.6.2 Indexing

The indexation operator in Python is the square bracket, `[` and `]`, and Python is zero-indexed (i.e., the first item in a list has the index `0`).

```
>>> numbers = [1, 3, 5, 7, 9, 11]

>>> numbers[2]
5
```

A negative integer index is calculated from the rightmost element of an iterable with an integer index (i.e., the entry with the highest ordinal index):

```
>>> numbers = [1, 3, 5, 7, 9, 11]

>>> numbers[-2]
7
```

The slicing operator (`:`) is applicable only to iterables (indexables that have an ordinal integer index). It retrieves values between two indices:

```
>>> numbers = [1, 3, 5, 7, 9, 11]

>>> numbers[2:4]
[5, 7]
```

Omitting one side of the slicing operator automatically inserts the lowest (or highest) value:

```
>>> numbers = [1, 3, 5, 7, 9, 11]

>>> numbers[2:]
[5, 7, 9, 11]
```

Calling for an index that is not defined (out of range) yields a `IndexError`.

A special form of nonindexed access is pushing and popping, distinct from ordinary indexing in that they alter the collection.

- Pushing, inserting an item at the "rightmost" end of a mutable indexable iterable, is accomplished by the `.append()` method of lists, taking the item to be pushed as its argument.
- Popping, using the `.pop()` method, returns the "rightmost" element and removes it from the iterable.

A.6.3 Tuples

Tuples are immutable, indexable iterables. That means none of their values can be changed once they are defined, although you are of course free to reassign a changed version of the tuple to the variable name. What you cannot, however, do is to change the object itself:

```
>>> numbers = (1, 2, 3)
>>> numbers[2] = 5
TypeError: 'tuple' object does not support item assignment
```

In the same vein, you cannot add new entries to, or remove entries from, a tuple.

By convention, we refer to tuples of a particular length as an n-ple (with "tuple" to rhyme with "topple" and "2-ple" to rhyme with "too-ple"). Because tuples are immutable, they are more memory-efficient: unlike for their mutable cousins, lists, Python does not need to allocate additional memory space to deal with potential expansion. This may also be a desirable feature: tuples throw a `TypeError` if the code seeks an assignment to an element or slice of the tuple. Thus tuples may be a good way to protect values from accidental overwriting as well.

A.6.4 Lists

A list is a mutable, indexable iterable. Lists are defined using square bracket notation:

```
>>> numbers = [1, 2, 3]
>>> numbers[2] = 5
>>> numbers[2]
5
```

A.6.5 Sets

A set is a mutable iterable with a uniqueness constraint, i.e., a set will have no more than one copy of the same value. Note that “same,” in this context, means identical in terms of Python: thus `{2, 2}` would not be a valid set, but `{2, "2"}` would be. Note also that from this perspective, a variable is the same as its value: if `x = 2`, then `set([2, x])` will result in the set `{2};the fact that we referred to \verb2|` by way of a variable does not make it distinct from its value.

Sets are created by passing an iterable, mutable or not, to the `set()` function.

A.6.6 Dictionaries

Dictionaries comprise values (the “right hand side”) indexed by hashable keys (the “left hand side”). Any data type that can be hashed can be a key. This includes strings and numeric formats, but also tuples. It does not, however, include lists, which are mutable and consequently not hashable.

```
>>> capitals = {"Germany": "Berlin",
                "Austria": "Vienna",
                "France": "Paris"}

>>> capitals["Austria"]
Vienna
```

Calling for a key that is not defined within the object yields a `KeyError`, rather than an `IndexError`. A dictionary’s keys can be listed using the `.keys()` method, whereas its values can be listed using the `.values()` method.

A.6.7 Comprehensions

A comprehension is a way to create an iterable or indexable object (i.e., a tuple, a list or a dictionary) using a generating expression and a range. For instance, we may describe the numbers 0 through 100 as `range(0, 100)`. If we wanted to obtain a list of the squares of all numbers in this range, we would write

```
squares = [i ** 2 for i in range(0, 100)]
```

This is equivalent to the set-builder notation

$$s = \left\{ \overbrace{x^2}^{\texttt{i ** 2}} \quad \underbrace{\,|\, x \in}_{\texttt{for i in in}} \quad \underbrace{[0, 100]}_{\texttt{range(0, 100)}} \right\}$$

A comprehension may include a conditional:

```
squares = [i ** 2 for i in range(0, 100) if i %% 3 == 0]
```

The above example only calculates the value if `i` is divisible by 3. This would be equivalent to the Iverson bracket [mod $_3 i = 0$].

A dictionary comprehension uses an index to set the keys of the dictionary, and then uses the value of that index to calculate the value assigned to it:

```
indexed_squares = {k: k ** 2 for k in range(0, 100)}
```

This builds a dictionary `{1: 1, 2: 4, ...}`.

This is not to be confused with the set comprehension, which also uses curly braces:

```
set_of_squares = {i ** 2 for i in range(-100, 100)}
```

This will build a set, `(0, 1, 2, ...)`.

A.6.8 Generators

A generator, unlike a comprehension, does not create an entire object, but rather evaluates every time a new item is called for. It is, in this sense, the lazy-evaluated cousin of a list comprehension. Thus a generator expression does not need to store a potentially large number of items in memory.

```
squares_generator = (i ** 2 for i in range(0, 100) if i %% 3 == 0)
```

Note that this is syntactically similar to the list comprehension, except for the use of parentheses instead of square brackets. Calling `squares` from the list comprehension example will return a list, whereas calling `squares_generator` returns a generator object.

When calling the `next()` function on a generator object, we get the next result:

```
>>> next(squares_generator)
0
>>> next(squares_generator)
9
>>> next(squares_generator)
36
```

Once the generator has "run out," it will return a `StopIteration` exception.

Note that you can generally turn a finite generator into a list by casting it as a list: `list(squares_generator)` will return the same object as `squares`. Note also that generators are stateful, i.e., they keep track of all the objects that have been called.

Generators are not indexable, but they are iterable. That means it is possible to run a function over the range of the generator (which is rather ill-advised if the generator is infinite). One drawback of iterating over a generator, rather than an indexable iterable, is that accessing elements other than the current value proves somewhat difficult. If that is necessary, finite generators can be cast as lists or tuples. Note that this largely obviates the greatest benefit of generators in terms of computational efficiency, which is that they are "lazily evaluated": instead of generating each value, the individual values are only calculated when requested. If this is an envisaged use, it might be easier to go straight to generating the entire list or tuple in advance.

A.6.9 Ranges and enumerations

Ranges and enumerations are special generators built into Python for convenience.

`range()` takes a value called the "stop value," and optionally, a start value and a step value, and returns a generator that at every step, counts in increments of the step value from zero or the start value (if provided) up to the stop value. There are two pitfalls worth noting when it comes to `range`:

- `range` returns a generator, not a list. If a list is desired, the generator must be cast into it (`list(range(4))`, for instance).
- Where a start value is provided, it precedes the stop value. This is somewhat counterintuitive, as typically, optional parameters come after required parameters, but also follows how we think of ranges of numbers ("one to ten").

For similar constructs, NumPy offers `np.linspace()`, which will return a given number of equally spaced values in the range and `np.arange()`, which does the same, except with a step size specification, rather than a number of samples.

Enumerations make our lives a little easier where we have to loop through an indexable iterable, while also keeping track of where we are. The common programming idiom of setting up a counter, then retrieving the n-th element and incrementing n until it reaches a certain value, is actually an antipattern in Python. Enumerations are generators that return a 2-ple consisting of an index and a value. It is common to see the idiom

```
for idx, val in enumerate(collection):
    ...
```

This stores the currently enumerated element's index as `idx` and its value as `val`, a highly efficient way that saves us having to iterate over the length of the array (which, incidentally, is not possible in some cases, e.g., infinite generators).

A.6.10 Itertools

`itertools` is a package that comes bundled with Python and creates various iterators, including infinite iterators. Though it comes with Python, it must be imported using the `import` keyword. Some of the most useful iterators provided by `itertools` (for our purposes at least) include:

- `accumulate()`: returns the cumulative sum of an element and all its predecessors in an iterable.
- `compress()`: takes two iterables of equal length, the data and the selector, and returns a sequence that includes the n-th element of the data if the n-th element of the selector evaluates to `True`.
- `product()`: returns the Cartesian product of all elements of the input iterables.
- `repeat()`: returns an iterator that returns the argument at every iteration. Unless a number of repetitions is set by an additional second argument, it will return an infinite iterator.

- `permutations()`: returns every tuple of a given length, of all possible orderings, without repetition. If the length of permutations is not specified, it is set to the length of the input iterable.
- `combinations()`: returns combination tuples of a specified length. Its counterpart, `combinations_with_replacement()`, repeats individual items. Both functions treat elements as unique by index, not by value. Thus if the input sequence is `("A", "B", "A", "C")`, then for 2-length combinations, `"AA"` will occur twice.

A.7 Arithmetics and basic operations

A.7.1 Arithmetic operations

Python provides the basic arithmetic operations (`+`, `-`, `*` and so on). The exponentiation operator in Python is `**`—the caret (`^`) is the bitwise XOR operator, which is quite different. Python also provides two division operators—`/` for simple division and `//` for floor division. The modulus operator is `%`.

For all of these operators, suffixing it with the assignment operator, `=` creates an assignment of the value to a variable of the operation's result on that variable. Thus `a **= 2` is equivalent to `a = a ** 2`, `a //= 3` is equivalent to `a = a // 3`, and so on.

A.7.2 Comparison operators

All comparisons in Python are objects of the `bool` (Boolean) type. In Python, the equality comparison operator is `==`; note that `=` is an assignment operator, not a comparison operator.

In comparisons with equalities, the comparison precedes the equal sign, e.g., `>=` for $\geq$, rather than `=>`.

A.7.3 Set and logical operators

Python provides two logical operators:

1. `is` and `is not` test for identity, not equality. Thus two distinct objects that are equal in value but not identical are not going to satisfy `is`. This is crucially the case where the value is the same, but the type is different. `2` and `"2"` are fundamentally different to Python; one is an integer, the other is a string, and to Python, that is actually enough to stop even considering their values, because objects of different types are generally inequal (with the exception that if a `float` can be cast to an `int`, and their values are equal, they are considered logically equal).
2. `and` and `or` chain comparisons. Note that they take precedence before the comparisons themselves. Comparisons are evaluated in series, and "short circuit" as soon as the first "truthy" value is found.

In addition, Python provides two logical operations on collections: `any()` and `all()`.

1. `any(x)` is `True` if any member of `x` is truthy.
2. `all(x)` is `True` if all members of `x` are truthy.

Finally, Python provides the membership operator (∈) as `in`. `a in b` is true for a collection `b` if `a` is an element of `b`. The negation of `in` is `not in`, e.g., `a not in b`. Note that for purposes of `in`, anything that can be indexed or iterated over will be treated as a collection, e.g., `''r'' in ''golden retriever''` is a perfectly valid (and true) statement. In addition, `in` is a strictly elementwise membership operator, that is, if a collection is a subset of another collection, but not an element within it on its own, `in` will return `False`. Thus `(1, 2, 4) in (1, 2, 3, 4)` is `False`.

A.8 Program structure

A.8.1 Imports

The `import` keyword allows you to bring an installed Python package's scope into your current scope. `import` imports the entire package, which will be available under its own name. For instance,

```
import numpy
```

imports the popular NumPy package.

For certain packages, you may wish to import a particular subpackage or function only. For instance, if one only intends to use the interpolation functions in Scipy, one can import only the `interpolate` subpackage by specifying it as the object to be imported:

```
from scipy import interpolate
```

You can go more specific, too: for instance, to import only the barycentric interpolator,

```
from scipy.interpolate import BarycentricInterpolator
```

Python imports the subject of the `import` keyword and all below it. So `from scipy import interpolate` makes the `scipy.interpolate` subpackage available as `interpolate`, rather than `scipy.interpolate`.

Imports allow you to alias the imported object to any (valid) name of your choice using the `as` keyword. This is very common with frequently used packages. For instance, NumPy is often imported to the alias `np`:

```
import numpy as np
```

It is possible to import every item at a level of a package. `from scipy.interpolate import *` imports every object in the `scipy.interpolate` subpackage, and brings them into your code's namespace. This is a bad idea for many reasons, most importantly because you might not be aware of name clashes, inadvertently overwriting a variable you have previously set. Wildcard imports are generally not good coding practice, but quite common nonetheless.

A.8.2 Main call

A common Python pattern is often seen in code that is intended to be directly called:

```
if __name__ == "__main__":
    ...
```

This is used to specify a particular behavior in case the file is called from the command line.

A.8.3 Environments

For reproducibility, it is often useful to have clearly specified environments. Python environment management goes beyond the scope of this appendix. However, it is worth briefly mentioning two main approaches.

Virtualenv allows you to keep multiple virtual environments on your computer, with a defined set of packages at different versions. Conda is another solution to creating reproducible environments, using the Anaconda or Miniconda Python distributions. Both of these are good choices for creating reproducible environments.

A.9 Object-oriented programming

In Python, just about everything is an object, so we have in theory been using object-oriented programming all the way up to now as well. However, when OOP is discussed in the context of Python, this typically means working with classes. A class in Python, and in OOP in general, is a prototype of an object: it is a blueprint that lays out the properties an object would have and the behaviors it would exhibit (the latter being implemented through methods). This is to be distinguished from an instance of a class, which is a concrete object that has been instantiated. To use the commonly deployed analogy, a class is a blueprint, an object is a finished building.

A.9.1 Initialization

Python classes are initialized with the `class` keyword, and the typical naming convention for classes is CamelCase:

```
class CompartmentalDiseaseModel:
    pass
```

By default, classes subclass the `object` class. If we wished to subclass a specific class, we would supply it as an argument to the new class's name:

```
class SIRModel(CompartmentalDiseaseModel):
    pass
```

The constructor class in Python is called `__init__()`. Whatever is passed onto a class upon creating an instance of it is passed as an argument to `__init__()`:

```
class SIRModel(CompartmentalDiseaseModel):
    def __init__(self, beta, gamma):
        ...
```

We can then construct a `SIRModel` object by calling the class with its parameters:

```
m = SIRModel(0.3, 0.2)
```

and access the model's parameters as attributes:

```
m.beta
```

```
>>> 0.3
```

```
m.gamma
```

```
>>> 0.2
```

Much in line with Python's general use, a class specification is a block of code. In short, anything indented under the `class ...:` line is part of that class. `def` thus declares not a stand-alone function but a method of the class (strictly speaking, in Python, methods are functions until the class is instantiated, which is when they become methods, but this is largely an internal process we do not engage with). Methods generally take `self` as their first argument, with some exceptions to be discussed below.

A.9.2 Attribute access and self

`self` is by far the most misunderstood feature of Python. In object-oriented programming, `self` allows us to refer from within the class to any object instantiated from that class. To put it less abstractly: when we use `self` in, say, the `SIRModel` class above, it means that any time we instantiate `SIRModel`, `self` will refer to that object. It is, for instance, common to assign certain parameters to be properties of the class during initialization:

```
class SIRModel(CompartmentalDiseaseModel):
    def __init__(self, beta, gamma):
        self.beta = beta
        self.gamma = gamma
```

This essentially means that any object instantiated from `SIRModel` will have the properties `beta` and `gamma`. An exception from the self-reference requirement in classes concerns static methods:

```
class SIRModel(CompartmentalDiseaseModel):
    ...
    @staticmethod
    def say_hi():
        print("I'm a static method. I can say hi without 'self'!")
```

A static method is a method that is inside the class "for organizational purposes": unlike a legitimate, fully-fledged method, it cannot read or alter the object's state. Static methods are useful for keeping functions that are logically tied to a class in the same place, but without the need to take a `self` argument.

A.9.3 Inheritance

Inheritance describes a class deriving from another class (the superclass). In the following example, `SIRModel` inherits from `CompartmentalDiseaseModel`, or, alternatively, it subclasses `CompartmentalDiseaseModel`:

```
class SIRModel(CompartmentalDiseaseModel):
    def __init__(self, beta, gamma):
        super().__init__()

        self.beta = beta
        self.gamma = gamma
```

The `super()` function returns a reference to the superclass of the current class; in this case, that would be `CompartmentalDiseaseModel`. This also means that we can pass arguments to the superclass constructor. Note that a superclass's methods and properties are inherited by the subclass (and its subclasses and so on, all the way down). Consequently, `SIRModel` will have every method a `CompartmentalDiseaseModel` object would have.

A.9.4 Multiple inheritance and the MRO

Python objects may inherit from any number of classes. This includes classes explicitly designed to be composed into a function, called mixins:

```
class DataFrameExportMixin:
    def __init__(self):
        ...

    def export_data(self):
        ...

class SIRModel(CompartmentalDiseaseModel, DataFrameExportMixin):
    def __init__(self, beta, gamma):
        super().__init__()
```

The sole purpose of `DataFrameExportMixin` is to provide the `export_data()` functionality. It is important, however, to put classes in the right order for multiple inheritance. When a method of an instance is called, Python will first look within the class. If the method is not defined in the class, Python will use its Method Resolution Order (MRO) to find the method. The MRO in Python is determined by the C3 algorithm, which can get fairly complicated as the number of classes grows. However, a simple approximation is as follows:

1. Python will first search in the leftmost superclass (in this case, that would be `CompartmentalDiseaseModel`).
2. If this fails to yield a result, it will search in the leftmost superclass's superclass.
3. If this fails to yield a result, it will move along the superclasses of the leftmost superclass, left to right.
4. If this fails to yield a result, it will search in the superclasses of the leftmost superclass of the leftmost superclass, then proceed left to right. This will continue until the entire depth is exhausted.
5. Then, Python will move on to the second-from-the-left superclass, and repeat the same procedure.

A.10 Ancillary tools

A.10.1 Linting

Chances are, whatever IDE you are using already has some linting functionality built in, highlighting syntax errors. If you want a stand-alone linter, `flake8`, PyLint, and `black`, a highly opinionated linter that requires virtually no configuration, are equally good options.

A.10.2 Type enforcement

Currently, `mypy` is the preferred type checker for Python. This can be obtained by installing it as a Python package (`pip install mypy`), and invoked by `mypy my_python_script.py`, substituting the name of the file to be checked for static typing. Many IDEs also support type checking, either through their own inherent logic or by invoking `mypy`.

A.11 Beyond the standard library

Much of Python's strength for data-driven applications comes not from the language itself but from the ecosystem built around it. We rely on some key packages that facilitate our work. Introducing each of these could easily fill an entire chapter. Therefore I shall confine myself to a very brief outline of what we use the particular package for. As well-maintained and widely used packages, all of these have been extensively documented, and many have a comfortable learning curve and an abundance of training materials and tutorials online. As such, their official websites and documentations ought to be the first port of call for readers wishing to learn about these packages in more depth.

A.11.1 NumPy

NumPy is our principal tool for representing multidimensional arrays. Canonically, `numpy` is imported as `np`, which is how we will refer to the package in what follow.

A.11.1.1 Arrays

The key object of NumPy is the `array`, which can accommodate up to 32 dimensions (which ought to be enough for most applications). In NumPy, matrices are provided as lists of rows, i.e., the matrix

$$\begin{pmatrix} 1 & 3 \\ 5 & 7 \end{pmatrix}$$

would be created as

```
np.array([[1, 3],
          [5, 7]])
```

There are certain special arrays that NumPy supports:

- `np.ones` creates an array where each value is `1`.
- `np.zeros` creates an array where each value is `0`.
- `np.full` takes an argument of a value and fills the array with it.
- `np.empty` creates an "empty" array. This is not, actually, "empty" but filled with whatever is at the memory address assigned to each element. `np.empty` differs from the previous approaches in that it does not overwrite the memory addresses assigned to each element with a value (e.g., a zero in `np.zeros`), but merely creates the mapping between array positions and memory addresses. This makes it faster, but somewhat unpredictable. Be sure not to use `np.empty` if there is a risk of confusion between initialized values and your data.

- `np.eye` creates an identity matrix, i.e., a matrix whose elements $m_{i,j}$ are 1 if $i = j$ else 0.

A.11.1.2 Indexing

An array is indexed in Numpy by a tuple of integers, the length of which corresponds to the number of dimensions (called axes in this context) of the array, encapsulated by the indexing operators (square brackets, `[ ]`) and separated by commas. Other than specific integers, the slice operator : can be used for indexing, which allows for a number of interesting features:

- Indexing an array with a range (slice) returns all elements within that range. Thus for the 1-dimensional array `arr`, `arr[5:10]` returns elements between the 5th and the 10th element. Like all indexing in Python, this is exclusive.
- Not specifying either side of the slice operator corresponds to the lowest (left) or highest (right) index, i.e., `arr[:5]` returns all entries from the beginning of the array to the 5th element.
- Leaving both sides of the slice operator corresponds to the entire range in that dimension. Thus, for a 2-dimensional array, `[3, :]` returns the entire row of index 3 (i.e., the 4th row of the array, since Python is 0-indexed).
- The slice operator accepts periodic slicing, using the same symbol. Thus `arr[1:10:2]` means "return every second element in the range `1:10`.'' This third argument is typically referred to as the "stride." A negative stride follows backwards from the highest-indexed element down, thus `arr[::-1]` returns the reverse of the array.

Missing indices are expanded to :, thus indexing an array with a tuple of a length lower than the array's dimensions corresponds to the specified index for the specified dimension(s) and : for all others. Thus `arr[2]` for a 3-dimensional array would be equivalent to `arr[2,:,:]`. In addition, indexing supports the ellipsis symbol, ..., which expands to as many slice operators as needed to completely index the array. For a 4-dimensional array,

- `arr[1, ...]` corresponds to `arr[1, :, :, :]`
- `arr[..., 2, 3]` corresponds to `arr[:, :, 2, 3]`
- `arr[1, ..., 4]` corresponds to `arr[1, :, :, 4]`

A special form of indexing is using a Boolean array of the same size as the array to be indexed. This may be created explicitly (as a `np.array` object) or implicitly. The truth value of an array is an array of the same shape with the elementwise result of the condition, i.e., `arr > 2` returns an array of the same dimensions as `arr`, where each element is `True` if the corresponding element of `arr` is greater than 2, and otherwise `False`. Thus indexing `arr` with this condition (`arr[arr > 2]`) returns all elements of `arr` greater than 2. Note that Boolean indexing will always return a 1-dimensional array, even if the results could make a coherent higher-dimensional array.

Numpy arrays support a range of methods and routines. The most commonly used are

- matrix transpose: `.T`

- shape: `.shape` returns the shape tuple of a matrix, in rows × columns form
- flattening: `.ravel` returns a 1-dimensional array from the input array.

A.11.1.3 Broadcasting

Broadcasting refers to a technique NumPy uses to deal with arithmetic operations on different-sized arrays. In general, most arithmetic operations are not defined for different-shaped matrices. Broadcasting, however, allows these operations to be performed if the shapes of the `np.array` objects are broadcastable. In this case, the smaller matrix is repeatedly "spread" across the larger one. A trivial example is the multiplication of an array by a scalar. In this case, the scalar is "spread" over every element of the array.

The fundamental rule of broadcasting is that two arrays are broadcastable if their specified axes, read from right to left, are either equal or one of them is one.

- A $3 \times 4 \times 1$ and a 4×2 array are broadcastable. The first axis pair is $1, 2$: this is broadcastable, because one of them is one. The second axis pair is $4, 4$: this is broadcastable, because they are equal. The result is a $3 \times 4 \times 2$ array.
- A $3 \times 4 \times 5$ and a 4×2 array are not broadcastable. The first axis pair is $5, 2$: since these are neither equal nor is either of them 1.
- A scalar is broadcastable to any array of any size.

A.11.1.4 Vectorization and universal functions

Where a function is iterated over a loop, a significant overhead is incurred every time. The cost of Python's "duck typing" is that every time a function is called on an object, the code needs to be "dispatched," that is, it needs to determine what the type of the argument or arguments is, and how to proceed therefrom. However, as we have noted, NumPy admits a relatively limited range of data types, and these are specified well in advance. Consequently, if we know we are dealing with a `np.int64` (64-bit integer) column, there is no real need for the interpreter to examine every single element and determine what their type is, and what code to execute: they are all the same type (if they had not been, NumPy would not have sanctioned their inclusion in a column with the set data type).

Vectorization is the ability to run a function over a collection (which may or may not be a vector from the mathematical perspective) in an efficient manner without incurring this overhead. For instance, given a NumPy array `cases`, `np.mean(cases * 0.8)` is more effective than writing an explicit loop that adds, for each element of cases, the respective value to an accumulator array, then takes the sum, then divides it by the length. This is because for both `cases * 0.8` and `np.mean(...)`, NumPy is aware of what type the arguments are, and, more importantly, that any collection is going to be typically a collection of the same type of object. It is usually a good general rule to consider iteration, rather than application of a vectorized function to be a last resort.

A.11.2 Pandas

Pandas builds on NumPy to create the tabular structure known as a `DataFrame`. In NumPy, we indexed items by the ordinal of the row, column, and higher-dimensional axes. In Pandas, these axes can take a list of names (the row index is generally referred to as the index, whereas the column index is referred to as column names), which makes referring to them much easier. Pandas also adds some crucial functionality, such as merging, function application, and some other useful routines.

A.11.2.1 Structure

Structurally, a `DataFrame` is made up of columns, which are each a `pd.Series` object. A `pd.Series` object is essentially a list with an index and a fixed data type. The overarching idea of a `DataFrame` is that the elements of the individual `Series` may share an index, which ties the entire construct together in a neat tabular layout.

In general, most common types can be elements of a Pandas index, as long as they are unique. In addition to simple indexing, Pandas provides the `MultiIndex` capability, which is sometimes (perhaps not entirely accurately) referred to as hierarchical indexing. A `MultiIndex` is essentially a tiered index, with the elements represented as an n-length tuple, where n is the number of levels of the `MultiIndex`, and the i-th element of the tuple is the index value for that level. For instance, a `MultiIndex` to report cases of a disease could be structured by country, then state, then county, and finally city. The corresponding index for Great Falls, Virginia, would be

```
("US", "VA", "Fairfax County", "Great Falls")
```

The advantage of a MultiIndex is that it represents a hierarchy, i.e., Great Falls is within Fairfax County, which in turn is in Virginia, and so on. This means that specifying each level filters down to that value. `df["US"]["VA"]["Fairfax County"]` thus gives us all entries for Fairfax County, VA. This is vastly more efficient than filtering by values, because the index essentially acts as a function that, given up to n values, returns the set of values, without having to compare or evaluate the values themselves.

The hierarchy of indexing is useful because, as we noted, index entries have to be unique. The issue of indexing counties illustrates this quite well; there are thirty-one Washington Counties in the US, 26 Jefferson Counties, and almost as many Franklin, Jackson, and Lincoln Counties, each. Over half of the counties in the United States share a name with another county in another state, which means we could not easily index by county; if we told Pandas we wanted all cases from Washington County, it would be rightly confused as to which one we are looking for. On the other hand, there is only one of each county name per state. By hierarchically indexing, the index for Washington County, Georgia (`("US", "GA", "Washington County")`) would be distinct from Washington County, Oregon (`("US", "OR", "Washington County")`), and thus indexable.

A special type of index that deserves brief mention is the `DateTimeIndex`, which can be created from anything that can be cast to a `datetime` object. The `DateTimeIndex` makes the values "time-sensitive," and opens up the plethora of tools

Pandas provides, such as resampling. The Pandas user guide's chapter on time series functionality is worth consulting if such a use case is considered.

A.11.2.2 Indexing

Pandas indexes elements by name, rather than by ordinal. There are, generally, three ways to access things in Pandas:

1. Attribute access: for a `DataFrame`, attribute access will retrieve the column with that name (if the name is capable of being an attribute: `"Total Cases"` is a perfectly fine column name, but will not work for attribute access) and for a `Series`, it will retrieve the element with that index. Thus `df.total_cases` will give the `total_cases` column for a data frame, and `df.total_cases.Virginia` will give the entry in that series with the index value of `Virginia`.
2. Label-based access: this is by far the most frequent way of obtaining items from a `DataFrame`; it uses the location indexer `df.loc`. Thus `df.loc["Virginia", "total_cases"]` retrieves the item with the index `Virginia` and column name `total_cases` (note the order) from `df`. Label-based access works with ranges, slices, and lists of values, e.g., `df.loc[("Virginia", "Ohio"), :]` would retrieve all columns for the rows indexed with `Virginia` and `Ohio`.
3. Position-based access: this is NumPy-style positional indexing, and uses the indexer `df.iloc`. Thus `df.iloc[3, 4]` will return the value from the 4th row and 5th column (since Python is zero-indexed), whatever the name of those columns.

Primarily, `df.total_cases` is equivalent to `df["total_cases"]`. To index multiple columns quickly, you may use double brackets:

```
df[["total_cases", "test_positivity_ratio"]]
```

It is worth noting that though there are many ways to access a value and chained indexing (`df["total_cases"]["Virginia"]`) is normally perfectly fine (although a little inefficient), there is one situation in which it is absolutely crucial to avoid it: assigning values.

New users to Pandas are bound to encounter, at some point, the dreaded `SettingWithCopyWarning`, together with what is a somewhat convoluted explanation of views and copies. The short(er) explanation is that a chained assignment consists of two separate operations, where neither knows much about the other. When you call `df["total_cases"]["Virginia"]`, two things happen in sequence:

- First, Pandas takes the entire `total_cases` column and returns it as a `Series`.
- Then, Pandas takes the element from that indexed as `Virginia`.

At the end, when Pandas returns you the value, it does so without any awareness that this is actually a particular element of the original data frame. If you then assign a new value, say,

```
df["total_cases"]["Virginia"] = 123
```

Pandas will take that `Series`, find the value indexed by `Virginia`, and change it. The problem is that Pandas may return values to you in one of two ways: either as a view or as a temporary, throw-away copy, and which of these will occur depends on what is going on behind the scenes, at the level of memory management. Thus when you operate on `df["total_cases"]`, you do not ordinarily know whether you are dealing with a temporary object or not, that is, there is no way to know if your assignment will "stick." On the other hand, `df.loc["Virginia", "total_cases"]` guarantees that it will change `df`. The same goes for `.iloc`. For this reason, you should never use chained indexing when assigning values.

A.11.2.3 Filtering

The most common idiom for filtering values in Pandas is subsetting it with a predicate. Consider us being interested only in states that have more than 100 total cases:

```
df.loc[df.total_cases > 100]
```

What happens here is worth describing in detail.

- `df.total_cases > 100` takes the `Series` called `total_cases` and returns an equal-length `Series` of Boolean values that will be `True` if and only if the predicate holds. In other words, the n-th element of the `Series` being returned is the truth value of `total_cases[n] > 100`.
- This Boolean array is used to retrieve indices, giving the `.loc` indexer a list of booleans, as long as the index returns the rows where the value is `True`.

The beauty of this is that predicates can be quite versatile, including operators provided by Pandas, such as `.isin()`, which returns `True` if a value is in a particular collection and `False` otherwise. In addition, it is possible to perform any logical operation on them. Thus if we are interested in cases that do not have more than 100 cases, we could write (for a less efficient take on `total_cases <= 100`)

```
df.loc[~(df.total_cases > 100)]
```

Here, ~ acts as the logical negation operator, inverting the result of the initial Boolean filter.

Boolean filters can also be joined, by putting them in braces and following them up with the desired logical operation:

```
df.loc[((df.total_cases > 100) &
        (df.test_positivity_ratio > 0.2)) |\
            (df.hospitalization_change > 25)]
```

This filters `df` for instances where either `total_cases` exceed 100 or `test_positivity_ratio` exceeds 0.2, or alternatively, where `hospitalization_change` exceeds 25. It is clear to see that even relatively complex filtering can be implemented with ease through Boolean indexing.

Table A.2 Windowing functions in Pandas. For `weighted` windows, `win_type` must be one of the functions in `scipy.signal.windows`. Exponentially weighted windows may be specified by their span, their half-life, center-of-mass, or the smoothing factor α.

Type	Method	Parameters
Rolling	`rolling`	`window` (window width)
Weighted	`rolling`	`window` (window width), `win_type` (window type)
Expanding	`expanding`	`min_periods` (minimum width)
Exponentially weighted	`ewm`	one of four parameters

A.11.2.4 Window functions

Window functions are enormously useful for performing a range of operations, such as rolling averages or expanding sums. In Pandas, a windowed operation is specified as follows:

```
df.<type of window>(<parameter of window>).<operation>()
```

For instance, an exponentially weighted mean with a half-life of 7 days would be specified as

```
df.ewm(halflife="7D").mean()
```

The window functions supported by Pandas are laid out in Table A.2. Note that not every window supports every aggregation function, and it is worth examining the documentation in detail for the most appropriate window in nontrivial cases.

One interesting feature of window functions is that if they operate over a `DateTimeIndex`, their parameters can accommodate different times of measurement. For instance,

```
df.total_cases.rolling(7).sum()
```

returns a 7-item rolling average, regardless of the time between them. If we have reliable daily reporting, for instance, then this approximates a 7-day moving average well enough. However, if data is sporadic, there are irregularities or—as is the case in the earliest days of an epidemic—there are not necessarily cases associated with every day, it is helpful to attach a `DateTimeIndex` to the data frame and specify the window width as a time-denominated quantity, e.g., `rolling("7D")` (where `D` stands, of course, for days).

A.11.2.5 Aggregation

Any reader familiar with SQL would certainly have fond memories of the `GROUPBY` keyword, which, as the name suggests, groups elements by a value and performs some aggregate function that typically yields a single scalar. Consider, for instance, a data frame in which we list hospitals by the county they are in, and their number of ICU beds. If we wanted to know the total number of ICU beds in each county, we would

1. first, group rows by county,
2. then, obtain an aggregation, in this case, a sum.

In Pandas, this looks as follows:

```
df.groupby(["county"])["icu_beds"].sum()
```

It is rather important to understand what is actually happening here. The initial `groupby` call creates a `GroupBy` object, which is largely intended to be machine-digestible and not really for human consumption (although `.get_group()` returns a single group). When the next operation is applied, Pandas uses our separation of the data set into multiple subsets in aggregating values. Thus instead of giving us the sum of the entire `icu_beds` column, it gives us the sums with respect to each distinct value of `county`.

Grouping can be over an arbitrary number of columns, as long as there is at least one value column left. Ordinarily, aggregation will result in the creation of an index (if grouping by a single column) or a `MultiIndex` (if grouping by several). If this behavior is not desired, passing the parameter `as_index=False` to `groupby` will simply return the values of the groups as ordinary columns.

There are, generally, two ways of calculating aggregations. If we have a single value column left, or if we want the same aggregation function for the entire data frame, and the function is one of the commonly provided functions by Pandas (e.g., `sum`, `mean`, `std`, and so on), we may write

```
df.groupby(["state", "county"]).mean()
```

or, if there are multiple columns,

```
df.groupby(["state", "county"])["cases"].sum()
```

Note that if you use this approach, all columns that are not grouped on must be amenable to the operation before you apply it. If you have a text column you are not aggregating, a call to `.mean()` will yield an error, since `mean()` is not defined for strings. One idiom is, as used above, to select one or more columns after grouping but before performing the aggregation calculation.

An often better and definitely tidier approach is the use of the `.agg()` method, which takes a dictionary of column names and how we want to aggregate them. This may be a function, a lambda, or a string representing any of the functions that are available on groups. Thus each of the following does the same, but to different columns:

```
df.groupby(["state", "county"]).agg({
    "beds": np.sum,
    "icu_beds": lambda x: np.sum(x),
    "occupied_beds": "sum"
})
```

A.11.2.6 Function application

It is, technically, possible to iterate over each element in a `DataFrame` and perform some operation. However, in general, this is considered inefficient and much of an antipattern. `df.iterrows()`, which creates an iterator over a data frame, is considered to be an antipattern for larger `DataFrames`, although they are generally accepted where the purpose is to create or operate on a different object (for an example, see Computational Note 3.4).

The preferred method is function application, performed by the `.apply()` method of a data frame. This method takes a function, and applies it along the specified axis (`0` is column-wise, `1` is row-wise application). Where the function takes arguments, these may be provided either as the parameter `args` or `kwargs` to the function, or lambdified:

```
df.apply(determine_school_term, axis=1, schedule="England")
```

```
df.apply(lambda x: determine_school_term(x, schedule="England), axis=1)
```

Both of the above are equivalent.

Writing functions that "apply well" (i.e., which run efficiently on application) is at times more an art than a science, and a good deal of experience writing vectorized functions and understanding the internals is required to write functions that apply 60–80 times faster than their nonvectorized counterparts. The `raw` parameter, when set to `True`, instructs `.apply()` to pass the row or column onto the function not as a `pd.Series` but as a `np.array`, which is rather more efficient if the receiving function is a NumPy vectorized function.

A.11.3 SciPy

SciPy is a "kitchen sink" of functionalities for scientific numerical methods. It contains a wealth of methods and techniques, implemented quite efficiently, such as ODE solvers, FFT, and other methods.

A.11.3.1 ODE solving

SciPy provides two ODE solving APIs—`odeint`, which is older, but is included in this book in some examples (principally in Chapter 2), because it is still quite widely used, and `solve_ivp`. Both are within the `scipy.integrate` subpackage.

`solve_ivp` and `odeint` both take a functional definition of a differential equation, a timespan of evaluation, and initial values. Based on this, they calculate the state of the system with respect to the quantities of evaluation over the timespan of evaluation.

```
def deriv(t, y, beta, gamma):
    S, I, R = y
    dSdt = -beta * S * I
    dIdt = beta * S * I - gamma * I
```

```
    dRdt = gamma * I
    return dSdt, dIdt, dRdt
```

The function above is a quite typical way of building a derivative function, and Chapter 2 explains the process in more detail. Note that one crucial difference between `odeint` and `solve_ivp` is the order of parameters; for `odeint`, the `y` parameter precedes the time parameter `t`.

```
res = solve_ivp(fun=deriv,
                t_span=(0, 100),
                y0=y_0,
                args=(beta, gamma),
                max_step=1)
```

This solves the IVP for the function `deriv` between `0` and `100` with the initial state vector `y_0` (which must be of the same length as the input vector, otherwise the destructuring assignment in the first line of the `deriv` function's body will not work).

The result object (which we here assign to `res`) is a bunch with a number of fields, accessible as attributes. These include `t`, the time series of times of evaluation, and `y`, the values at each of the time steps. Where the derivative function returns a vector of values (as our example indeed does), this will be a two-dimensional `array`.

A.11.3.2 Minimization

SciPy's `optimize` subpackage contains a number of tools for minimization, including a wide range of commonly used solvers, both deterministic and stochastic. The common way of using `minimize` is to define an objective function, provide some useful guesses, and specify constraints, if any. This is a useful tool for a range of simple optimization tasks.

```
res_opt = minimize(loss,
                   x0=(0.5, 1.3),
                   bounds=((0, 2), (1, 3)))
```

This function minimizes the arguments of a loss function `loss` with two parameters, bounded by `0` and `2` for the first parameter and `1` and `3` for the second. It is essentially equivalent to

$$\underset{\substack{0\leq x_1\leq 2,\\ 1\leq x_2\leq 3}}{\arg\min}\ \mathrm{loss}(x_1, x_2). \tag{A.1}$$

More complex bounds can be provided by supplying a `LinearConstraint` and/or a `NonlinearConstraint` to the `minimize` function's `constraints` parameter. These can define fairly complex relationships, including trigonometric relations, but are not supported by every solver.

A.11.3.3 Curve fitting

Strictly speaking, curve fitting is a minimization problem: given x and $\hat{y} = f(x, \theta)$, find the values of θ so as to minimize $\delta(y, \hat{y})$ (where δ is some metric of fit, e.g., sum of squares). However, since we use it all the time, it has its separate function in SciPy's `optimize` subpackage, `curve_fit`.

In what ought to emerge as somewhat of a pattern at this point, we begin by defining our objective function, which takes x and each element of θ and returns $\hat{y}$. Say, for instance, that we wanted to fit an exponential model, as we would do in the earliest phases of an epidemic:

$$\hat{y} = y_0 \, 2^{\frac{x}{T_d}}.$$

This formulation relates the modeled number of cases $\hat{y}$, to the initial number of cases at time 0 (y_0) and T_d, the doubling time (the time it takes to get from n to $2n$ cases). In this case, we have a simple univariate fit problem:

$$\underset{T_d \in \mathbb{R}_{>0}}{\arg\min} \, \delta(y, y_0 2^{\frac{x}{T_d}}),$$

where y_0 is a known quantity (the first element of y), and x is the span of evaluation. We would code this as the objective function

```
y_0 = y[0]

def objective_fx(x, T_d):
    return y_0 * 2 ** (x/T_d)
```

We can then call the optimization as

```
popt, pcov = optimize.curve_fit(objective_fx,
                                x,
                                y,
                                p0=(5))
```

This yields the optimized parameters (`popt`) and the covariance matrix, `pcov`.

You might have noticed we were not exactly asked how we wished the distance to be determined between y and $\hat{y}$. By default, `curve_fit()` uses a nonlinear least squares fit. Out of the box, this is a Levenberg–Marquardt algorithm if no bounds are provided, and the Trust region reflective approach, where there are bounds specified. For a different curve fitting method, we can always rewrite the problem as a minimization problem.

A.12 Where to find help

If you are stuck, the first place to check out is often the official Python documentation, or the documentation of the package that is causing problems. Alternatively, you can also search for your issue on StackOverflow, a question-answer site for software developers. There is a vivid Python community on StackOverflow, with over 1.8 m questions answered at the time of writing.

If you are migrating to Python from another programming language, Rosetta Code may be a useful learning tool. Rosetta Code is a programming chrestomathy website, which lists solutions of relatively common programming tasks in a variety of languages. Python solutions exist for almost all of the over 1100 tasks, and comparing the Python way of solving problems to other languages, you become familiar with what can be a helpful learning aid.

Codecademy and freeCodeCamp both offer Python courses free of charge (at the time of writing). MOOCs provided by sites such as edX and Coursera feature courses both on Python and on other subjects that use Python as the computational tool of choice. Unlike Codecademy and freeCodeCamp, these are less interactive, and thus less immersive experiences, and more akin to conventional classroom instruction.

A number of books have been written on Python. Zed Shaw's *Learn Python the Hard Way* and Allen B. Downey's *Think Python* are both great first texts on Python, but both of these are a few years old and the rapid evolution of Python renders some of their contents out of date. Beginners may find that online courses, MOOCs, and independent study of the official Python documentation (especially the API reference for the basic library) provides a better experience.

References

[1] Laurie Garrett, The Coming Plague: Newly Emerging Diseases in a World Out of Balance, ISBN 0140250913, 1995.

[2] George E.P. Box, Science and statistics, Journal of the American Statistical Association 71 (356) (Dec. 1976) 791, https://doi.org/10.2307/2286841, ISSN: 01621459.

[3] Hesiod, Works and Days, Dover Publications, Mineola, NY, ISBN 0486452182, Aug. 2006.

[4] Jaime Cerda Lorca, Gonzalo Valdivia, John Snow, the cholera epidemic and the foundation of modern epidemiology, Revista Chilena de Infectologia 24 (4) (2007), ISSN: 0716-1018.

[5] S.W.B. Newsom, Pioneers in infection control: John Snow, Henry Whitehead, the Broad Street pump, and the beginnings of geographical epidemiology, https://doi.org/10.1016/j.jhin.2006.05.020, 2006.

[6] Stephen Wolfram, A New Kind of Science, Wolfram Media, Champaign, IL, 2002.

[7] Isaac Newton, Letter from Sir Isaac Newton to Robert Hooke, 1676.

[8] Chris A. MacK, Fifty years of Moore's law, IEEE Transactions on Semiconductor Manufacturing 24 (2) (2011), https://doi.org/10.1109/TSM.2010.2096437.

[9] Fred Brauer, Mathematical epidemiology: past, present, and future, https://doi.org/10.1016/j.idm.2017.02.001, 2017.

[10] Daniel Bernoulli, Essai d'une nouvelle analyse de la mortalité causée par la petite vérole, et des advantage de l'inoculation pour la prévenir, in: Die Werke von Daniel Bernoulli, Bd. 2 Analysis und Wahrscheinlichkeitsrechnung, 1766.

[11] Thomas Archibald, Craig Fraser, Ivor Grattan-Guinness, The history of differential equations, 1670–1950, Oberwolfach Reports (2009), https://doi.org/10.4171/owr/2004/51.

[12] Ronald Ross, Some a priori pathometric equations, British Medical Journal 1 (2830) (1915), https://doi.org/10.1136/bmj.1.2830.546, ISSN: 00071447.

[13] William Ogilvy Kermack, Anderson G. McKendrick, A contribution to the mathematical theory of epidemics, Proceedings of the Royal society of London. Series A, Containing Papers of a Mathematical and Physical Character 115 (772) (1927) 700–721.

[14] Numerical analysis: historical developments in the 20th century, https://doi.org/10.1016/c2009-0-10776-6, 2001.

[15] J.C. Butcher, G. Wanner, Runge-Kutta methods: some historical notes, Applied Numerical Mathematics 22 (1–3) (1996), https://doi.org/10.1016/s0168-9274(96)00048-7, SPEC. ISS., ISSN: 01689274.

[16] John Brooks, Dreadnought gunnery and the battle of Jutland: the question of fire control, https://doi.org/10.4324/9780203316207, 2005.

[17] Michael Tremblay, Bombsights and adding machines: translating wartime technology into peacetime sales, Bulletin of Science, Technology & Society 30 (3) (2010), https://doi.org/10.1177/0270467610371713, ISSN: 0270-4676.

[18] Jan Van Der Spiegel, et al., The ENIAC: history, operation and reconstruction in VLSI, in: The First Computers: History and Architectures, 2002.

[19] Thomas C. Schelling, Dynamic models of segregation, The Journal of Mathematical Sociology 1 (2) (1971), https://doi.org/10.1080/0022250X.1971.9989794, ISSN: 15455874.
[20] Itziar de Lecuona, María Villalobos-Quesada, European perspectives on big data applied to health: the case of biobanks and human databases, Developing World Bioethics 18 (3) (2018), https://doi.org/10.1111/dewb.12208, ISSN: 14718847.
[21] Victoria Dawe, John Hardie, The rise of the big data doctor: recent advances and warnings in digital public health technology, Electronic Journal of General Medicine 18 (3) (2021), https://doi.org/10.29333/ejgm/9763, ISSN: 25163507.
[22] Javier Andreu-Perez, et al., Big data for health, IEEE Journal of Biomedical and Health Informatics 19 (4) (2015), https://doi.org/10.1109/JBHI.2015.2450362, ISSN: 21682208.
[23] Allyson L. Byrd, Julia A. Segre, Adapting Koch's postulates, Science 351 (6270) (2016) 224–226.
[24] Chengcheng Shao, et al., Anatomy of an online misinformation network, PLoS ONE 13 (4) (2018), https://doi.org/10.1371/journa.pone.0196087, ISSN: 19326203.
[25] Petter Törnberg, Echo chambers and viral misinformation: modeling fake news as complex contagion, PLoS ONE 13 (9) (2018), https://doi.org/10.1371/journa.pone.0203958, ISSN: 19326203.
[26] Aaron Lynch, Thought contagion: how belief spreads through society: the new science of memes, The Kluwer International Series in Engineering & Computer Science (1996).
[27] Mick Roberts, Hans Heesterbeek, Bluff your way in epidemic models, Trends in Microbiology 1 (9) (1993), https://doi.org/10.1016/0966-842X(93)90075-3, ISSN: 0966842X.
[28] Oliver Sacks, Awakenings, Duckworth & Co., London, 1973.
[29] Fabio Sánchez, et al., Drinking as an epidemic-a simple mathematical model with recovery and relapse, in: Therapist's Guide to Evidence-Based Relapse Prevention, 2007.
[30] Manuel Plantegenest, Christophe Le May, Frédéric Fabre, Landscape epidemiology of plant diseases, https://doi.org/10.1098/rsif.2007.1114, 2007.
[31] C.L. Campbell, L.V. Madden, Introduction to Plant Disease Epidemiology, Wiley-Interscience, New York, 1990.
[32] Shady S. Atallah, et al., A plant-level, spatial, bioeconomic model of plant disease diffusion and control: grapevine leafroll disease, American Journal of Agricultural Economics 97 (1) (2015), https://doi.org/10.1093/ajae/aau032, ISSN: 14678276.
[33] Curtis L. Meinert, Clinical Trials Handbook: Design and Conduct, Wiley, Hoboken, NY, 2012.
[34] W.B. Carter, et al., Developing and testing a decision model for predicting influenza vaccination compliance, Health Services Research 20 (6 Pt 2) (Feb. 1986) 897–932, ISSN: 0017-9124, http://www.pubmedcentra.nih.gov/articerender.fcgi?artid=1068913&too=pmcentrez&rendertype=abstract.
[35] Guosheng Yin, Clinical Trial Design: Bayesian and Frequentist Adaptive Methods, Wiley, New York, NY, ISBN 0470581719, 2011.
[36] Michel Tibayrenc, Genetics and evolution of infectious diseases, in: Michel Tibayrenc (ed.), 2nd ed., ISBN 9780127999425, Elsevier, Boston, MA, 2017.
[37] Geoffrey S. Ginsburg, et al., Genomic and Precision Medicine: Infectious and Inflammatory Disease, Academic Press, London, UK, ISBN 9780128014967, 2019.
[38] Jordan Dominic, Peter Smith, Mathematical Techniques: An Introduction for the Engineering, Physical, and Mathematical Science, 4th ed., Oxford University Press, Oxford, UK, ISBN 0199282013, 2008.
[39] Matt J. Keeling, Pejman Rohani, Modeling Infectious Diseases in Humans and Animals, Princeton University Press, Princeton, NJ, ISBN 9781400841035, 2008.

[40] Roberto Therón Sánchez, Data visualization: pathways between art and science in the production and consumption of images, Fonseca Journal of Communication 23 (2021), https://doi.org/10.14201/FJC2021233960, ISSN: 21729077.
[41] Michael Y. Li, An Introduction to Mathematical Modeling of Infectious Diseases, vol. 2, Springer, 2018.
[42] Antonella Colonna Vilasi, The intelligence cycle, Open Journal of Political Science 08 (01) (2018), https://doi.org/10.4236/ojps.2018.81003, ISSN: 2164-0505.
[43] John Mantle, D.A.J. Tyrrell, An epidemic of influenza on Tristan da Cunha, Epidemiology and Infection 71 (1) (1973) 89–95.
[44] Giulia Giordano, et al., Modelling the Covid-19 epidemic and implementation of population-wide interventions in Italy, Nature Medicine 26 (6) (2020), https://doi.org/10.1038/s41591-020-0883-7, ISSN: 1546170X.
[45] Fiona M. Guerra, et al., The basic reproduction number (R0) of measles: a systematic review, Lancet. Infectious Diseases 17 (12) (2017), https://doi.org/10.1016/S1473-3099(17)30307-9, ISSN: 14744457.
[46] Shelley L. Deeks, et al., An assessment of mumps vaccine effectiveness by dose during an outbreak in Canada, CMAJ. Canadian Medical Association Journal 183 (9) (2011), https://doi.org/10.1503/cmaj.101371, ISSN: 14882329.
[47] Valentina Costantino, et al., How valid are assumptions about re-emerging smallpox? A systematic review of parameters used in smallpox mathematical models, https://doi.org/10.1093/mimed/usx092, 2018.
[48] Nishiura Hiroshi, Correcting the actual reproduction number: a simple method to estimate R0 from early epidemic growth data, International Journal of Environmental Research and Public Health 7 (1) (2010), https://doi.org/10.3390/ijerph7010291, ISSN: 16604601.
[49] Ying Liu, Joacim Rocklöv, The reproductive number of the Delta variant of Sars-CoV-2 is far higher compared to the ancestral Sars-CoV-2 virus, Journal of Travel Medicine 28 (7) (2021), https://doi.org/10.1093/jtm/taab124, ISSN: 17088305.
[50] Md Arif Billah, Md Mamun Miah, Md Nuruzzaman Khan, Reproductive number of coronavirus: a systematic review and meta-analysis based on global level evidence, PLoS ONE 15 (11) (November 2020), https://doi.org/10.1371/journa.pone.0242128, ISSN: 19326203.
[51] Abhishek Mallela, et al., Bayesian inference of state-level Covid-19 basic reproduction numbers across the United States, Viruses 14 (1) (2022), https://doi.org/10.3390/v14010157, ISSN: 19994915.
[52] Adnan Khan, et al., Estimating the basic reproductive ratio for the Ebola outbreak in Liberia and Sierra leone, Infectious Diseases of Poverty 4 (1) (2015), https://doi.org/10.1186/s40249-015-0043-3, ISSN: 20499957.
[53] Brian J. Coburn, Bradley G. Wagner, Sally Blower, Modeling influenza epidemics and pandemics: insights into the future of swine flu (H1N1), BMC Medicine 7 (2009), https://doi.org/10.1186/1741-7015-7-30, ISSN: 17417015.
[54] Katherine Royce, Feng Fu, Mathematically modeling spillovers of an emerging infectious zoonosis with an intermediate host, PLoS ONE 15 (8) (2020), https://doi.org/10.1371/journa.pone.0237780, ISSN: 19326203.
[55] Phrutsamon Wongnak, et al., A 'what-if' scenario: Nipah virus attacks pig trade chains in Thailand, BMC Veterinary Research 16 (1) (2020), https://doi.org/10.1186/s12917-020-02502-4, ISSN: 17466148.
[56] P.C. Cross, et al., Female elk contacts are neither frequency nor density dependent, Ecology 94 (9) (2013), https://doi.org/10.1890/12-2086.1, ISSN: 00129658.

[57] J.A.P. Heesterbeek, K. Dietz, The concept of R 0 in epidemic theory, Statistica Neerlandica 50 (1) (Mar. 1996) 89–110, https://doi.org/10.1111/j.1467-9574.1996.tb01482.x, ISSN: 0039-0402.

[58] John C. Forbes, et al., A national review of vertical HIV transmission, AIDS 26 (6) (2012), https://doi.org/10.1097/QAD.0b013e328350995c, ISSN: 02699370.

[59] Ziyad Al-Aly, Yan Xie, Benjamin Bowe, High-dimensional characterization of post-acute sequelae of Covid-19, Nature 594 (7862) (2021), https://doi.org/10.1038/s41586-021-03553-9, ISSN: 14764687.

[60] Kevin Wing, et al., Surviving Ebola: a historical cohort study of Ebola mortality and survival in Sierra leone 2014-2015, PLoS ONE 13 (12) (2018), https://doi.org/10.1371/journa.pone.0209655, ISSN: 19326203.

[61] Aaron D. Blackwell, et al., Helminth infection, fecundity, and age of first pregnancy in women, Science 350 (6263) (2015), https://doi.org/10.1126/science.aac7902, ISSN: 10959203.

[62] Changyun Hu, et al., Infections with immunogenic trypanosomes reduce tsetse reproductive fitness: potential impact of different parasite strains on vector population structure, PLoS Neglected Tropical Diseases 2 (3) (2008), https://doi.org/10.1371/journa.pntd.0000192, ISSN: 1935-2735.

[63] Barry S. Hewlett, Richard P. Amolat, Cultural contexts of Ebola in Northern Uganda, Emerging Infectious Diseases 9 (10) (2003), https://doi.org/10.3201/eid0910.020493, ISSN: 10806040.

[64] Amanda Tiffany, et al., Estimating the number of secondary Ebola cases resulting from an unsafe burial and risk factors for transmission during the West Africa Ebola epidemic, PLoS Neglected Tropical Diseases 11 (6) (2017), https://doi.org/10.1371/journa.pntd.0005491, ISSN: 19352735.

[65] J. Legrand, et al., Understanding the dynamics of Ebola epidemics, Epidemiology and Infection 135 (4) (2007), https://doi.org/10.1017/S0950268806007217, ISSN: 14694409.

[66] Chulwoo Park, Traditional funeral and burial rituals and Ebola outbreaks in West Africa: a narrative review of causes and strategy interventions, Journal of Health and Social Sciences 5 (1) (2020), https://doi.org/10.19204/2020/trdt8.

[67] Kerton R. Victory, et al., Ebola transmission linked to a single traditional funeral ceremony – Kissidougou, Guinea, December, 2014-January 2015, Morbidity and Mortality Weekly Report 64 (14) (2015), ISSN: 1545-861X.

[68] WHO, How to conduct safe and dignified burial of a patient who has died from suspected or confirmed Ebola virus disease, in: WHO/EVD/Guidance/Burials/14.2, October 2014.

[69] Carrie F. Nielsen, et al., Improving burial practices and cemetery management during an Ebola virus disease epidemic – Sierra Leone, 2014, Morbidity and Mortality Weekly Report 64 (1) (2015), ISSN: 1545-861X.

[70] Nicholas C. Grassly, Christophe Fraser, Geoffrey P. Garnett, Host immunity and synchronized epidemics of syphilis across the United States, Nature 433 (7024) (2005), https://doi.org/10.1038/nature03072, ISSN: 00280836.

[71] Anthony M. Carter, Comparative placentation, in: Encyclopedia of Reproduction, 2018, https://doi.org/10.1016/B978-0-12-809633-8.20546-3.

[72] Kathryn M. Edwards, Maternal antibodies and infant immune responses to vaccines, Vaccine 33 (47) (2015), https://doi.org/10.1016/j.vaccine.2015.07.085, ISSN: 18732518.

[73] Filio Marineli, et al., Mary Mallon (1869–1938) and the history of typhoid fever, Annals of Gastroenterology 26 (2) (2013), ISSN: 11087471.

[74] Kirsten S. Adriani, The uncomfortable history of Mary Mallon and typhoid fever, Nederlandsch Tijdschrift Voor Geneeskunde 163 (2019), ISSN: 1876-8784.

[75] Ran Canetti, et al., Privacy-preserving automated exposure notification, https://eprint.iacr.org/2020/863, 2020.
[76] Yousef Alimohamadi, Maryam Taghdir, Mojtaba Sepandi, Estimate of the basic reproduction number for Covid-19: a systematic review and meta-analysis, https://doi.org/10.3961/JPMPH.20.076, 2020.
[77] J.J. Pandit, Managing the R0 of Covid-19: mathematics fights back, Anaesthesia 75 (12) (2020), https://doi.org/10.1111/anae.15151, ISSN: 13652044.
[78] David M. Auerbach, et al., Cluster of cases of the Acquired Immune Deficiency Syndrome. Patients linked by sexual contact, The American Journal of Medicine 76 (3) (1984), https://doi.org/10.1016/0002-9343(84)90668-5, ISSN: 00029343.
[79] J. Wallinga, M. Lipsitch, How generation intervals shape the relationship between growth rates and reproductive numbers, Proceedings of the Royal Society of London. Series B, Biological Sciences 274 (1609) (2007), https://doi.org/10.1098/rspb.2006.3754, ISSN: 14712970.
[80] John Griffin, et al., Rapid review of available evidence on the serial interval and generation time of Covid-19, BMJ Open 10 (11) (2020), https://doi.org/10.1136/bmjopen-2020-040263, ISSN: 20446055.
[81] Denis Mollison, The structure of epidemic models, in: Epidemic Models: Their Structure and Relation to Data, 1995, Section 3.
[82] Klaus Dietz, Dieter Schenzle, Proportionate mixing models for age-dependent infection transmission, Journal of Mathematical Biology 22 (1) (1985), https://doi.org/10.1007/BF00276550, ISSN: 03036812.
[83] Andrew Harvey, Paul Kattuman, A farewell to R: time-series models for tracking and forecasting epidemics, Journal of the Royal Society Interface 18 (182) (2021), https://doi.org/10.1098/rsif.2021.017, ISSN: 17425662.
[84] Hannah Ritchie, et al., Coronavirus pandemic (Covid-19), in: Our World in Data, 2020, https://ourwordindata.org/coronavirus.
[85] Tore Huag, Kjell Tormod Nilssen, Ecological implications of harp seal phoca groenlandica invasions in northern Norway, Developments in Marine Biology 4.C (1995), https://doi.org/10.1016/S0163-6995(06)80053-3, ISSN: 01636995.
[86] Chester A. Alper, et al., Genetic prediction of nonresponse to hepatitis B vaccine, The New England Journal of Medicine 321 (11) (1989), https://doi.org/10.1056/nejm198909143211103, ISSN: 0028-4793.
[87] David A. Nace, et al., Antibody responses after mRNA-based Covid-19 vaccination in residential older adults: implications for reopening, Journal of the American Medical Directors Association 22 (8) (2021), https://doi.org/10.1016/j.jamda.2021.06.006, ISSN: 15389375.
[88] Jens Wrammert, et al., Human B cell responses to cholera infection, The Journal of Immunology 196 (1 Supplement) (2016).
[89] Vladimir Igorevich Arnol'd, Catastrophe Theory, Springer Berlin Heidelberg, Berlin, Heidelberg, ISBN 978-3-540-16199-8, 1986.
[90] Sabra L. Klein, Andrea Hodgson, Dionne P. Robinson, Mechanisms of sex disparities in influenza pathogenesis, Journal of Leukocyte Biology 92 (1) (2012), https://doi.org/10.1189/jb.0811427, ISSN: 0741-5400.
[91] Margaret Hellard, et al., The elimination of hepatitis C as a public health threat, Cold Spring Harbor Perspectives in Medicine 10 (4) (2020), https://doi.org/10.1101/cshperspect.a036939, ISSN: 21571422.
[92] Mitchell L. Shiffman, The next wave of hepatitis C virus: the epidemic of intravenous drug use, https://doi.org/10.1111/iv.13647, 2018.

[93] J.O. Lloyd-Smith, et al., Superspreading and the effect of individual variation on disease emergence, Nature 438 (7066) (2005), https://doi.org/10.1038/nature04153, ISSN: 14764687.

[94] M.E.J. Woolhouse, et al., Heterogeneities in the transmission of infectious agents: implications for the design of control programs, Proceedings of the National Academy of Sciences of the United States of America 94 (1) (1997), https://doi.org/10.1073/pnas.94.1.338, ISSN: 00278424.

[95] Richard A. McKay, "Patient Zero": The absence of a patient's view of the early North American AIDS epidemic, Bulletin of the History of Medicine 88 (1) (2014), https://doi.org/10.1353/bhm.2014.0005, ISSN: 00075140.

[96] Christian Janke, et al., Beyond Ebola treatment units: severe infection temporary treatment units as an essential element of Ebola case management during an outbreak, https://doi.org/10.1186/s12879-017-2235-x, 2017.

[97] Skylar Marvel, et al., The Covid-19 pandemic vulnerability index (PVI) dashboard: monitoring county level vulnerability, medRxiv: the preprint server for health sciences, https://doi.org/10.1101/2020.08.10.20169649, 2020.

[98] Luis E.C. Rocha, Fredrik Liljeros, Petter Holme, Simulated epidemics in an empirical spatiotemporal network of 50,185 sexual contacts, PLoS Computational Biology 7 (3) (2011), https://doi.org/10.1371/journa.pcbi.1001109, ISSN: 1553734X.

[99] Michael Worobey, et al., 1970s and 'Patient 0' HIV-1 genomes illuminate early HIV/AIDS history in North America, Nature 539 (7627) (2016), https://doi.org/10.1038/nature19827, ISSN: 14764687.

[100] Robert F. Garry, et al., Documentation of an AIDS virus infection in the United States in 1968, Journal of the American Medical Association 260 (14) (1988) 15383598, https://doi.org/10.1001/jama.1988.03410140097031, ISSN: 15383598.

[101] Joël Mossong, et al., Social contacts and mixing patterns relevant to the spread of infectious diseases, PLoS Medicine 5 (3) (2008), https://doi.org/10.1371/journa.pmed.0050074, ISSN: 15491277.

[102] Thang Hoang, et al., A systematic review of social contact surveys to inform transmission models of close-contact infections, Epidemiology 30 (5) (2019), https://doi.org/10.1097/EDE.0000000000001047, ISSN: 15315487.

[103] Kiesha Prem, Alex R. Cook, Mark Jit, Projecting social contact matrices in 152 countries using contact surveys and demographic data, PLoS Computational Biology 13 (9) (2017), https://doi.org/10.1371/journa.pcbi.1005697, ISSN: 15537358.

[104] Dina Mistry, et al., Inferring high-resolution human mixing patterns for disease modeling, Nature Communications 12 (1) (Dec. 2021) 323, https://doi.org/10.1038/s41467-020-20544-y, ISSN: 2041-1723.

[105] Abigail I. Leibowitz, et al., Association between prison crowding and Covid-19 incidence rates in Massachusetts prisons, April 2020-January 2021, JAMA Internal Medicine 181 (10) (2021), https://doi.org/10.1001/jamainternmed.2021.4392, ISSN: 21686114.

[106] Steven Ruggles, et al., IPUMS USA: Version 12.0, Minneapolis, MN, 2022.

[107] Virginia Department of Education, Enrollment and demographics: fall membership reports, https://www.doe.virginia.gov/statistics_reports/enroment/index.shtm, 2021.

[108] US Census Bureau, CB2000CBP All Sectors: County Business Patterns, including ZIP Code Business Patterns, by Legal Form of Organization and Employment Size Class for the U.S., States, and Selected Geographies, Tech. Rep., US Census Bureau, 2020, https://data.census.gov/cedsci/tabe?q=CB2000CBP%3A%20A%20Sectors%3A%20County%20Business%20Patterns,%20incuding%20ZIP%20Code%20Business%20Patterns,%20by%20Lega%20Form%20of%20Organization%20and%20Empoyment%20Size%

20Cass%20for%20the%20U.S.,%20States,%20and%20Seected%20Geographies%3A%202020&tid=CBP2020.CB2000CBP.
[109] US Census Bureau, EEOALL1R EEO 1-Year Detailed Census Occupation by Sex and Race/Ethnicity for Residence Geography, 2018, 5-year estimates, Tech. Rep., US Census Bureau, 2018, https://data.census.gov/cedsci/tabe?q=EEOALL1R&tid=ACSEEO5Y2018.EEOALL1R.
[110] R.M. Anderson, R.M. May, Age-related changes in the rate of disease transmission: implications for the design of vaccination programmes, Journal of Hygiene 94 (3) (1985), https://doi.org/10.1017/S002217240006160X, ISSN: 00221724.
[111] David Quammen, Spillover: Animal Infections and the Next Human Pandemic, Bodley Head, London, ISBN 9781847920102, 2012.
[112] Maureen Moyo, et al., Tuberculosis patients at the human-animal interface: potential zooanthroponotic and zoonotic transmission, One Health 13 (2021), https://doi.org/10.1016/j.oneht.2021.100319, ISSN: 23527714.
[113] Ronald Ross, On some peculiar pigmented cells found in two mosquitos fed on malarial blood, British Medical Journal 2 (1929) (1897), https://doi.org/10.1136/bmj.2.1929.1786, ISSN: 00071447.
[114] G.C. Cook, An illustrated history of malaria, Transactions of the Royal Society of Tropical Medicine and Hygiene 94 (2) (2000), https://doi.org/10.1016/s0035-9203(00)90294-3, ISSN: 00359203.
[115] S.R. Christophers, History of malaria, BMJ. British Medical Journal 1 (4711) (1951), https://doi.org/10.1136/bmj.1.4711.865-a, ISSN: 0959-8138.
[116] Thomas E. Wellems, Karen Hayton, Rick M. Fairhurst, The impact of malaria parasitism: from corpuscles to communities, https://doi.org/10.1172/JCI38307, 2009.
[117] Anthony James Wilson, et al., What is a vector?, Philosophical Transactions of the Royal Society of London. Series B, Biological Sciences 372 (1719) (2017), https://doi.org/10.1098/rstb.2016.0085, ISSN: 14712970.
[118] Tony Schountz, et al., Immunological control of viral infections in bats and the emergence of viruses highly pathogenic to humans, Frontiers in Immunology 8 (2017), https://doi.org/10.3389/fimmu.2017.01098, ISSN: 16643224.
[119] Eric M. Leroy, et al., Fruit bats as reservoirs of Ebola virus, Nature 438 (7068) (2005), https://doi.org/10.1038/438575a, ISSN: 14764687.
[120] Sandip Mandal, Ram Sarkar, Somdatta Sinha, Mathematical models of malaria – a review, https://doi.org/10.1186/1475-2875-10-202, 2011.
[121] Asha Ram Meena, Review on surveillance and bionomics of (Aedes mosquitoes) Dengue vectors, International Journal of Entomology Research 5 (1) (2020).
[122] Azael Che-Mendoza, et al., Abundance and Seasonality of Aedes aegypti (Diptera: Culicidae) in Two Suburban Localities of South Mexico, with Implications for Wolbachia (Rickettsiales: Rickettsiaceae)-Carrying Male Releases for Population Suppression, Journal of Medical Entomology 58 (4) (2021), https://doi.org/10.1093/jme/tjab052, ISSN: 19382928.
[123] Daniel Foreman-Mackey, et al., emcee: the MCMC Hammer, Publications of the Astronomical Society of the Pacific 125 (925) (Mar. 2013) 306–312, https://doi.org/10.1086/670067, ISSN: 00046280.
[124] Jonathan Goodman, Jonathan Weare, Ensemble samplers with affine invariance, Communications in Applied Mathematics and Computational Science 5 (1) (2010), https://doi.org/10.2140/camcos.2010.5.65, ISSN: 21575452.
[125] E. Paul J. Gibbs, The evolution of one health: a decade of progress and challenges for the future, Veterinary Record 174 (2014) 4, https://doi.org/10.1136/vr.g143, ISSN: 00424900.

[126] Kim Gruetzmacher, et al., The Berlin principles on one health – bridging global health and conservation, Science of the Total Environment 764 (2021), https://doi.org/10.1016/j.scitotenv.2020.142919, ISSN: 18791026.

[127] Michelle L. Baker, Mary Tachedjian, Lin Fa Wang, Immunoglobulin heavy chain diversity in Pteropid bats: evidence for a diverse and highly specific antigen binding repertoire, Immunogenetics 62 (3) (2010), https://doi.org/10.1007/s00251-010-0425-4, ISSN: 00937711.

[128] J.G. Breman, et al., The epidemiology of Ebola Haemorrhagic Fever in Zaire, 1976. Ebola Virus Haemorrhagic Fever, 1978.

[129] Daniel Crowley, et al., Identifying suspect bat reservoirs of emerging infections, Vaccines 8 (2) (2020), https://doi.org/10.3390/vaccines8020228, ISSN: 2076393X.

[130] Almudena Marí Saéz, et al., Investigating the zoonotic origin of the West African Ebola epidemic, EMBO Molecular Medicine 7 (1) (2015), https://doi.org/10.15252/emmm.201404792, ISSN: 1757-4676.

[131] Luke Nyakarahuka, et al., Isolated case of Marburg virus disease, Kampala, Uganda, 2014, Emerging Infectious Diseases 23 (6) (2017), https://doi.org/10.3201/eid2306.170047, ISSN: 10806059.

[132] Daniel G. Bausch, et al., Marburg hemorrhagic fever associated with multiple genetic lineages of virus, The New England Journal of Medicine 355 (9) (2006), https://doi.org/10.1056/nejmoa051465, ISSN: 0028-4793.

[133] Janusz T. Paweska, et al., Marburg virus infection in Egyptian rousette bats, South Africa, 2013–2014, Emerging Infectious Diseases 24 (6) (2018), https://doi.org/10.3201/eid2406.17216, ISSN: 10806059.

[134] Brian R. Amman, et al., Isolation of Angola-like Marburg virus from Egyptian rousette bats from West Africa, Nature Communications 11 (1) (2020), https://doi.org/10.1038/s41467-020-14327-8, ISSN: 20411723.

[135] David J. Wolking, et al., Boma to banda – a disease sentinel concept for reduction of diarrhoea, Pastoralism 6 (1) (2016), https://doi.org/10.1186/s13570-016-0059-8, ISSN: 20417136.

[136] Isabelle Anne Bisson, Benard J. Ssebide, Peter P. Marra, Early detection of emerging zoonotic diseases with animal morbidity and mortality monitoring, EcoHealth 12 (1) (2015), https://doi.org/10.1007/s10393-014-0988-x, ISSN: 16129210.

[137] Laura Amato, et al., Integrated early warning surveillance: Achilles' heel of One Health?, Microorganisms 8 (1) (2020), https://doi.org/10.3390/microorganisms8010084, ISSN: 20762607.

[138] Joost van Herten, et al., Ethical decision-making in zoonotic disease control, Journal of Agricultural and Environmental Ethics 33 (2) (2020), https://doi.org/10.1007/s10806-020-09828-x, ISSN: 1187-7863.

[139] Laith Hussain-Alkhateeb, et al., Early warning systems (EWSs) for Chikungunya, Dengue, malaria, yellow fever, and Zika outbreaks: what is the evidence? A scoping review, PLoS Neglected Tropical Diseases 15 (9) (2021), https://doi.org/10.1371/journa.pntd.0009686, ISSN: 19352735.

[140] Janet L. Gardner, Winter flocking behaviour of speckled warblers and the Allee effect, Biological Conservation 118 (2) (2004), https://doi.org/10.1016/j.biocon.2003.08.018, ISSN: 00063207.

[141] Leah C. Wilson, James L. Goodson, Marcy A. Kingsbury, Seasonal variation in group size is related to seasonal variation in neuropeptide receptor density, Brain, Behavior and Evolution 88 (2) (2016), https://doi.org/10.1159/000448372, ISSN: 14219743.

[142] David T.S. Hayman, Biannual birth pulses allow filoviruses to persist in bat populations, Proceedings of the Royal Society of London. Series B, Biological Sciences 282 (1803) (2015), https://doi.org/10.1098/rspb.2014.259, ISSN: 14712954.
[143] Wolfgang Zimmermann, Robert Kammerer, The immune-modulating pregnancy-specific glycoproteins evolve rapidly and their presence correlates with hemochorial placentation in primates, BMC Genomics 22 (1) (2021), https://doi.org/10.1186/s12864-021-07413-8, ISSN: 14712164.
[144] John M. Barry, The Great Influenza: the Epic Story of the Deadliest Plague in History, Penguin Books, New York, NY, ISBN 0143036491, 2005.
[145] Gabriel Ånestad, Svein Arne Nordbø, Virus interference. did rhinoviruses activity hamper the progress of the 2009 influenza A (H1N1) pandemic in Norway?, Medical Hypotheses 77 (6) (2011), https://doi.org/10.1016/j.mehy.2011.09.021, ISSN: 03069877.
[146] Abdul H. Mohsen, et al., Hepatitis C and HIV-1 coinfection, Gut 51 (4) (2002), https://doi.org/10.1136/gut.51.4.601, ISSN: 00175749.
[147] Bruno Guy, et al., From research to phase III: preclinical, industrial and clinical development of the Sanofi Pasteur tetravalent Dengue vaccine, Vaccine 29 (42) (2011), https://doi.org/10.1016/j.vaccine.2011.06.094, ISSN: 0264410X.
[148] Saranya Sridhar, et al., Effect of Dengue serostatus on Dengue vaccine safety and efficacy, The New England Journal of Medicine (2018), https://doi.org/10.1056/nejmoa1800820, ISSN: 0028-4793.
[149] Scott B. Halstead, Dengvaxia sensitizes seronegatives to vaccine enhanced disease regardless of age, Vaccine 35 (47) (2017), https://doi.org/10.1016/j.vaccine.2017.09.08, ISSN: 18732518.
[150] Ronald U. Mendoza, et al., Public trust and the Covid-19 vaccination campaign: lessons from the Philippines as it emerges from the Dengvaxia controversy, International Journal of Health Planning and Management 36 (6) (2021), https://doi.org/10.1002/hpm.3297, ISSN: 10991751.
[151] Won Eui Hong, Robert L. Pego, Exclusion and multiplicity for stable communities in Lotka–Volterra systems, Journal of Mathematical Biology 83 (2) (2021), https://doi.org/10.1007/s00285-021-01638-7, ISSN: 14321416.
[152] Neil Ferguson, Roy Anderson, Sunetra Gupta, The effect of antibody-dependent enhancement on the transmission dynamics and persistence of multiple-strain pathogens, Proceedings of the National Academy of Sciences of the United States of America 96 (2) (1999), https://doi.org/10.1073/pnas.96.2.790, ISSN: 00278424.
[153] Nessma Adil Mahmoud Yousif, et al., The impact of Covid-19 vaccination campaigns accounting for antibody-dependent enhancement, PLoS ONE 16 (4) (2021), https://doi.org/10.1371/journa.pone.0245417, ISSN: 19326203.
[154] Nikhra Vinod, Stages in Covid-19 vaccine development: the Nemesis, the Hubris and the Elpis, International Journal of Clinical Virology 4 (1) (2020), https://doi.org/10.29328/journa.ijcv.1001028.
[155] Wajihul Hasan Khan, et al., Covid-19 pandemic and vaccines: update on challenges and resolutions, Frontiers in Cellular and Infection Microbiology 11 (2021), https://doi.org/10.3389/fcimb.2021.690621, ISSN: 22352988.
[156] Shabeena Tawar, Parnika Chandola, Sougat Ray, Is antibody-dependent enhancement a cause for COVID vaccine hesitancy, Journal of Marine Medical Society 23 (1) (2021), https://doi.org/10.4103/jmms.jmms_93_21, ISSN: 0975-3605.
[157] Daniele Focosi, et al., Previous humoral immunity to the endemic seasonal alphacoronaviruses NL63 and 229E is associated with worse clinical outcome in Covid-19 and suggests original antigenic sin, Life 11 (4) (2021), https://doi.org/10.3390/ife11040298, ISSN: 20751729.

[158] Susana V. Bardina, et al., Enhancement of Zika virus pathogenesis by preexisting anti-flavivirus immunity, Science 356 (6334) (2017), https://doi.org/10.1126/science.aa4365, ISSN: 10959203.

[159] Jesse J. Waggoner, et al., Antibody-dependent enhancement of severe disease is mediated by serum viral load in pediatric Dengue virus infections, Journal of Infectious Diseases 221 (11) (2020), https://doi.org/10.1093/INFDIS/JIZ618, ISSN: 15376613.

[160] Leah C. Katzelnick, et al., Antibody-dependent enhancement of severe Dengue disease in humans, Science 358 (6365) (2017), https://doi.org/10.1126/science.aan6836, ISSN: 10959203.

[161] Erida Gjini, Patricia H. Brito, Integrating antimicrobial therapy with host immunity to fight drug-resistant infections: classical vs. adaptive treatment, PLoS Computational Biology 12 (4) (2016), https://doi.org/10.1371/journa.pcbi.1004857, ISSN: 15537358.

[162] Quang Huy Nguyen, et al., Insights into the processes that drive the evolution of drug resistance in Mycobacterium tuberculosis, Evolutionary Applications 11 (9) (2018), https://doi.org/10.1111/eva.12654, ISSN: 17524571.

[163] Andreas Handel, Roland R. Regoes, Rustom Antia, The role of compensatory mutations in the emergence of drug resistance, PLoS Computational Biology 2 (10) (2006), https://doi.org/10.1371/journa.pcbi.0020137, ISSN: 15537358.

[164] Amel Kevin Alame Emane, et al., Drug resistance, fitness and compensatory mutations in mycobacterium tuberculosis, Tuberculosis 129 (2021), https://doi.org/10.1016/j.tube.2021.102091, ISSN: 1873281X.

[165] Barbara Hellriegel, Immunoepidemiology – bridging the gap between immunology and epidemiology, Trends in Parasitology 17 (2) (2001), https://doi.org/10.1016/S1471-4922(00)01767-0, ISSN: 14714922.

[166] Jacques Monod, The growth of bacterial cultures, Annual Review of Microbiology 3 (1949) 1, https://doi.org/10.1146/annurev.mi.03.100149.002103, ISSN: 0066-4227.

[167] Joseph M. Benoun, Jasmine C. Labuda, Stephen J. McSorley, Collateral damage: detrimental effect of antibiotics on the development of protective immune memory, mBio 7 (6) (2016), https://doi.org/10.1128/mBio.01520-16, ISSN: 21507511.

[168] Sharon L. Messenger, Ian J. Molineux, J.J. Bull, Virulence evolution in a virus obeys a trade-off, Proceedings of the Royal Society of London. Series B, Biological Sciences 266 (1417) (1999), https://doi.org/10.1098/rspb.1999.0651, ISSN: 14712970.

[169] Élodie Chapuis, Audrey Arnal, Jean Baptiste Ferdy, Trade-offs shape the evolution of the vector-borne insect pathogen xenorhabdus nematophila, Proceedings of the Royal Society of London. Series B, Biological Sciences 279 (1738) (2012), https://doi.org/10.1098/rspb.2012.0228, ISSN: 14712954.

[170] P.D. Williams, et al., Evidence of trade-offs shaping virulence evolution in an emerging wildlife pathogen, Journal of Evolutionary Biology 27 (6) (2014), https://doi.org/10.1111/jeb.12379, ISSN: 14209101.

[171] Samuel Alizon, Minus van Baalen, Transmission-virulence trade-offs in vector-borne diseases, Theoretical Population Biology 74 (1) (2008), https://doi.org/10.1016/j.tpb.2008.04.003, ISSN: 00405809.

[172] S. Alizon, et al., Virulence evolution and the trade-off hypothesis: history, current state of affairs and the future, Journal of Evolutionary Biology 22 (2) (2009), https://doi.org/10.1111/j.1420-9101.2008.01658.x, ISSN: 1010061X.

[173] Tobias E. Hector, Isobel Booksmythe, Digest: little evidence exists for a virulence-transmission trade-off, Evolution 73 (4) (2019), https://doi.org/10.1111/evo.13724, ISSN: 15585646.

[174] C.H. Wirsing von Koenig, Nicole Guiso, Global burden of pertussis: signs of hope but need for accurate data, Lancet. Infectious Diseases 17 (9) (2017), https://doi.org/10.1016/S1473-3099(17)30357-2, ISSN: 14744457.
[175] Pedro Plans-Rubió, Evaluation of the establishment of herd immunity in the population by means of serological surveys and vaccination coverage, Human Vaccines and Immunotherapeutics 8 (2) (2012), https://doi.org/10.4161/hv.18444, ISSN: 2164554X.
[176] Annika B. Wilder-Smith, Kaveri Qureshi, Resurgence of measles in Europe: a systematic review on parental attitudes and beliefs of measles vaccine, Journal of Epidemiology and Global Health 10 (1) (2020), https://doi.org/10.2991/JEGH.K.191117.001, ISSN: 22106014.
[177] Dennis K. Flaherty, The vaccine-autism connection: a public health crisis caused by unethical medical practices and fraudulent science, Annals of Pharmacotherapy 45 (10) (2011), https://doi.org/10.1345/aph.1Q318, ISSN: 10600280.
[178] Stephen Coleman, The historical association between measles and pertussis: a case of immune suppression?, SAGE Open Medicine 3 (2015), https://doi.org/10.1177/2050312115621315, ISSN: 2050-3121.
[179] Brigitta M. Laksono, et al., Studies into the mechanism of measles-associated immune suppression during a measles outbreak in the Netherlands, Nature Communications 9 (1) (2018), https://doi.org/10.1038/s41467-018-07515-0, ISSN: 20411723.
[180] Guillermo B. Morales, Miguel A. Muñoz, Immune amnesia induced by measles and its effects on concurrent epidemics, Journal of the Royal Society Interface 18 (179) (2021), https://doi.org/10.1098/rsif.2021.0153, ISSN: 17425662.
[181] Chandini Raina MacIntyre, Valentina Costantino, David J. Heslop, The potential impact of a recent measles epidemic on Covid-19 in Samoa, BMC Infectious Diseases 20 (1) (2020), https://doi.org/10.1186/s12879-020-05469-7, ISSN: 14712334.
[182] Sandeepa M. Eswarappa, Sylvie Estrela, Sam P. Brown, Within-host dynamics of multi-species infections: facilitation, competition and virulence, PLoS ONE 7 (6) (2012), https://doi.org/10.1371/journa.pone.0038730, ISSN: 19326203.
[183] D. Goletti, et al., Effect of Mycobacterium tuberculosis on HIV replication. Role of immune activation, Journal of Immunology 157 (3) (1996) 1271–1278, ISSN: 0022-1767.
[184] M. Best, A. Katamba, D. Neuhauser, Making the right decision: Benjamin Franklin's son dies of smallpox in 1736, Quality and Safety in Health Care 16 (6) (2007), https://doi.org/10.1136/qshc.2007.023465, ISSN: 14753898.
[185] José Esparza, Andreas Nitsche, Clarissa R. Damaso, Investigations on the historical origin and evolution of the smallpox vaccine, Gaceta Medica de Caracas 128 (1) (2020), https://doi.org/10.47307/GMC.2020.128.s1.11, ISSN: 03674762.
[186] Boying Ma, Smallpox remedies and variolation in ancient China, A History of Medicine in Chinese Culture (2020), https://doi.org/10.1142/9789813237995_0014.
[187] Yuki Furuse, Akira Suzuki, Hitoshi Oshitani, Origin of measles virus: divergence from rinderpest virus between the 11th and 12th centuries, Virology Journal 7 (2010), https://doi.org/10.1186/1743-422X-7-52, ISSN: 1743422X.
[188] Normile Dennis, Driven to extinction, Science 319 (5870) (2008), https://doi.org/10.1126/science.319.5870.1606, ISSN: 00368075.
[189] Maimuna S. Majumder, et al., Substandard vaccination compliance and the 2015 measles outbreak, JAMA Pediatrics 169 (5) (May 2015), https://doi.org/10.1001/jamapediatrics.2015.0384, ISSN: 2168-6203.
[190] Amy A. Parker, et al., Implications of a 2005 measles outbreak in Indiana for sustained elimination of measles in the United States (5), The New England Journal of Medicine 355 (Aug. 2006), https://doi.org/10.1056/NEJMoa060775, ISSN: 0028-4793.

[191] M. McCarthy, Measles cases exceed 100 in US outbreak, BMJ. British Medical Journal 350 (feb03) (Feb. 2015) 13, https://doi.org/10.1136/bmj.h622, ISSN: 1756-1833.

[192] Vittorio Demicheli, et al., Vaccines for measles, Mumps and rubella in children, Cochrane Database of Systematic Reviews (Feb. 2012), https://doi.org/10.1002/14651858.CD004407.pub3, ISSN: 14651858.

[193] D. James Nokes, Jonathan Swinton, The control of childhood viral infections by pulse vaccination, Mathematical Medicine and Biology 12 (1) (1995), https://doi.org/10.1093/imammb/12.1.29, ISSN: 14778599.

[194] Mamta Barik, Sudipa Chauhan, Sumit Kaur Bhatia, Efficacy of pulse vaccination over constant vaccination in Covid-19: a dynamical analysis, Communications in Mathematical Biology and Neuroscience (2020) 2020, https://doi.org/10.28919/cmbn/5187, ISSN: 20522541.

[195] Boris Shulgin, Lewi Stone, Zvia Agur, Pulse vaccination strategy in the SIR epidemic model, Bulletin of Mathematical Biology 60 (6) (1998), https://doi.org/10.1016/S0092-8240(98)90005-2, ISSN: 00928240.

[196] Anna K. Lugnér, et al., Cost-effectiveness of targeted vaccination to protect newborns against pertussis: comparing neonatal, maternal, and cocooning vaccination strategies, Vaccine 31 (46) (2013), https://doi.org/10.1016/j.vaccine.2013.09.028, ISSN: 0264410X.

[197] Geeta K. Swamy, Sarahn M. Wheeler, Neonatal pertussis, cocooning and maternal immunization, Expert Review of Vaccines 13 (9) (2014), https://doi.org/10.1586/14760584.2014.944509, ISSN: 17448395.

[198] Jingzhou Liu, et al., Epidemiological game-theory dynamics of chickenpox vaccination in the USA and Israel, Journal of the Royal Society Interface 9 (66) (2012), https://doi.org/10.1098/rsif.2011.0001, ISSN: 17425662.

[199] Chris T. Bauch, David J.D. Earn, Vaccination and the theory of games, Proceedings of the National Academy of Sciences of the United States of America 101 (36) (2004), https://doi.org/10.1073/pnas.0403823101, ISSN: 00278424.

[200] Heidi J. Larson, Kenneth Hartigan-Go, Alexandre de Figueiredo, Vaccine confidence plummets in the Philippines following Dengue vaccine scare: why it matters to pandemic preparedness, Human Vaccines and Immunotherapeutics 15 (3) (2019), https://doi.org/10.1080/21645515.2018.1522468, ISSN: 2164554X.

[201] Philippe Monteyne, Francis E. André, Is there a causal link between hepatitis B vaccination and multiple sclerosis?, Vaccine 18 (19) (2000), https://doi.org/10.1016/S0264-410X(99)00533-2, ISSN: 0264410X.

[202] Luca D. Bertzbach, et al., Latest insights into Marek's disease virus pathogenesis and tumorigenesis, Cancers 12 (3) (2020), https://doi.org/10.3390/cancers12030647, ISSN: 20726694.

[203] Sanjay M. Reddy, Yoshihiro Izumiya, Blanca Lupiani, Marek's disease vaccines: current status, and strategies for improvement and development of vector vaccines, Veterinary Microbiology 206 (2017), https://doi.org/10.1016/j.vetmic.2016.11.024, ISSN: 18732542.

[204] Gerrit Ansmann, Efficiently and easily integrating differential equations with JiTCODE, JiTCDDE, and JiTCSDE, Chaos 28 (4) (2018), https://doi.org/10.1063/1.5019320, ISSN: 10541500.

[205] David A. Kennedy, Andrew F. Read, Why does drug resistance readily evolve but vaccine resistance does not?, Proceedings of the Royal Society of London. Series B, Biological Sciences 284 (1851) (2017), https://doi.org/10.1098/rspb.2016.2562, ISSN: 14712954.

[206] Anita H. Melnyk, Alex Wong, Rees Kassen, The fitness costs of antibiotic resistance mutations, Evolutionary Applications 8 (3) (2015), https://doi.org/10.1111/eva.12196, ISSN: 17524571.
[207] M.E. Alexander, et al., Modelling the effect of a booster vaccination on disease epidemiology, https://doi.org/10.1007/s00285-005-0356-0, 2006.
[208] Zoltan Kiss, et al., Nationwide effectiveness of first and second Sars-CoV2 booster vaccines during the delta and omicron pandemic waves in Hungary (HUN-VE 2 study), Frontiers in Immunology (2022), https://doi.org/10.3389/fimmu.2022.905585.
[209] Rose Marie Carlsson, et al., Modeling the waning and boosting of immunity from infection or vaccination, Journal of Theoretical Biology 497 (2020), https://doi.org/10.1016/j.jtbi.2020.110265, ISSN: 10958541.
[210] Peter Oliver Okin, The yellow flag of quarantine: an analysis of the historical and prospective impacts of socio-legal controls over contagions, Dissertation Abstracts International: Section B: The Sciences and Engineering 73 (8-B(E)) (2013), ISSN: 0419-4217.
[211] Gian Franco Gensini, Magdi H. Yacoub, Andrea A. Conti, The concept of quarantine in history: from plague to Sars, Journal of Infection 49 (4) (2004), https://doi.org/10.1016/j.jinf.2004.03.002, ISSN: 01634453.
[212] Moritz E. Wigand, Thomas Becker, Florian Steger, Psychosocial reactions to plagues in the cultural history of medicine: a medical humanities approach, The Journal of Nervous and Mental Disease 208 (6) (2020), https://doi.org/10.1097/NMD.0000000000001200, ISSN: 1539736X.
[213] Benjamin Barber, Post-pandemic literature: force and moral obligation to the other in Boccaccio's Decameron and Camus' The Plague, in: COVID-19 Pandemic, Crisis Responses and the Changing World, 2021.
[214] Rowena D. Jones, et al., Quantitative risk assessment of rabies entering Great Britain from North America via cats and dogs, Risk Analysis 25 (3) (2005), https://doi.org/10.1111/j.1539-6924.2005.00613.x, ISSN: 02724332.
[215] J.A. Baker, Singapore's circuit breaker and beyond: Timeline of the Covid-19 reality, 2020.
[216] David R. Lawrence, John Harris, Red herrings, circuit-breakers and ageism in the Covid-19 debate, Journal of Medical Ethics 47 (9) (2021), https://doi.org/10.1136/medethics-2020-107115, ISSN: 14734257.
[217] American Medical Association, Code of Medical Ethics Opinion 8.4: Ethical Use of Quarantine & Isolation, Tech. Rep., American Medical Association, Chicago, IL, 2016.
[218] Alice Desclaux, et al., Accepted monitoring or endured quarantine? Ebola contacts' perceptions in Senegal, Social Science and Medicine 178 (2017), https://doi.org/10.1016/j.socscimed.2017.02.009, ISSN: 18735347.
[219] Marlies Hesselman, et al., Energy poverty in the Covid-19 era: mapping global responses in light of momentum for the right to energy, Energy Research and Social Science 81 (2021), https://doi.org/10.1016/j.erss.2021.102246, ISSN: 22146296.
[220] Bridget J. Crawford, Emily Gold Waldman, Period poverty in a pandemic: harnessing law to achieve menstrual equity, Washington University Law Review 98 (2021) 5.
[221] Ranie Ahmed, et al., Racial equity in the fight against Covid-19: a qualitative study examining the importance of collecting race-based data in the Canadian context, Tropical Diseases, Travel Medicine and Vaccines 7 (1) (2021), https://doi.org/10.1186/s40794-021-00138-2, ISSN: 20550936.
[222] Gregory Phillips, et al., Addressing the disproportionate impacts of the Covid-19 pandemic on sexual and gender minority populations in the United States: actions to-

ward equity, LGBT Health 7 (6) (2020), https://doi.org/10.1089/gbt.2020.0187, ISSN: 23258306.

[223] Ishaan Sachdeva, et al., The disparities faced by the LGBTQ+ community in times of Covid-19, Psychiatry Research 297 (2021), https://doi.org/10.1016/j.psychres.2021.113725, ISSN: 18727123.

[224] Hagai Rossman, et al., Hospital load and increased Covid-19 related mortality in Israel, Nature Communications 12 (1) (2021), https://doi.org/10.1038/s41467-021-22214-z, ISSN: 20411723.

[225] Kristoffer Strålin, et al., Mortality in hospitalized Covid-19 patients was associated with the Covid-19 admission rate during the first year of the pandemic in Sweden, Infectious Diseases 54 (2) (2022), https://doi.org/10.1080/23744235.2021.1983643, ISSN: 23744235.

[226] Alejandro E. Maciás, et al., Real-world evidence of Dengue burden on hospitals in Mexico: insights from the automated subsystem of hospital discharges (Saeh) database, Revista de Investigacion Clinica. 71 (3) (2019), https://doi.org/10.24875/RIC.18002681.

[227] Herbert Kayiga, et al., Improving the quality of obstetric care for women with obstructed labour in the national referral hospital in Uganda: lessons learnt from criteria based audit, BMC Pregnancy and Childbirth 16 (1) (2016), https://doi.org/10.1186/s12884-016-0949-1, ISSN: 14712393.

[228] Marie Crandall, et al., Effects of closure of an urban level I trauma centre on adjacent hospitals and local injury mortality: a retrospective, observational study, BMJ Open 6 (5) (2016), https://doi.org/10.1136/bmjopen-2016-011700, ISSN: 20446055.

[229] Jacinta I. Pei Chen, et al., Covid-19 and Singapore: from early response to circuit breaker, Annals of the Academy of Medicine, Singapore 49 (8) (2020), https://doi.org/10.47102/annas-acadmedsg.2020239, ISSN: 03044602.

[230] Walid Jumblatt Abdullah, Soojin Kim, Singapore's responses to the Covid-19 outbreak: a critical assessment, American Review of Public Administration 50 (6–7) (2020), https://doi.org/10.1177/0275074020942454, ISSN: 15523357.

[231] Andrew Hall, et al., Cost and benefit of military quarantine policies, Preventive Medicine 143 (2021), https://doi.org/10.1016/j.ypmed.2020.106371, ISSN: 10960260.

[232] Cliff C. Kerr, et al., Controlling Covid-19 via test-trace-quarantine, Nature Communications 12 (1) (2021), https://doi.org/10.1038/s41467-021-23276-9, ISSN: 20411723.

[233] Souvik Dubey, et al., Psychosocial impact of Covid-19, Diabetes and Metabolic Syndrome. Clinical Research and Reviews 14 (5) (2020), https://doi.org/10.1016/j.dsx.2020.05.035, ISSN: 18780334.

[234] Yuchang Jin, et al., Mass quarantine and mental health during Covid-19: a meta-analysis, Journal of Affective Disorders 295 (2021), https://doi.org/10.1016/j.jad.2021.08.067, ISSN: 15732517.

[235] Andrew William Byrne, et al., Inferred duration of infectious period of Sars-CoV-2: rapid scoping review and analysis of available evidence for asymptomatic and symptomatic Covid-19 cases, BMJ Open 10 (8) (2020), https://doi.org/10.1136/bmjopen-2020-039856, ISSN: 20446055.

[236] John P.A. Ioannidis, Precision shielding for Covid-19: metrics of assessment and feasibility of deployment, BMJ Global Health 6 (1) (2021), https://doi.org/10.1136/bmjgh-2020-004614, ISSN: 20597908.

[237] George Davey Smith, David Spiegelhalter, Shielding from Covid-19 should be stratified by risk, https://doi.org/10.1136/bmj.m2063, 2020.

[238] Thornton Wilder, The Eighth Day, ISBN 0060088915, 1967.

[239] Thana C. De Campos, The traditional definition of pandemics, its moral conflations, and its practical implications: a defense of conceptual clarity in global health laws and policies, Cambridge Quarterly of Healthcare Ethics 29 (2) (2020), https://doi.org/10.1017/S0963180119001002, ISSN: 14692147.

[240] Peter Doshi, The elusive definition of pandemic influenza, Bulletin of the World Health Organization 89 (7) (2011), https://doi.org/10.2471/bt.11.086173, ISSN: 0042-9686.

[241] Benjamin J. Singer, Robin N. Thompson, Michael B. Bonsall, The effect of the definition of 'pandemic' on quantitative assessments of infectious disease outbreak risk, Scientific Reports 11 (1) (2021), https://doi.org/10.1038/s41598-021-81814-, ISSN: 20452322.

[242] Victoria A. Catenacci, James O. Hill, Holly R. Wyatt, The obesity epidemic, Clinics in Chest Medicine 30 (3) (2009), https://doi.org/10.1016/j.ccm.2009.05.001, ISSN: 02725231.

[243] Michael J. Reilly, Keon M. Parsa, Matthew Biel, Is there a selfie epidemic?, JAMA Facial Plastic Surgery 21 (5) (2019), https://doi.org/10.1001/jamafacia.2019.0419, ISSN: 21686076.

[244] L.J. Donaldson, J. Cavanagh, J. Rankin, The dancing plague: a public health conundrum, Public Health 111 (4) (1997), ISSN: 0033-3506.

[245] Gregor Rohmann, Gregor Rohmann, I. Einführung: Chorea, Veitstanz, Tanzwut – zwischen Religion und Medizin, Tanzwut (2012), https://doi.org/10.13109/9783666367212.15.

[246] Carlos Castillo-Chavez, Baojun Song, Dynamical models of tuberculosis and their applications, Mathematical Biosciences and Engineering 1 (2) (2004) 361–404, https://doi.org/10.3934/mbe.2004.1.361, ISSN 1551-0018.

[247] E.H. Elbasha, C.N. Podder, A.B. Gumel, Analyzing the dynamics of an SIRS vaccination model with waning natural and vaccine-induced immunity, Nonlinear Analysis: Real World Applications 12 (5) (2011), https://doi.org/10.1016/j.nonrwa.2011.03.015, ISSN: 14681218.

[248] A.B. Gumel, Causes of backward bifurcations in some epidemiological models, Journal of Mathematical Analysis and Applications 395 (1) (2012), https://doi.org/10.1016/j.jmaa.2012.04.077, ISSN: 0022247X.

[249] David Greenhalgh, Martin Griffiths, Backward bifurcation, equilibrium and stability phenomena in a three-stage extended BRSV epidemic model, Journal of Mathematical Biology 59 (1) (2009), https://doi.org/10.1007/s00285-008-0206-y, ISSN: 03036812.

[250] Roy M. Anderson, Robert M. May, Infectious Diseases of Humans: Dynamics and Control, OUP, 1991.

[251] William Farr, Causes of death in England and Wales, Tech. Rep., The Registrar General of Births, Deaths and Marriages in England, London, 1840, pp. 69–98, https://hd.hande.net/2027/njp.32101064041955.

[252] Petter Holme, Extinction times of epidemic outbreaks in networks, PLoS ONE 8 (12) (2013), https://doi.org/10.1371/journa.pone.0084429, ISSN: 19326203.

[253] W.P. Cleveland, G.C. Tiao, Decomposition of seasonal time series: a model for the census X11 program, Journal of the American Statistical Association 71 (355) (1976), https://doi.org/10.1080/01621459.1976.10481532, ISSN: 1537274X.

[254] Robert B. Cleveland, et al., STL: a seasonal-trend decomposition procedure based on loess (with discussion), Journal of Official Statistics 6 (1990).

[255] Estela Bee Dagum, Silvia Bianconcini, Time series components, in: Seasonal Adjustment Methods and Real Time Trend-Cycle Estimation, Springer, Cham, Switzerland, ISBN 978-3-319-31820-2, 2016, pp. 29–57, Chap. 2.

[256] Romulus Breban, et al., Is there any evidence that syphilis epidemics cycle?, Lancet. Infectious Diseases 8 (9) (2008), https://doi.org/10.1016/S1473-3099(08)70203-2, ISSN: 14733099.

[257] Christopher Torrence, Gilbert P. Compo, A practical guide to wavelet analysis, Bulletin of the American Meteorological Society 79 (1) (1998), https://doi.org/10.1175/1520-0477(1998)079<0061:APGTWA>2.0.CO;2, ISSN: 00030007.

[258] S.D. Meyers, B.G. Kelly, J.J. O'Brien, An introduction of wavelet analysis in oceanography and meteorology: with application to the dispersion of Yanai waves, Monthly Weather Review 121 (10) (1993), https://doi.org/10.1175/1520-0493(1993)121<2858:AITWAI>2.0.CO;2, ISSN: 00270644.

[259] A.B. Christie, Poliomyelitis: clinical features, Public Health 61 (C) (1947), https://doi.org/10.1016/S0033-3506(47)80090-3, ISSN: 00333506.

[260] N. Nathanson, Epidemiologic aspects of poliomyelitis eradication, Reviews of Infectious Diseases 6 (Suppl 2) (1984), https://doi.org/10.1093/cinids/6.suppement_2.s308, ISSN: 01620886.

[261] Stephen L. Cochi, et al., The global polio eradication initiative: progress, lessons learned, and polio legacy transition planning, Health Affairs 35 (2) (2016), https://doi.org/10.1377/hthaff.2015.1104, ISSN: 15445208.

[262] Sunil Bahl, et al., Global polio eradication – way ahead, Indian Journal of Pediatrics 85 (2) (2018), https://doi.org/10.1007/s12098-017-2586-8, ISSN: 09737693.

[263] Micaela Elvira Martinez, The calendar of epidemics: seasonal cycles of infectious diseases, PLoS Pathogens 14 (11) (2018), https://doi.org/10.1371/journa.ppat.1007327, ISSN: 15537374.

[264] S.F. Dowell, Seasonal variation in host susceptibility and cycles of certain infectious diseases, Emerging Infectious Diseases 7 (3) (2001), https://doi.org/10.3201/eid0703.017301, ISSN: 10806040.

[265] Benjamin Sultan, et al., Climate drives the meningitis epidemics onset in West Africa, PLoS Medicine 2 (2005), https://doi.org/10.1371/journa.pmed.0020006, ISSN: 15491277.

[266] Matt J. Keeling, Pejman Rohani, Bryan T. Grenfell, Seasonally forced disease dynamics explored as switching between attractors, Physica D: Nonlinear Phenomena 148 (3–4) (2001), https://doi.org/10.1016/S0167-2789(00)00187-1, ISSN: 01672789.

[267] Liang Mao, et al., Modeling monthly flows of global air travel passengers: an open-access data resource, Journal of Transport Geography 48 (2015), https://doi.org/10.1016/j.jtrangeo.2015.08.017, ISSN: 09666923.

[268] Shengjie Lai, et al., Global holiday datasets for understanding seasonal human mobility and population dynamics, Scientific Data 9 (1) (2022), https://doi.org/10.1038/s41597-022-01120-z, ISSN: 20524463.

[269] Jens Halvorsrud, Ivar Örstavik, An epidemic of rotavirus-associated gastroenteritis in a nursing home for the elderly, Scandinavian Journal of Infectious Diseases 12 (3) (1980), https://doi.org/10.3109/inf.1980.12.issue-3.01, ISSN: 16511980.

[270] Stephen Davis, et al., Predictive thresholds for plague in Kazakhstan, Science 304 (5671) (2004), https://doi.org/10.1126/science.1095854, ISSN: 00368075.

[271] Ljupco Kocarev, et al., Discrete chaos – I: theory, IEEE Transactions on Circuits and Systems I: Regular Papers 53 (6) (2006), https://doi.org/10.1109/TCSI.2006.874181, ISSN: 10577122.

[272] Jorge Duarte, et al., Chaos analysis and explicit series solutions to the seasonally forced SIR epidemic model, Journal of Mathematical Biology 78 (7) (2019), https://doi.org/10.1007/s00285-019-01342-7, ISSN: 14321416.

[273] Suzanne M. O'Regan, et al., Chaos in a seasonally perturbed SIR model: avian influenza in a seabird colony as a paradigm, Journal of Mathematical Biology 67 (2) (2013), https://doi.org/10.1007/s00285-012-0550-9, ISSN: 03036812.
[274] Olivia P. Judson, The rise of the individual-based model in ecology, Trends in Ecology and Evolution 9 (1) (1994), https://doi.org/10.1016/0169-5347(94)90225-9, ISSN: 01695347.
[275] Colin Sparrow, The Lorenz Equations: Bifurcations, Chaos, and Strange Attractors, Springer, New York, NY, ISBN 978-1-4612-5767-7, 1982.
[276] André F. Steklain, Ahmed Al-Ghamdi, Euaggelos E. Zotos, Using chaos indicators to determine vaccine influence on epidemic stabilization, Physical Review E 103 (3) (2021), https://doi.org/10.1103/PhysRevE.103.032212, ISSN: 24700053.
[277] Rebecca Solnit, The Encyclopedia of Trouble and Spaciousness, Trinity University Press, San Antonio, TX, 2014.
[278] Rockefeller in charge, Gunnison, CO, https://www.cooradohistoricnewspapers.org/?a=d&d=GNC19181101-01.2.11, Nov. 1918.
[279] Howard Markel, et al., Nonpharmaceutical influenza mitigation strategies, US communities, 1918–1920 pandemic, Emerging Infectious Diseases 12 (12) (2006), https://doi.org/10.3201/eid1212.060506, ISSN: 10806059.
[280] Flu gets us at last, Gunnison, CO, https://www.cooradohistoricnewspapers.org/?a=d&d=GNC19190314-01.2.2, Mar. 1919.
[281] W.R. Tobler, A computer movie simulating urban growth in the Detroit region, Economic Geography 46 (1970), https://doi.org/10.2307/143141, ISSN: 00130095.
[282] Benjamin D. Dalziel, et al., Urbanization and humidity shape the intensity of influenza epidemics in U.S. cities, Science 362 (6410) (2018), https://doi.org/10.1126/science.aat6030, ISSN: 10959203.
[283] Veaceslav Mir, Post-pandemic city: historical context for new urban design, Transylvanian Review of Administrative Sciences 2020 (special issue) (2020), https://doi.org/10.24193/tras.SI2020.6.
[284] Nelson Gouveia, Claudio Kanai, Pandemics, cities and public health, Ambiente e Sociedade 23 (2020), https://doi.org/10.1590/1809-4422ASOC20200120VU2020L3ID, ISSN, 1414753X.
[285] Carlos Moreno, et al., Introducing the "15-minute city": sustainability, resilience and place identity in future post-pandemic cities, Smart Cities 4 (1) (2021), https://doi.org/10.3390/smartcities4010006, ISSN: 26246511.
[286] Nuño Mardones Fernández de Valderrama, José Luque-Valdivia, Izaskun Aseguinolaza-Braga, The 15 minutes-city, a sustainable solution for postCOVID19 cities?, Ciudad y Territorio Estudios Territoriales 52 (205) (2020), https://doi.org/10.37230/CyTET.2020.205.13, ISSN: 11334762.
[287] Maria Nicola Buonocore Mattia De Martino, Chiara Ferro, Digital transformation and cities: how Covid-19 has boosted a new evolution of urban spaces, Journal of Urban Regeneration and Renewal 15 (1) (2021), ISSN: 17529646.
[288] Rahul Jaiswal, Anshul Agarwal, Richa Negi, Smart solution for reducing the Covid-19 risk using smart city technology, IET Smart Cities 2 (2) (2020), https://doi.org/10.1049/iet-smc.2020.004, ISSN: 26317680.
[289] Roy Dibyendu, et al., Utilizing smart city cyber-physical infrastructure for tracking and monitoring pandemics like Covid-19 with the ICCC as the nerve centre, Digital Government: Research and Practice 2 (1) (2021), https://doi.org/10.1145/3428124, ISSN: 26390175.
[290] Lucas Melgaço, Rosamunde van Brakel, Smart cities as surveillance theatre, Surveillance and Society 19 (2) (2021), https://doi.org/10.24908/ss.v19i2.14321, ISSN: 14777487.

[291] J. Olejnik, et al., Differences in cytokine responses during infection with high and low pathogenic Ebola viruses, Cytokine 63 (3) (2013), ISSN: 1043-4666.
[292] P.B. Jahrling, et al., Preliminary report: isolation of Ebola virus from monkeys imported to USA, The Lancet 335 (8688) (1990), https://doi.org/10.1016/0140-6736(90)90737-P, ISSN: 01406736.
[293] Geoff Boeing, OSMnx: a Python package to work with graph-theoretic OpenStreetMap street networks, The Journal of Open Source Software 2 (2017) 12, https://doi.org/10.21105/joss.00215.
[294] Geoff Boeing, OSMnx: new methods for acquiring, constructing, analyzing, and visualizing complex street networks, Computers, Environment and Urban Systems 65 (2017), https://doi.org/10.1016/j.compenvurbsys.2017.05.004, ISSN: 01989715.
[295] B. Castelnuovo, A review of compliance to anti tuberculosis treatment and risk factors for defaulting treatment in Sub Saharan Africa, African Health Sciences 10 (4) (2010), ISSN: 16806905.
[296] S. Govender, R. Mash, What are the reasons for patients not adhering to their anti-TB treatment in a South African district hospital?, South African Family Practice 51 (6) (2009), https://doi.org/10.1080/20786204.2009.10873916, ISSN: 20786204.
[297] Sawsan Elbireer, et al., Tuberculosis treatment default among HIV-TB co-infected patients in urban Uganda, Tropical Medicine and International Health 16 (8) (2011), https://doi.org/10.1111/j.1365-3156.2011.02800.x, ISSN: 13602276.
[298] Steen M. Hansen, et al., Association between driving distance from nearest fire station and survival of out-of-hospital cardiac arrest, Journal of the American Heart Association 7 (21) (2018), https://doi.org/10.1161/JAHA.118.008771, ISSN: 20479980.
[299] Michael T. Cudnik, et al., A geospatial assessment of transport distance and survival to discharge in out of hospital cardiac arrest patients: implications for resuscitation centers, Resuscitation 81 (5) (2010), https://doi.org/10.1016/j.resuscitation.2009.12.030, ISSN: 03009572.
[300] Charlotte Kelly, et al., Are differences in travel time or distance to healthcare for adults in global North countries associated with an impact on health outcomes? A systematic review, BMJ Open 6 (11) (2016), https://doi.org/10.1136/bmjopen-2016-01305, ISSN: 20446055.
[301] Olufemi Samuel Amoo, Care seeking behavior of citizens during pandemics: a case study of Covid-19 in Nigeria, Acta Scientific Microbiology 3 (2020) 9, https://doi.org/10.31080/asmi.2020.03.0689.
[302] Julie H. Hernandez, et al., Is Covid-19 community level testing effective in reaching at-risk populations? Evidence from spatial analysis of new orleans patient data at walk-up sites, BMC Public Health 21 (1) (2021), https://doi.org/10.1186/s12889-021-10717-9, ISSN: 14712458.
[303] Aida Rosalia Guhlincozzi, Aynaz Lotfata, Travel distance to flu and Covid-19 vaccination sites for people with disabilities and age 65 and older, Chicago metropolitan area, Journal of Health Research (2021), https://doi.org/10.1108/JHR-03-2021-0196, ISSN: 2586940X.
[304] S.L. Hakimi, Optimum distribution of switching centers in a communication network and some related graph theoretic problems, Operations Research 13 (3) (1965), https://doi.org/10.1287/opre.13.3.462, ISSN: 0030-364X.
[305] S. O'Mahony, Medicine and the McNamara fallacy, Journal of the Royal College of Physicians of Edinburgh 47 (3) (2017), https://doi.org/10.4997/JRCPE.2017.315, ISSN: 14782715.
[306] V. Latora, M. Marchiori, A measure of centrality based on network efficiency, New Journal of Physics 9 (2007), https://doi.org/10.1088/1367-2630/9/6/188, ISSN: 13672630.

[307] Yiping Chen, et al., Finding a better immunization strategy, Physical Review Letters 101 (5) (2008), https://doi.org/10.1103/PhysRevLett.101.058701, ISSN: 00319007.

[308] Andrew Crooks, et al., Agent-Based Modeling and the City: A Gallery of Applications, Urban Book Series, 2021.

[309] Sasha Azad, Chris Martens, Little computer people: a survey and taxonomy of simulated models of social interaction, Proceedings of the ACM on Human-Computer Interaction 5 (CHIPLAY) (2021), https://doi.org/10.1145/3474672.

[310] Mariusz Maziarz, Martin Zach, Assessing the quality of evidence from epidemiological agent-based models for the Covid-19 pandemic, History and Philosophy of the Life Sciences 43 (1) (2021), https://doi.org/10.1007/s40656-020-00357-4, ISSN: 17426316.

[311] Srinivasan Venkatramanan, et al., Using data-driven agent-based models for forecasting emerging infectious diseases, Epidemics 22 (2018), https://doi.org/10.1016/j.epidem.2017.02.010, ISSN: 18780067.

[312] Brandon D.L. Marshall, Sandro Galea, Formalizing the role of agent-based modeling in causal inference and epidemiology, American Journal of Epidemiology 181 (2) (2015), https://doi.org/10.1093/aje/kwu274, ISSN: 14766256.

[313] Marco A. Janssen, Calvin Pritchard, Allen Lee, On code sharing and model documentation of published individual and agent-based models, Environmental Modelling and Software 134 (2020), https://doi.org/10.1016/j.envsoft.2020.104873, ISSN: 13648152.

[314] Eric Silverman, et al., Situating agent-based modelling in population health research, Emerging Themes in Epidemiology 18 (1) (2021), https://doi.org/10.1186/s12982-021-00102-7, ISSN: 17427622.

[315] Zahraa Marafie, et al., Autocoach: an intelligent driver behavior feedback agent with personality-based driver models, Electronics (Switzerland) 10 (11) (2021), https://doi.org/10.3390/eectronics10111361, ISSN: 20799292.

[316] Laura Burbach, et al., Who shares fake news in online social networks? An agent-based model of different personality models and behaviors in social networks, in: ACM UMAP 2019 – Proceedings of the 27th ACM Conference on User Modeling, Adaptation and Personalization, 2019.

[317] Philippe J. Giabbanelli, Boone Tison, James Keith, The application of modeling and simulation to public health: assessing the quality of agent-based models for obesity, Simulation Modelling Practice and Theory 108 (2021), https://doi.org/10.1016/j.simpat.2020.102268, ISSN: 1569190X.

[318] Christopher R. Williams, Armin R. Mikler, Incorporating disgust as disease-avoidant behavior in an agent-based epidemic model, in: Lecture Notes in Computer Science (Including Subseries Lecture Notes in Artificial Intelligence and Lecture Notes in Bioinformatics), vol. 9708, 2016.

[319] Altuntas Fevzi, et al., Covid-19 in hematopoietic cell transplant recipients, Bone Marrow Transplantation 56 (4) (2021), https://doi.org/10.1038/s41409-020-01084-x-, ISSN: 14765365.

[320] Jure Leskovec, Andrej Krevl, SNAP Datasets: Stanford Large Network Dataset Collection, 2014.

[321] Laura Temime, et al., NosoSim: an agent-based model of nosocomial pathogens circulation in hospitals, Procedia Computer Science 1 (1) (2010), https://doi.org/10.1016/j.procs.2010.04.251.

[322] Thi Mui Pham, et al., Interventions to control nosocomial transmission of Sars-CoV-2: a modelling study, BMC Medicine 19 (1) (2021), https://doi.org/10.1186/s12916-021-02060-y, ISSN: 17417015.

[323] Tracey Elizabeth Claire Jones-Konneh, et al., Agent-based modeling and simulation of nosocomial infection among healthcare workers during Ebola virus disease outbreak in Sierra leone, Tohoku Journal of Experimental Medicine 245 (4) (2018), https://doi.org/10.1620/tjem.245.231, ISSN: 13493329.
[324] D.W. Tarry, Dicrocoelium dendriticum: the life cycle in Britain, Journal of Helminthology 43 (3–4) (1969), https://doi.org/10.1017/S0022149X00004971, ISSN: 14752697.
[325] Gözde Gürelli, Importance of land snails in dicrocoeliosis epidemiology, Turkiye Parazitolojii Dergisi 41 (3) (2017), https://doi.org/10.5152/tpd.2017.5177, ISSN: 21463077.
[326] Bradley J. Van Paridon, et al., Life cycle, host utilization, and ecological fitting for invasive lancet liver fluke, dicrocoelium dendriticum, emerging in Southern Alberta, Canada, Journal of Parasitology 103 (3) (2017), https://doi.org/10.1645/16-140, ISSN: 19372345.
[327] Melissa A. Beck, Cameron P. Goater, Douglas D. Colwell, Comparative recruitment, morphology and reproduction of a generalist trematode, dicrocoelium dendriticum, in three species of host, Parasitology 142 (10) (2015), https://doi.org/10.1017/S0031182015000621, ISSN: 14698161.
[328] P.T. Tverdokhlebov, Prevention and control of dicrocoeliasis, Veterinariia 3 (Mar. 1971) 69–70, ISSN: 0042-4846.
[329] Melissa Pagaoa, et al., Trends in nationally notifiable sexually transmitted disease case reports during the US Covid-19 pandemic, January to December 2020, Sexually Transmitted Diseases 48 (10) (2021), https://doi.org/10.1097/OLQ.0000000000001506, ISSN: 15374521.
[330] Filippo Bonato, et al., Syphilis and the Covid-19 pandemic: did the lockdown stop risky sexual behavior?, Clinics in Dermatology 39 (4) (2021), https://doi.org/10.1016/j.cindermato.2020.11.006, ISSN: 18791131.
[331] Kelly A. Johnson, et al., Measuring the impact of the Covid-19 pandemic on sexually transmitted diseases public health surveillance and program operations in the state of California, Sexually Transmitted Diseases 48 (8) (2021), https://doi.org/10.1097/OLQ.0000000000001441, ISSN: 15374521.
[332] Amollo Otiende, Human rights and ethics in HIV and AIDS control programmes, Discovery and Innovation 17 (SPEC. ISS.) (2005), https://doi.org/10.4314/dai.v17i2.15697, ISSN: 1015079X.
[333] Rzeszutek Marcin, et al., HIV/AIDS stigma and psychological well-being after 40 years of HIV/AIDS: a systematic review and meta-analysis, European Journal of Psychotraumatology 12 (1) (2021), https://doi.org/10.1080/20008198.2021.1990527, ISSN: 20008066.
[334] Polyxeni Potter, Jan Steen (c. 1625–1679). Beware of Luxury (c. 1665), Emerging Infectious Diseases 9 (8) (2003), https://doi.org/10.3201/eid0908.ac0908, ISSN: 1080-6040.
[335] Kim Yi Dionne, Fulya Felicity Turkmen, The politics of pandemic othering: putting Covid-19 in global and historical context, International Organization (2020), https://doi.org/10.1017/S0020818320000405, ISSN: 15315088.
[336] Annamaria Silvana De Rosa, Terri Mannarini, The "invisible other": social representations of Covid-19 pandemic in media and institutional discourse, Papers on Social Representations 29 (2) (2020), ISSN: 18193978.
[337] Julia Goodall, et al., HIV and AIDS related stigma: a necessary protective mechanism for children in high exposure areas?, South African Journal of Psychology 41 (2011) 2, https://doi.org/10.1177/008124631104100207, ISSN: 00812463.
[338] Marina Economou, Social distance in Covid-19: drawing the line between protective behavior and stigma manifestation, Psychiatrike = Psychiatriki 32 (3) (2021), https://doi.org/10.22365/jpsych.2021.025, ISSN: 11052333.

[339] Emma Ohlsson-Nevo, Jan Karlsson, Impact of health-related stigma on psychosocial functioning in the general population: construct validity and Swedish reference data for the stigma-related social problems scale (SSP), Research in Nursing and Health 42 (1) (2019), https://doi.org/10.1002/nur.21924, ISSN: 1098240X.

[340] Ida Viktoria Kolte, et al., The contribution of stigma to the transmission and treatment of tuberculosis in a hyperendemic indigenous population in Brazil, PLoS ONE 15 (12) (2020), https://doi.org/10.1371/journa.pone.0243988, ISSN: 19326203.

[341] Mariam Davtyan, Brandon Brown, Morenike Oluwatoyin Folayan, Addressing Ebola-related stigma: lessons learned from HIV/AIDS, Global Health Action 7 (1) (2014), https://doi.org/10.3402/gha.v7.26058, ISSN: 16549880.

[342] Manuela Colombini, et al., The risks of partner violence following HIV status disclosure, and health service responses: narratives of women attending reproductive health services in Kenya, Journal of the International AIDS Society 19 (1) (2016), https://doi.org/10.7448/IAS.19.1.20766, ISSN: 17582652.

[343] Fatmah Almoayad, et al., Stigmatisation of Covid-19 in Riyadh, Saudi Arabia: a cross-sectional study, Sultan Qaboos University Medical Journal 21 (4) (2021), https://doi.org/10.18295/squmj.4.2021.044, ISSN: 20750528.

[344] E. Fee, N. Krieger, Understanding AIDS: historical interpretations and the limits of biomedical individualism, American Journal of Public Health 83 (10) (1993), https://doi.org/10.2105/AJPH.83.10.1477, ISSN: 00900036.

[345] Thomas L. Long, AIDS and American Apocalypticism: The Cultural Semiotics of an Epidemic, SUNY Press, Albany, NY, ISBN 9780791461679, 2004.

[346] Judith Bruchfeld, Margarida Correia-Neves, Gunilla Kallenius, Tuberculosis and HIV coinfection, Cold Spring Harbor Perspectives in Medicine 5 (7) (2015), https://doi.org/10.1101/cshperspect.a017871, ISSN: 21571422.

[347] Samaa T. Gobran, Petronela Ancuta, Naglaa H. Shoukry, A tale of two viruses: immunological insights into HCV/HIV coinfection, Frontiers in Immunology 12 (2021), https://doi.org/10.3389/fimmu.2021.726419, ISSN: 16643224.

[348] Megan Oaten, Richard J. Stevenson, Trevor I. Case, Disgust as a disease-avoidance mechanism, Psychological Bulletin 135 (2) (2009), https://doi.org/10.1037/a0014823, ISSN: 00332909.

[349] Pelin Gul, et al., Disease avoidance motives trade-off against social motives, especially mate-seeking, to predict social distancing: evidence from the Covid-19 pandemic, Social Psychological and Personality Science (2021), https://doi.org/10.1177/19485506211046462, ISSN: 19485514.

[350] Pavel A. Kislyakov, Elena A. Shmeleva, Strategies of prosocial behavior during the Covid-19 pandemic, The Open Psychology Journal 14 (1) (2021), https://doi.org/10.2174/1874350102114010266, ISSN: 1874-3501.

[351] Nicola Grignoli, et al., Influence of empathy disposition and risk perception on the psychological impact of lockdown during the coronavirus disease pandemic outbreak, Frontiers in Public Health 8 (2021), https://doi.org/10.3389/fpubh.2020.567337, ISSN: 22962565.

[352] Michael F. Parry, et al., Precipitous fall in common respiratory viral infections during Covid-19, Open Forum Infectious Diseases 7 (11) (2020), https://doi.org/10.1093/ofid/ofaa511, ISSN: 23288957.

[353] Sonja J. Olsen, et al., Decreased influenza activity during the Covid-19 pandemic – United States, Australia, Chile, and South Africa, 2020, Morbidity and Mortality Weekly Report 69 (37) (2020), https://doi.org/10.15585/mmwr.mm6937a6, ISSN: 0149-2195.

[354] Jovana Milan Pavlovic, Dragica Petar Pesut, Maja Borivoje Stosic, Influence of the Covid-19 pandemic on the incidence of tuberculosis and influenza, Revista do Instituto de Medicina Tropical de Sao Paulo 63 (2021), https://doi.org/10.1590/S1678-9946202163053, ISSN: 16789946.
[355] Grant Young, et al., Rapid decline of seasonal influenza during the outbreak of Covid-19, ERJ Open Research 6 (3) (2020), https://doi.org/10.1183/23120541.00296-2020.
[356] H.E. Romo, J.L. Kenney, A.C. Brault, Decreased transmission of Zika virus in Aedes aegypti mosquitoes co-inoculated with an insectspecific flavivirus, American Journal of Tropical Medicine and Hygiene 97 (5) (2017), ISSN: 0002-9637.
[357] Louise Rodrigues, et al., Superinfection with woodchuck hepatitis virus strain WHVNY of livers chronically infected with strain WHV7, Journal of Virology 89 (1) (2015), https://doi.org/10.1128/jvi.02361-14, ISSN: 0022-538X.
[358] Francesca Lipari, This is how we do it: how social norms and social identity shape decision making under uncertainty, Games 9 (4) (2018), https://doi.org/10.3390/g9040099, ISSN: 20734336.
[359] Abe Taiga, et al., The toilet paper issue during the Covid-19 crisis, in: Proceedings – 2020 IEEE/WIC/ACM International Joint Conference on Web Intelligence and Intelligent Agent Technology, WI-IAT 2020, 2020.
[360] Yongling Lin, et al., The hierarchical sensitivity to social misalignment during decision-making under uncertainty, Social Cognitive and Affective Neuroscience 16 (6) (2021), https://doi.org/10.1093/scan/nsab022, ISSN: 17495024.
[361] Rex N. Ali, Harvey Rubin, Saswati Sarkar, Countering the potential re-emergence of a deadly infectious disease-information warfare, identifying strategic threats, launching countermeasures, PLoS ONE 16 (August 2021) 8, https://doi.org/10.1371/journa.pone.0256014, ISSN: 19326203.
[362] Yang Ye, et al., Effect of heterogeneous risk perception on information di-fusion, behavior change, and disease transmission, Physical Review E 102 (4) (2020), ISSN: 24700053 https://doi.org/10.1103/PhysRevE.102.042314.
[363] P. Clifford, A. Sudbury, A model for spatial conflict, Biometrika 60 (1973) 3, https://doi.org/10.2307/2335008, ISSN: 00063444.
[364] Pooja Agarwal, Mackenzie Simper, Rick Durrett, The q-voter model on the torus, Electronic Journal of Probability 26 (2021), https://doi.org/10.1214/21-EJP682, ISSN: 10836489.
[365] André C.R. Martins, Continuous opinions and discrete actions in opinion dynamics problems, International Journal of Modern Physics C 19 (4) (2008), https://doi.org/10.1142/S0129183108012339, ISSN: 01291831.
[366] Tomasz M. Gwizdalla, Randomized Sznajd model, International Journal of Modern Physics C 17 (12) (2006), https://doi.org/10.1142/S0129183106010170, ISSN: 01291831.
[367] Jimit R. Majmudar, et al., Voter models and external influence, Journal of Mathematical Sociology 44 (1) (2020), https://doi.org/10.1080/0022250X.2019.1625349, ISSN: 15455874.
[368] Rainer Hegselmann, Ulrich Krause, Opinion dynamics and bounded confidence: models, analysis and simulation, JASSS 5 (3) (2002), ISSN: 14607425.
[369] Seyed Rasoul Etesami, Tamer Basar, Game-theoretic analysis of the Hegselmann-Krause model for opinion dynamics in finite dimensions, IEEE Transactions on Automatic Control 60 (7) (2015), https://doi.org/10.1109/TAC.2015.2394954, ISSN: 00189286.
[370] Antonio F. Peralta, et al., Effect of algorithmic bias and network structure on coexistence, consensus, and polarization of opinions, Physical Review E 104 (4) (2021), https://doi.org/10.1103/PhysRevE.104.044312, ISSN: 24700053.

[371] Antonio F. Peralta, János Kertész, Gerardo Iñiguez, Opinion formation on social networks with algorithmic bias: dynamics and bias imbalance, Journal of Physics: Complexity 2 (4) (2021), https://doi.org/10.1088/2632-072X/ac340f, ISSN, 2632072X.
[372] Michelle M. Mello, Vaccine misinformation and the first amendment—the price of free speech, JAMA Health Forum 3 (3) (2022), https://doi.org/10.1001/jamaheathforum.2022.0732.
[373] Matthew J. Berryman, Neil F. Johnson, Derek Abbott, Contagions across networks: colds and markets, Proceedings of the SPIE 6039 (2005), https://doi.org/10.1117/12.638311.
[374] Sam Yeaman, Alana Schick, Laurent Lehmann, Social network architecture and the maintenance of deleterious cultural traits, Journal of the Royal Society Interface 9 (70) (2012), https://doi.org/10.1098/rsif.2011.0555, ISSN: 17425662.
[375] Raazesh Sainudiin, David Welch, The transmission process: a combinatorial stochastic process for the evolution of transmission trees over networks, Journal of Theoretical Biology 410 (2016), https://doi.org/10.1016/j.jtbi.2016.07.038, ISSN: 10958541.
[376] Soodeh Hosseini, Mohammad Abdollahi Azgomi, Adel Torkaman Rahmani, Dynamics of a rumor-spreading model with diversity of configurations in scale-free networks, International Journal of Communication Systems 28 (18) (2015), https://doi.org/10.1002/dac.3016, ISSN: 10991131.
[377] Wang Hongmei, et al., Susceptible-infectious-critical-recovered rumor spreading model in social networks, International Journal of Modern Physics C (2021), https://doi.org/10.1142/S012918312250053X, ISSN: 01291831.
[378] Shangde Gao, Yan Wang, Lisa Sundahl Platt, Modeling U.S. health agencies' message dissemination on Twitter and users' exposure to vaccine-related misinformation using system dynamics, Proceedings of the International ISCRAM Conference 2021-May (2021).
[379] Jin-Hee Cho, et al., Uncertainty-based false information propagation in social networks, ACM Transactions on Social Computing 2 (2) (2019), https://doi.org/10.1145/3311091, ISSN: 2469-7818.
[380] Jiajun Xian, et al., The optimal edge for containing the spreading of SIS model, Journal of Statistical Mechanics: Theory and Experiment 2020 (4) (2020), https://doi.org/10.1088/1742-5468/ab780d, ISSN: 17425468.
[381] Akanksha Mathur, Chandra Prakash Gupta, Dynamic SEIZ in online social networks: epidemiological modeling of untrue information, International Journal of Advanced Computer Science and Applications 11 (7) (2020), https://doi.org/10.14569/IJACSA.2020.0110771, ISSN: 21565570.
[382] Azhar Hussain, et al., The anti-vaccination movement: a regression in modern medicine, Cureus (2018), https://doi.org/10.7759/cureus.2919, ISSN: 2168-8184.
[383] Seth C. Kalichman, et al., Faster than warp speed: early attention to COVD-19 by anti-vaccine groups on Facebook, Journal of Public Health 44 (1) (2022), https://doi.org/10.1093/pubmed/fdab093, ISSN: 17413850.
[384] Christopher J. Mckinley, Fanny Lauby, Anti-Vaccine Beliefs and COVID-19 Information Seeking on Social Media: Examining Processes Influencing COVID-19 Beliefs and Preventative Actions, Tech. Rep., 2021.
[385] Amy E. Leader, et al., Understanding the messages and motivation of vaccine hesitant or refusing social media influencers, Vaccine 39 (2) (2021), https://doi.org/10.1016/j.vaccine.2020.11.058, ISSN: 18732518.
[386] Dorothy M. Horstmann, Robert W. McCollum, Poliomyelitis virus in human blood during the "Minor Illness" and the asymptomatic infection, Proceedings of the Society for Experimental Biology and Medicine 82 (3) (1953), https://doi.org/10.3181/00379727-82-20139, ISSN: 15353699.

[387] Joseph W.S. Timothy, et al., Early transmission and case fatality of Ebola virus at the index site of the 2013–16 West African Ebola outbreak: a cross-sectional seroprevalence survey, Lancet. Infectious Diseases 19 (4) (2019), https://doi.org/10.1016/S1473-3099(18)30791-6, ISSN: 14744457.

[388] Adnan I. Qureshi, et al., Study of Ebola virus disease survivors in Guinea, Clinical Infectious Diseases 61 (7) (2015), https://doi.org/10.1093/cid/civ453, ISSN: 15376591.

[389] Judith R. Glynn, et al., Asymptomatic infection and unrecognised Ebola virus disease in Ebola-affected households in Sierra leone: a cross-sectional study using a new non-invasive assay for antibodies to Ebola virus, Lancet. Infectious Diseases 17 (6) (2017), https://doi.org/10.1016/S1473-3099(17)30111-1, ISSN: 14744457.

[390] Placide Mbala, et al., Evaluating the frequency of asymptomatic Ebola virus infection, Philosophical Transactions of the Royal Society B: Biological Sciences 372 (1721) (2017), https://doi.org/10.1098/rstb.2016.0303, ISSN: 14712970.

[391] Longzhao Liu, et al., Modeling confirmation bias and peer pressure in opinion dynamics, Frontiers in Physics 9 (2021), https://doi.org/10.3389/fphy.2021.649852, ISSN: 2296424X.

[392] James Dickson Murray, Mathematical Biology II: Spatial Models and Biomedical Applications, 3rd, 2003.

[393] Alberto Giubilini, Julian Savulescu, Dominic Wilkinson, Queue questions: ethics of Covid-19 vaccine prioritization, Bioethics 35 (4) (2021), https://doi.org/10.1111/bioe.12858, ISSN: 14678519.

[394] Maxwell J. Smith, Why we should not 'just use age' for Covid-19 vaccine prioritisation, Journal of Medical Ethics (2021), https://doi.org/10.1136/medethics-2021-107443, ISSN: 14734257.

[395] Hajime Sato, Countermeasures and vaccination against terrorism using smallpox: pre-event and post-event smallpox vaccination and its contraindications, Environmental Health and Preventive Medicine 16 (5) (2011), https://doi.org/10.1007/s12199-010-0200-z, ISSN: 1342078X.

[396] Travis C. Porco, et al., Logistics of community smallpox control through contact tracing and ring vaccination: a stochastic network model, BMC Public Health 4 (2004), https://doi.org/10.1186/1471-2458-4-34, ISSN: 14712458.

[397] Richard D. Turner, Graham H. Bothamley, Cough and the transmission of tuberculosis, The Journal of Infectious Diseases 211 (9) (May 2015) 1367–1372, https://doi.org/10.1093/infdis/jiu625, ISSN: 0022-1899.

[398] Daniel Havlichek, et al., Age-related hepatitis B seroconversion rates in health care workers, American Journal of Infection Control 25 (5) (1997), https://doi.org/10.1016/S0196-6553(97)90090-0, ISSN: 01966553.

[399] P.F.M. Teunis, J.C.H. van Eijkeren, Estimation of seroconversion rates for infectious diseases: effects of age and noise, Statistics in Medicine 39 (21) (2020), https://doi.org/10.1002/sim.8578, ISSN: 10970258.

[400] Grace Sembajwe, et al., Patterns of negative seroconversion in ongoing surveys of SARS-CoV-2 antibodies among workers in New York's largest healthcare system, Occupational and Environmental Medicine 78 (11) (2021), https://doi.org/10.1136/oemed-2021-107382, ISSN: 14707926.

[401] Kristien Cloots, et al., Male predominance in reported visceral leishmaniasis cases: nature or nurture? A comparison of population-based with health facility-reported data, PLoS Neglected Tropical Diseases 14 (1) (2020), https://doi.org/10.1371/JOURNAL.PNTD.0007995, ISSN: 19352735.

[402] Tinuade A. Ogunlesi, Durotoye M. Olanrewaju, Socio-demographic factors and appropriate health care-seeking behavior for childhood illnesses, Journal of Tropical Pediatrics 56 (6) (2010), https://doi.org/10.1093/tropej/fmq009, ISSN: 01426338.

[403] Khadija Said, et al., Diagnostic delay and associated factors among patients with pulmonary tuberculosis in Dar es Salaam, Tanzania, Infectious Diseases of Poverty 6 (1) (2017), https://doi.org/10.1186/s40249-017-0276-4, ISSN: 20499957.

[404] S. Sarrassat, et al., Distance to care, care seeking and child mortality in rural Burkina Faso: findings from a population-based cross-sectional survey, Tropical Medicine and International Health 24 (1) (2019), https://doi.org/10.1111/tmi.13170, ISSN: 13653156.

[405] Daniel C. Oshi, et al., Gender-related factors influencing women's health seeking for tuberculosis care in Ebonyi state, Nigeria, Journal of Biosocial Science 48 (1) (2016), https://doi.org/10.1017/S0021932014000534, ISSN: 14697599.

[406] Moumita Das, et al., The gendered experience with respect to health-seeking behaviour in an urban slum of Kolkata, India, International Journal for Equity in Health 17 (1) (2018), https://doi.org/10.1186/s12939-018-0738-8, ISSN: 14759276.

[407] Emily Walton, Denise L. Anthony, Do you want to see a doctor for that? Contextualizing racial and ethnic differences in care-seeking, Research in the Sociology of Health Care 35 (2017), https://doi.org/10.1108/S0275-495920170000035013, ISSN: 02754959.

[408] Nishita Padmanabhan, S. Poornima, A study to assess the stigma related to tuberculosis among directly observed treatment short-course (DOTS) providers and patients on DOTS therapy attending DOTS centres of Mandya City, Karnataka, India, International Journal of Community Medicine and Public Health (2016), https://doi.org/10.18203/2394-6040.ijcmph20163367, ISSN: 2394-6032.

[409] P. Malfertheiner, Compliance, adverse events and antibiotic resistance in helicobacter pylori treatment, Scandinavian Journal of Gastroenterology 28 (S196) (1993), https://doi.org/10.3109/00365529309098341, ISSN: 00365521.

[410] Lorna Marie West, Maria Cordina, Educational intervention to enhance adherence to short-term use of antibiotics, Research in Social and Administrative Pharmacy 15 (2) (2019), https://doi.org/10.1016/j.sapharm.2018.04.011, ISSN: 15517411.

[411] M.V. Shestakova, et al., Cognitive impairment and compliance in chronic heart failure, Zhurnal Nevrologii i Psihiatrii imeni S.S. Korsakova 117 (6) (2017), https://doi.org/10.17116/jnevro20171176253-57, ISSN: 23094729.

[412] Leanne E. Unicomb, et al., Motivating antibiotic stewardship in Bangladesh: identifying audiences and target behaviours using the behaviour change wheel, BMC Public Health 21 (1) (2021), https://doi.org/10.1186/s12889-021-10973-9, ISSN: 14712458.

[413] Amnir Hadachi, Mozhgan Pourmoradnasseri, Kaveh Khoshkhah, Unveiling large-scale commuting patterns based on mobile phone cellular network data, Journal of Transport Geography 89 (2020), https://doi.org/10.1016/j.jtrangeo.2020.102871, ISSN: 09666923.

[414] Marco Mamei, et al., Evaluating origin–destination matrices obtained from CDR data, Sensors (Switzerland) 19 (20) (2019), https://doi.org/10.3390/s19204470, ISSN: 14248220.

[415] Georg Heiler, et al., Country-wide mobility changes observed using mobile phone data during Covid-19 pandemic, in: Proceedings – 2020 IEEE International Conference on Big Data, Big Data 2020, 2020.

[416] Nishant Kishore, et al., Measuring mobility to monitor travel and physical distancing interventions: a common framework for mobile phone data analysis, The Lancet Digital Health (2020), https://doi.org/10.1016/S2589-7500(20)30193-X.

[417] Nazanin Sabri, Ridhi Kashyap, Ingmar Weber, Examining global mobile diffusion and mobile gender gaps through Facebook's advertising data, in: HT 2021 – Proceedings of the 32nd ACM Conference on Hypertext and Social Media, 2021.

[418] Nuhu Diraso Gapsiso, Rahila Jibrin, Women and Nigerian ICT policy: the inevitability of gender mainstreaming, in: Overcoming Gender Inequalities Through Technology Integration, 2016.

[419] Masoomali Fatehkia, Ridhi Kashyap, Ingmar Weber, Using Facebook ad data to track the global digital gender gap, World Development 107 (2018), https://doi.org/10.1016/j.worddev.2018.03.007, ISSN: 18735991.

[420] Akihiro Fujihara, Hiroyoshi Miwa, Homesick Lévy walk: a mobility model having ichi-go ichi-e and scale-free properties of human encounters, in: Proceedings – International Computer Software and Applications Conference, 2014.

[421] Koichiro Sugihara, Naohiro Hayashibara, Target exploration by Nomadic Lévy walk on unit disk graphs, International Journal of Grid and Utility Computing 11 (2) (2020), https://doi.org/10.1504/IJGUC.2020.105536, ISSN: 17418488.

Index

S

Printed in the United States
by Baker & Taylor Publisher Services